A Monstrous Commotion

Also by Gareth Williams

Textbook of Diabetes, with John Pickup, editions 1–3 (1991–2002)

Handbook of Diabetes, with John Pickup, editions 1–3 (1992–2004)

Obesity: Science to Practice, with Gema Frühbeck (2009)

Angel of Death: The Story of Smallpox (2010)

Paralysed with Fear: The Story of Polio (2013)

Critical acclaim for *Angel of Death*:

'Wonderfully researched, vividly written . . . an example of medical history at its absolute best.'   *Michael Neve, Wellcome Book Prize Panel, 2010*

'Williams recounts the history of smallpox in a breezy, accessible style. And what a history it is.'   *Clive Anderson, New Scientist Big Read, 2010*

'An engaging narrative . . . medical history is interwoven with social history and reflections on contemporary issues.'   *BBC History Magazine, 2010*

'A meticulously researched story with pace and flair . . . Both the history and the science are terrific.'   *Medical Journalists' Association, 2011*

Critical acclaim for *Paralysed with Fear*:

'His splendid book, riveting from beginning to end, is a model of its kind . . . a consistently fascinating account.'   *Literary Review, 2013*

'An incredible story told by a great storyteller.'   *Lancet, 2013*

'A wonderful biography of polio and a revealing story of the development of 20th-century medicine, warts and all.'   *BBC History Magazine, 2013*

'A human story, humanely told – authoritative and insightful, warm in tone and compulsively readable.'   *British Polio Fellowship Bulletin, 2013*

'A detailed, science-rich account.'   *Nature, 2013*

A Monstrous Commotion

The Mysteries of Loch Ness

GARETH WILLIAMS

This edition first published in Great Britain in 2015 by
Orion Books
An imprint of the Orion Publishing Group Ltd
Carmelite House, 50 Victoria Embankment,
London EC4Y 0DZ
An Hachette Livre UK Company

1 3 5 7 9 10 8 6 4 2

A CIP catalogue record for this book is available
from the British Library.

ISBN: 978 1 4091 5873 8

Typeset by Input Data Services Ltd, Bridgwater, Somerset

Printed and bound by CPI Group (UK) Ltd, Croydon, CR0 4YY

The Orion Publishing Group's policy is to use papers that are natural,
renewable and recyclable and made from wood grown in sustainable
forests. The logging and manufacturing processes are expected to conform
to the environmental regulations of the country of origin.

Every effort has been made to fulfil requirements with regard to
reproducing copyright material. The author and publisher will be glad to
rectify any omissions at the earliest opportunity.

www.orionbooks.co.uk

To Caroline, Tim, Jo and Tessa
And in memory of my parents, Alwyn and Joan

Rachainn a thaomadh na fairge dha nan iarradh e orm.
I would go to drain the sea for him if he asked me.

Gaelic proverb

Contents

Chronology

The modern era (1930–) of the Loch Ness Monster

AAS = Academy of Applied Science; LN(P)IB = Loch Ness (Phenomenon) Investigation Bureau.

1930   27 August: Three young men fishing report a massive wake off Tor Point

1933   28 April: John and Aldie Mackay report a large animal like a whale near Abriachan Pier

       22 July: Mr and Mrs George Spicer report a large 'prehistoric' animal crossing the road north of Foyers

       17 October: Philip Stalker writes an article about the Loch Ness Monster for the *Scotsman*, entitled 'The Plesiosaurus Theory'

       12 November: Hugh Gray photographs a 40-foot 'serpent' off Foyers – the first photograph of the Monster

       14 November: Lieutenant Commander Rupert Gould begins his two-week investigation at Loch Ness

       19 December: *Daily Mail* expedition, led by Marmaduke Wetherell, arrives at Loch Ness; they discover a trail of footprints south of Foyers on 22 December

1934   3 January: Wetherell's 'Monster' footprints are identified by the Natural History Museum as the left hind foot of a hippopotamus

       5 January: Arthur Grant encounters a large animal by the roadside near Abriachan

       March: Publication of *The Home of the Loch Ness Monster* by Colonel W. H. Lane, suggesting that the Monster is a giant salamander

       19 April: Robert K. Wilson takes the 'Surgeon's Photograph', showing a curved neck and head, near Invermoriston

May: Publication of *The Loch Ness Monster and Others*, by Rupert Gould, arguing that the Monster was related to the sea serpent

July: Sir Edward Mountain's Expedition to Loch Ness, with observers trained in still and cine-photography

1945 Publication of *Loch Ness and Its Monster*, by Father Aloysius Carruth

1951 14 July: Lachlan Stuart photographs a three-humped animal below Whitefield

1954 2 December: echo sounder on the trawler *Rival III* detects a massive sonar contact, 500 feet below the surface of Loch Ness

1955 29 July: Peter Macnab photographs a 50-foot, two-humped animal in Urquhart Bay

1957 Publication of *More Than a Legend* by Constance Whyte, suggesting that the Monster is a surviving plesiosaur

1959 22 March: Dr Denys Tucker, Curator of Fishes at the Natural History Museum, watches a humped animal in the Loch off Glendoe and later identifies it as a plesiosaur

1960 28 February: Torquil MacLeod observes a 50-foot animal at the water's edge below the Horseshoe Scree

14 March: Tim Dinsdale begins a lengthy correspondence with Peter Scott

23 April: Tim Dinsdale films a single-humped object off Foyers; the film is shown on BBC Television's *Panorama* on 13 June

31 July: Denys Tucker is sacked by the Natural History Museum for refusing to retract his belief in the Loch Ness Monster

7 August: the yacht *Finola* encounters a large aquatic animal near Dores; Torquil MacLeod corroborates the sighting from the shore

19 October: Peter Scott convenes the Loch Ness Study Group of zoological experts to consider Dinsdale's film and other evidence

1961 Publication of *Loch Ness Monster* by Tim Dinsdale, arguing that the Monster is a surviving plesiosaur

Publication of *The Elusive Monster* by Maurice Burton, suggesting that sightings of a 'Monster' are explained by otters or vegetation mats

July: David James and Peter Scott found the Loch Ness Phenomena Investigation Bureau (LNPIB)

23 November: *The Field* publishes 'Time to meet the Monster' by David James MP

1962 July: Peter Baker leads the Oxford–Cambridge Expedition to Loch Ness, with surface observation and an underwater sonar sweep

24 August: F. W. (Ted) Holiday sees a huge slug-like animal off Foyers

October: the First David James Expedition (LNPIB) to Loch Ness

November: David James's panel of experts concludes that Loch Ness harbours a 'large animate object' worthy of further study

1963 24 February: David James and Peter Scott appear in Associated Television's *Report on the Loch Ness Monster*

June: the Second David James Expedition to Loch Ness; Peter Scott and others survey the Loch from gliders

1964 Summer: the Third David James Expedition to Loch Ness

15 October: David James loses his Brighton seat in the General Election

1965 September: Professor Roy Mackal, University of Chicago, visits Loch Ness and the Fourth David James Expedition, and is appointed Scientific Director of the LNPIB

1966 January: David James visits the Adventurers' Club in Chicago to raise funds for the LNPIB

February: the Joint Air Reconnaissance Intelligence Centre (JARIC) concludes that the object filmed by Tim Dinsdale in 1960 was not a boat and was 'probably animate'

Publication of *The Leviathans* by Tim Dinsdale

9 December: BBC Television broadcasts Peter Scott's *Look* programme, 'On the track of unknown animals', from Loch Ness

1967   13 June: Dick Raynor, LNPIB volunteer, films a long wake off Dores Bay

1968   Publication of *The Great Orm of Loch Ness* by F. W. Holiday, arguing that the Monster is a giant invertebrate descended from the extinct species *Tullimonstrum*

Summer: the Seventh David James Expedition to Loch Ness

Sonar survey, by Professor Gordon Tucker and Dr Hugh Braithwaite of Birmingham University, identifies large sonar contacts moving fast at depth

1969   Summer: submarines are deployed to search for the Monster during the Eighth LNPIB Expedition to the Loch

Frank Searle, photographer and hoaxer, takes up residence on the shore of Loch Ness

1970   The LNPIB becomes the LNIB; the Ninth LNIB Expedition to Loch Ness takes place

Summer: Elizabeth Montgomery Campbell and David Solomon lead the first expedition to Loch Morar to hunt for 'Morag', another large aquatic creature

September: Robert Rines and the Academy of Applied Science make their first trip to Loch Ness with various 'attractants' to lure the Monster

1971   23 June: Robert Rines sees a large, single-humped animal in Urquhart Bay

1972   Summer: Nicholas Witchell spends six months watching the Loch before starting university

8 August: the underwater 'flipper' photograph is taken in Urquhart Bay by Robert Rines and the AAS

October: the LNIB is forced to close down due to lack of funds

1974   Summer: Adrian Shine begins the exploration of Loch Morar in *Machan*, his home-built submersible observation hide

Publication of *The Loch Ness Story* by Nicholas Witchell

1975   Early June: Robert Rines meets Peter Scott at Slimbridge

20 June: the underwater 'whole body' and 'gargoyle head' photographs are taken in Urquhart Bay by Robert Rines and the AAS

November: Peter Scott coins a formal scientific name for the Monster: *Nessiteras rhombopteryx*, meaning 'the wonder from Ness with the diamond-shaped fin'

10 December: *Nature* publishes 'Naming the Loch Ness Monster' by Peter Scott and Robert Rines

11 December: presentation of the Rines photographs and other evidence at a Conference on the Loch Ness Monster, held at the House of Commons

1976   Publication of *The Monsters of Loch Ness* by Roy Mackal, in which he suggests that the Monster is a giant amphibian

Publication of *Nessie: Seven Years in Search of the Monster* by Frank Searle, containing numerous faked photographs

March: *Wildlife* publishes 'Why I believe in Nessie' by Sir Peter Scott

Summer: rival American expeditions investigate Loch Ness – the AAS/*New York Times* (Robert Rines's seventh visit) and *National Geographic*

1979   Formation of Loch Ness & Morar Project, directed by Adrian Shine

1981   Adrian Shine begins the Loch Ness Project, a systematic geographical and ecological survey of the Loch, using sonar and underwater television

1984   Frank Searle disappears from Loch Ness after 15 years

1987   July: Symposium on the Loch Ness Monster, organised by the International Society of Cryptozoology and the Society for the History of Natural History, is held in Edinburgh

October: Adrian Shine organises Operation Deepscan, a comprehensive sonar sweep of the Loch

1991   Steve Feltham, Monster hunter, arrives at Loch Ness

1992   Nicholas Witchell organises Project Urquhart, a sonar and ecological survey of the Loch that involves the Natural History Museum and the Royal Geographical Society

2008 Robert Rines's thirtieth and last visit to Loch Ness

2015 Steve Feltham completes 24 years of continuous observation at Loch Ness

Key players in the story

Baker, Peter F (1939–87). As a graduate student, organised the Oxford–Cambridge Expedition to Loch Ness in 1960 and an end-to-end sonar sweep of the Loch in 1962. Subsequently Fellow of the Royal Society and Professor of Physiology at King's College, London.

Burton, Dr Maurice (1898–1992). Curator of Sponges and later Deputy Keeper of Zoology at the Natural History Museum, London. Wrote several books including *Living Fossils* (1956) and *The Elusive Monster* (1961). Initially believed that the Monster was a plesiosaur; later, that sightings were of otters or masses of decaying vegetation.

Campbell, Alex (1901–83). Water bailiff for the Ness Fishery Board, based at Fort Augustus. Part-time correspondent for the *Inverness Courier* and other local papers for over 60 years. Filed many reports about the Monster, notably the index sighting by the Mackays in May 1933; saw the Monster himself on at least 17 occasions.

Crowley, Aleister (1875–1947). Occultist, high priest of his own religion ('Thelema'), poet and sexual endurance athlete. Owned Boleskine House, overlooking the Loch, from 1899 to 1913. Invented the 'Koloo Mavlick' to intimidate the locals, and later claimed that this gave rise to the legend of the Loch Ness Monster.

Dinsdale, Tim (1924–87). Aeronautical engineer who became a global celebrity after filming the Monster off Foyers in April 1960; in 1966, JARIC experts confirmed that the object filmed was 'probably animate' and not a boat. Wrote several best-selling books, beginning with *The Loch Ness Monster* (1961).

Edgerton, Harold 'Doc'. Professor of Engineering at Massachusetts Institute of Technology and inventor of ultra-high-speed exposure systems and the 'strobe' flash which revolutionised underwater photography. Nicknamed 'Papa Flash' by Jacques Cousteau.

Gould, Rupert T. (1890–1948). Lieutenant Commander (Retired) of the Royal Navy. Cartographer, restorer of John Harrison's historic marine chronometers, and author of books about natural enigmas including *The Case for the Sea-Serpent* (1930). Broadcaster and 'the man who knew (almost) everything' on the BBC's *Brains Trust* during the 1930s. Visited Loch Ness to investigate the Monster in November 1933 and wrote *The Loch Ness Monster and Others* (1934).

Grant, Arthur. Veterinary student who encountered a Monster on the Inverness–Drumnadrochit Road near Abriachan in the small hours of 5 January 1934.

Gray, Hugh. Employee of the British Aluminium plant at Foyers, who took the first acknowledged photograph of a Monster (a 40-foot 'serpent') near Foyers on 4 November 1933.

Holiday, F. W. 'Ted' (1920–79). Angling correspondent and writer who first saw the Monster off Foyers in August 1962. Became convinced that the Monster was, firstly, a giant slug-like invertebrate and, later, a multidimensional entity with telepathic powers. Wrote *The Great Orm of Loch Ness* (1968) and *The Dragon and the Disc* (1973).

James, David (1918–86). Adventurer, writer, politician and Laird of Torosay Castle on the Isle of Mull. Famous for a daring escape from a German prisoner-of-war camp in 1941. With Peter Scott, founded the Loch Ness Phenomena Investigation Bureau in 1961 and organised ten expeditions to Loch Ness between 1962 and 1971. MP for Brighton Kempton (1959–64) and North Dorset (1970–79).

Lane, Colonel W. H. (1874–1946). Veteran of the British Army in India and Burma, who also excavated the ruins of Babylon. Wrote the first book on the Monster, *The Home of the Loch Ness Monster* (1934), arguing that the creature was a giant salamander.

Mackal, Roy (1925–2013). Assistant Professor of Biochemistry at the University of Chicago, who visited Loch Ness in 1965 and became Scientific

Director of the LNPIB in 1966. Believed that the Monster was a giant amphibian and wrote *The Monsters of Loch Ness* (1975). Also wrote scientific papers on molecular virology and *In Search of Mokele-Mbembe* (1982), about his hunt for a surviving dinosaur in Africa.

Mackay, John and Donaldina (Aldie). Proprietors of the Drumnadrochit Hotel whose sighting of a large, whale-like animal off Abriachan Pier triggered the modern era of interest in the Monster. This 'Strange Spectacle' was reported by Alex Campbell in the *Inverness Courier* on 2 May 1933.

Macnab, Peter. Bank manager and town councillor from Ayrshire, who photographed a two-humped creature, estimated to be over 50 feet long, in Urquhart Bay in July 1955. His photograph persuaded Sir Alister Hardy, Professor of Zoology at Oxford, that the Monster existed.

Mountain, Sir Edward. Chairman of the Eagle Star Insurance Company, who once declined to insure the *Titanic*. In 1934, he financed and led the first organised expedition to photograph and film the Monster – an exercise he described as 'a success from the start'.

Murray, Sir John (1841–1914). Marine biologist, explorer and cartographer; veteran of the *Challenger* expedition to Antarctica during the 1880s. Led the massive Bathymetrical Survey (1899–1906) which catalogued the geography and ecology of Loch Ness and 561 other freshwater lakes in Scotland.

Nessiteras rhombopteryx. Formal scientific name which Sir Peter Scott proposed for the Loch Ness Monster in an article published by *Nature* in December 1975. Derived from classical Greek, the name means 'the wonder from Ness with the diamond-shaped fin'.

O'Connor, Peter. Fireman and member of the Northern Naturalists' Society, Newcastle. In May 1960, claimed to have photographed a long-necked, plesiosaur-like creature at close range with a Brownie camera; named the Monster *Nessisaurus o'connori*. Later set up business in Luton as 'The Taxidermist of Europe'.

Rines, Dr Robert H. (1922–2009). Boston-based patent lawyer who first read engineering at Massachusetts Institute of Technology and filed several patents for microwave and radar devices. Founded the Academy of Applied Science (1964), which funded numerous trips to investigate the

Monster. The AAS expeditions took famous underwater photographs, later published in *Nature*, which showed the Monster's diamond-shaped flipper (1972) and its body, neck and head (1975). Rines made his thirtieth and last visit to Loch Ness in 2008, aged 86.

Scott, Sir Peter Markham (1909–89). 'Painter and naturalist'; also broadcaster, writer, dinghy sailor (Olympic medallist, 1936) and British gliding champion. International conservationist and founder (first President) of the World Wildlife Fund. With David James, founded the Loch Ness Phenomena Investigation Bureau (1961); senior author, with Robert Rines, of 'Naming the Loch Ness Monster', published by *Nature* in December 1975.

Searle, Frank (1921–2005). Former soldier who spent 15 years from 1969 camping on the shores of Loch Ness, often with female companions whom he called his 'Girls Friday'; known as 'the man who took his camera to bed'. Claimed to have seen the Monster many times; many of his photographs were published by the popular press but were later dismissed as fakes. Ran the 'Frank Searle Loch Ness Investigation Centre' after the LNIB closed down in 1972, and wrote *Nessie: Seven Years in Search of the Monster* (1976). Disappeared suddenly from Loch Ness in 1984.

Shine, Adrian (born 1949). Self-taught ecologist who has explored Lochs Morar and Ness since the late 1970s, initially in a self-built submersible called *Machan*. Organised and led the Loch Ness Project and several systematic investigations of the Loch, including the comprehensive sonar sweep, Operation Deepscan (1987). Published numerous papers on the ecology and topography of Loch Ness and the search for the Monster. Designed the permanent Loch Ness Exhibition in Drumnadrochit.

Spicer, George. London-based businessman who reported seeing a 'prehistoric' animal crossing the road near Foyers in July 1933 – the first documented sighting of the Monster on land.

Stalker, Philip. Senior reporter with the *Scotsman* during the 1930s. In October 1933, wrote a highly influential article on the emerging mystery of the Monster, entitled 'The plesiosaurus theory'.

Stuart, Lachlan. While working as a woodsman in July 1951, took a photograph of a three-humped creature in the shallows below Whitefield. This was one of the classic photographs of the Monster, and key evidence in Constance Whyte's book, *More Than a Legend* (1957).

Tucker, Dr Denys W. (1921–2009). Expert on eels and deep-sea fish, and Curator of Fishes at the Natural History Museum, London, during the 1950s; nicknamed 'Eel Man' by the popular press. Saw the Monster in 1959 and identified it as a plesiosaur. Ignored warnings from the Archbishop of Canterbury and was sacked by the Museum in 1960 when he refused to retract his belief in the Monster.

Tucker, Professor D. Gordon (1914–90). Professor of Electronic and Electrical Engineering at the University of Birmingham who, with Dr Hugh Braithwaite, developed a novel underwater sonar apparatus during the mid-1960s. In August 1968, recorded large objects diving and rising at high speeds (too fast to be fish) in Loch Ness.

Wetherell, Marmaduke (1888–1938). Film director, scriptwriter and star from the silent era; self-styled big-game hunter and adventurer, and Fellow of the Geographical Society and the Zoological Society. Led the *Daily Mail* expedition to find the Monster in December 1933.

Whyte, Constance (1902–82). Former general practitioner; wife of the manager of the Caledonian Canal from the mid-1930s to the late 1950s. After the Second World War, followed up accounts of the Loch Ness Monster and wrote *More Than a Legend* (1957). Foundation Trustee of the Loch Ness Phenomenon Investigation Bureau (1961); resigned in 1966.

Wilson, Colonel Robert K. (1899–1969). Gynaecologist and surgeon who practised in the West End of London between the wars. In April 1934, took the 'Surgeon's Photograph', which became the iconic image of the Monster. Later, served with the Special Operations Executive in 1944–45 and wrote a classic textbook on automatic pistols.

Witchell, Nicholas (born 1953). Aged 19, spent several months in a wooden hut overlooking the Loch before beginning university to read Law. While still a student, wrote *The Loch Ness Story* (1974), which included the underwater photographs taken during the AAS expeditions led by Robert Rines. Became a journalist, correspondent and newsreader at the BBC. Led the multidisciplinary Project Urquhart study of Loch Ness in 1996.

List of Illustrations

Plates

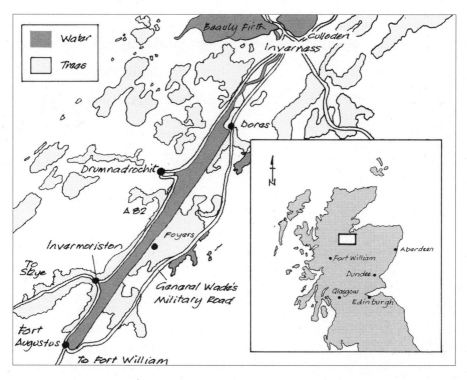

Map 1. Loch Ness and the north-eastern end of the Great Glen.
Map drawn by Ray Loadman.

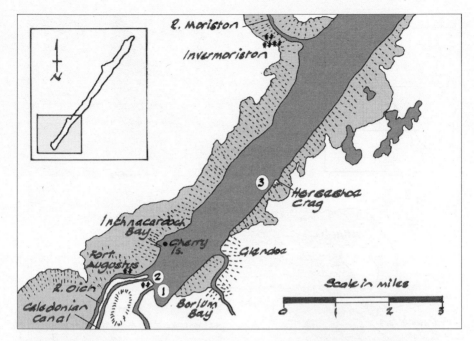

Map 2. Southern end of Loch Ness, from Fort Augustus to Invermoriston.
Classic sightings are indicated by the numbers: 1 – Margaret Munro, shore of
Borlum Bay, 5 June 1934. 2 – Alex Campbell, beyond the Abbey Boathouse in
Borlum Bay, 22 September 1933. 3 – Torquil MacLeod, foot of Horseshoe Crag,
28 February 1960. Map drawn by Ray Loadman.

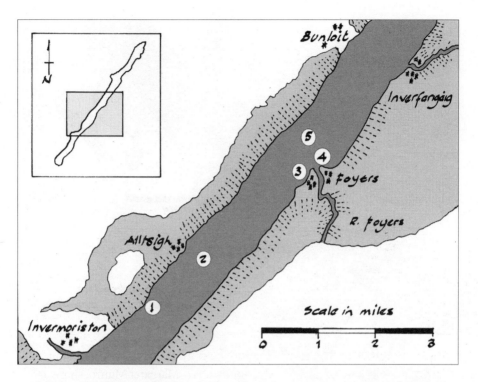

Map 3. Middle section of Loch Ness, from Invermoriston to Inverfarigaig.
Classic sightings indicated by the numbers: 1 – the 'Surgeon's Photograph, taken
by R.K. Wilson, 19 April 1934. 2 – Janet Fraser and others, the Halfway House tea
room, Alltsigh, 22 September 1933. 3 – the 'Great Orm' seen by F.W. Holiday, 26
August 1962. 4 – the first photograph taken of the Monster by Hugh Gray, Foyers,
12 November 1933. 5 – the film taken by Tim Dinsdale, 23 April 1960.
Map drawn by Ray Loadman.

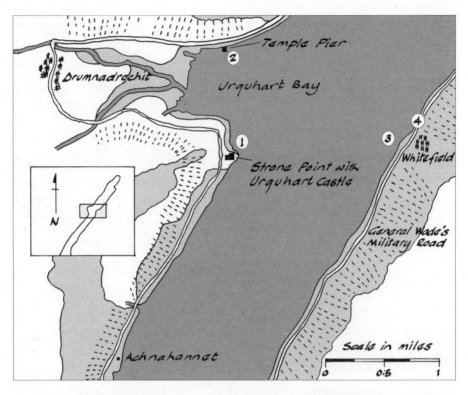

Map 4. Area of Loch Ness around Urquhart Bay and Drumnadrochit. Note that the scale is larger than for the other area maps. Classic sightings indicated by the numbers: 1 – photograph taken by Peter Macnab, near Urquhart Castle, 29 July 1955. 2 – 'Grid Sector 2903' near Temple Pier, where the 'flipper', 'whole body' and 'gargoyle' underwater photographs were taken by the Academy of Applied Science expeditions on 8 August 1972 and 20 June 1975. 3 – photograph taken by Lachlan Stuart on 14 July 1951. 4 – Mr and Mrs G. Spicer, 22 July 1933.
Map drawn by Ray Loadman.

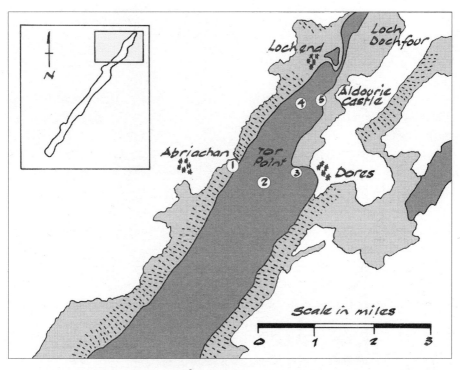

Map 5. North-eastern end of Loch Ness, including Abriachan, Dores and Lochend. Classic sightings indicated by the numbers: 1 – Arthur Grant, 4 January 1934. 2 – crew of the yacht *Finola*, also observed by Torquil MacLeod and others, 7 August 1960. 3 – the 'three young anglers' off Tor Point, 27 August 1930. 4 – the 'Strange Spectacle' seen by John and Aldie Mackay, 28 April 1933. 5 – Mrs Greta Finlay and her son, 18 August 1952. Map drawn by Ray Loadman.

Preface: Sneak preview

On a cold grey morning in November 1975, the postman delivered his usual bulky mailbag to 3 Little Essex Street, London WC2. The entrance looked unremarkable but the institution that lay behind it was famous throughout the world. It also stirred up strong feelings. Over the years, various unhappy people had threatened to kill the man in charge or set fire to themselves on the pavement outside. The malcontents were scientists who been pushed beyond reason by a polite note explaining that, unfortunately, it would not be possible to publish their research paper. They were in plentiful company, because rejection was the fate which awaited over 95 per cent of all the research papers submitted to this address. No 3 Little Essex Street housed the editorial office of *Nature*, the world's most prestigious scientific journal.

One of the manuscripts delivered that morning provoked a unique reaction in this temple of scientific excellence; many of those who saw it laughed out loud. A new animal species had been identified and, as was customary, a formal scientific name was proposed for the creature. Unusually, no specimens had been captured and no physical remains had ever been found. The key evidence consisted of two grainy underwater photographs, one showing a sturdy, diamond-shaped flipper attached to a rough surface, and the other a bulbous body drawn out into a long, graceful neck.

The lead author of the article was not a research zoologist but Sir Peter Scott, renowned as an ornithologist, painter, television personality and President of the World Wildlife Fund. Scott had produced a pen-and-ink drawing of the animal, which looked rather like a reconstruction of the ocean-going plesiosaurs that had supposedly died out with the dinosaurs over 60 million years ago. This one, however, had recently been photographed in the peat-stained waters of a deep lake in north-eastern Scotland. The formal name which Scott had coined for the new species was *Nessiteras rhombopteryx*, meaning 'the marvel from Ness with the diamond-shaped fin', and the paper was entitled 'Naming the Loch Ness Monster'.

The article was greeted with incredulity at every stage of its progress through *Nature*'s notoriously ruthless review process. Yet the issue which appeared on 11 December 1975 devoted three pages to Scott's paper, and the underwater photograph of the diamond-shaped flipper was plastered across its cover. The message was plain for all to see. The Loch Ness Monster, which had been mired in controversy ever since a string of sightings and photographs brought it international prominence during the early 1930s, was not a hoax or a trick of the light but a flesh-and-blood animal. And, incredibly, it appeared to be a 'living fossil' which had somehow sidestepped extinction and could provide a unique insight back to the Age of the Dinosaurs.

It had taken 15 years to convince Scott that the Monster existed; the decisive piece of evidence was the photograph that now filled the cover of the top scientific journal on the planet. For Scott, the creature held even greater significance. It was not just an astonishing relic of prehistory, but a powerful icon for wildlife conservation. Instantly recognisable and clinging to existence in its unlikely habitat, the Monster could spearhead his campaign to protect endangered animals across the world. To that end, Scott had already drafted artwork for the World Wildlife Fund's 'Save One Species' appeal, featuring a drawing of *Nessiteras rhombopteryx*.

At the same time, Scott knew that he was putting his reputation on the line by openly declaring his belief in the Monster. The vast majority of mainstream zoologists – all of whom would have been proud to see their work published in *Nature* – were still unpersuaded that the Monster was real. They were entrenched in their scepticism, just as the scientific establishment had been since the 1930s. To them, the Monster was not the greatest zoological coup of the century, but a preposterous insult to the intelligence of anyone who understood the basic rules of science. As soon as Scott's article appeared, experts on zoology, evolution and plesiosaurs began firing off furious letters to the editor of *Nature*. They did not mince their words. By publishing this nonsense, the editor had endangered the reputation of British science and had dragged this once-noble journal down to the level of the gutter press.

Over Christmas, Peter Scott was also swamped with letters, some critical and some offering support. One that reached him early in the New Year came from a professional zoologist who knew all about the narrow-mindedness and bigotry of the scientific mafia. Dr Denys Tucker was an expert on eels and deep-sea fish, whose career had been unstoppable until the fateful day in March 1959 when he visited Loch Ness and watched a large humped object travelling across the water. He had never encountered anything like it, even while editing the encyclopaedic *Freshwater Fishes of the World*. After reviewing all the evidence, Tucker was forced to conclude that the animal he had seen was a plesiosaur. The most likely candidate was a long-necked elasmosaurus, broadly similar to the creature which Peter Scott later drew for his *Nature* article. Having made up his mind, Tucker then did what every scientist does on discovering something new and exciting: he began telling his peers.

Tucker's superiors were dismayed to hear that he had visited scientific groups in London and Cambridge to give seminars about the plesiosaur in Loch Ness. Tucker held an influential position as Curator of Fishes at the Natural History Museum in London – an institution that had always denied the existence of the Loch Ness Monster. The Museum's Trustees, who included the Archbishop of Canterbury as well as top-class scientists, offered him the chance to recant. When he refused, they sacked him. Tucker, generally acknowledged to be one of the Museum's brightest young scientists, was given three days to clear his desk and was then barred permanently from the premises.

Tucker's letter reached Scott on 3 January 1976. It began by offering congratulations on the *Nature* paper and ended with a warning. 'The jackal pack are sniffing around again,' Tucker wrote, 'and this time they are after you.'

And so 1976 began on an ominous note for Sir Peter Scott. But there was worse to come.

1

A landscape made for a mystery

Type the coordinates 57° 20' 4.66" North and 4° 26' 36.90" West into Google Earth and look down from 3,000 feet. You will find yourself suspended above a bay scooped out of the western side of a body of dark, rather sullen water. To the north-east, the water stretches out into a long thin lake, hemmed in on both sides by the rough grey and brown of mountains. From its end, a river snakes away towards the sprawl of a large town, some 15 miles away. The farthest limit of the town is marked by the slender arc of a bridge which forms the gateway to the open sea.

In the opposite direction, the steep-sided valley runs away to the south-west with the dark ribbon of the lake laid along its floor. Beyond it lies a narrow strip of water, then another longer lake and, 30 miles away, the apex of a fjord which opens into a sea studded with irregular islands. The valley, its lakes and the fjord lie along a line that could have been drawn on a map with a ruler, running diagonally from north-east to south-west.

Now drop down towards the water from your starting point above the bay. As you lose altitude, you notice a road zigzagging up from the south, houses emerging from the camouflage of fields and trees, and the wakes of boats frozen in the instant when the satellite and its camera swept overhead, 120 miles up.

The picture coarsens 100 feet above the surface and freezes as you meet the water. Continuing your descent, you enter a space that rapidly distances itself from the world above. Ten feet down, the light is dull yellow and you might wonder what it feels like to be trapped in amber. Twenty feet below that, you can no longer make out your fingers with your hand held 12 inches from your face. At 45 feet, the water is like obsidian, black and impenetrable.

Point a powerful lamp straight down and you will see that you are hanging above a plain of grey sediment, sloping away from the shore. The gradient steepens about 30 feet further out then plunges into nothing – or, more accurately, 200 feet of water. A few hundred yards beyond that, the lake is nearly 600 feet deep. This is a cold, permanently dark underworld, completely invisible from the surface.

You are in Urquhart Bay, on the western shore of the largest body of fresh water in the British Isles and one of the deepest lakes in Europe. The undistinguished spot where you are suspended was immortalised by something strange that happened here in the small hours of 20 June 1975. An underwater camera, synchronised with an electronic flash that fired every 12 seconds, had been lowered from a boat several hours earlier. The spot had been carefully chosen, but hundreds of thousands of photographs already taken had shown only blank water with the occasional fish.

The same was true for over 2,400 photographs taken that day, but the frame exposed just after 4.32 a.m. had captured an extraordinary object. It was hanging in the water about 20 feet away, starkly lit from below by the flash: a bulbous body, carrying a fin-shaped projection that stuck out towards the camera, and drawn out into a graceful curve like a long neck.[1]

That image was the vindication of years of searching by a team that included world experts in underwater photography from one of America's great research institutions. It was to become a crucial weapon in a 40-year battle against cynics in the scientific establishment who refused to accept that Loch Ness harboured a large unknown animal – and that it looked like an aquatic reptile which should have been extinct for over 65 million years.

Drifter

Like much of the Scottish Highlands, Loch Ness is both beautiful and haunting. It is also an ideal place for nature to hide away something to mystify and tantalise. And, as the stage on which this story is played out, it provides a complex backdrop with special effects that enliven the performance.

Any impression that Loch Ness – or indeed Scotland – has been there forever is misleading. Hints of an exotic past are locked away in Scotland's rocks: sandstones that were once dunes in a desert, and fossilised tree ferns that flourished in a tropical swamp. These are not errors in the geological record, but mementos of a 500-million-year voyage that began near the South Pole. Dancing to the dead-slow tune of continental drift, the land that became Scotland crossed the Equator about 300 million years ago and headed for its current position, two-thirds of the way up the northern hemisphere. Hard though it is to believe when fighting through freezing Caledonian rain, the embryonic Scotland spent 250 million years basking in tropical sunshine. By then, the Highlands had been thrown up as huge folds in the Earth's crust, as high as today's Himalayas. During the Ice Ages, all but a few thousand feet of those virgin peaks were ground away by gla-ciers which were up to 4,000 feet thick.

The Great Glen, the 80-mile-long valley which contains Loch Ness, di-vides the Highlands above from the Grampian Mountains below. Curiously straight and pointing like a compass needle to the north-east, the Great Glen has had a violent past. Rocks on opposite sides of Loch Ness, barely a mile apart, can differ in age by up to 200 million years. A dark granite near Foyers on the eastern shore is unique locally but matches an outcrop at Strontian, 65 miles to the south-west. This is because the Great Glen marks a fault line that cracked open the Earth's crust for hundreds of miles, and the northern block of the Highlands was then pushed those 65 miles along the split. The movement left a belt of weakened rock along the fault line, which was later gouged away by glaciers to form the deep, steep-sided valley that became the Great Glen. When the ice eventually melted and the trillions of tons of its weight fell away, the land actually rose, like a mattress relieved of the weight of a sleeping body. The recesses of the Great Glen then filled with water, creating a chain of three lakes.[2]

Loch Ness, occupying its north-eastern end, is the largest and drains into the North Sea through the River Ness. It came into being about 12,000 years ago. Little Loch Oich, just five miles long, lies near the mid-point of the Glen. Last in line is ten-mile-long Loch Lochy, from which the river of the same name runs into Loch Linnhe, the funnel-shaped fjord that opens into the sea opposite the Isle of Mull.

Vital statistics

Loch Ness is a substantial body of water. At 263,000 million cubic feet, its volume is the greatest of any lake in Britain and twenty times that of Lake Windermere, the pride of England's Lake District. The Loch is long, thin and deep. It is mostly just a mile across, but a motor boat will take a couple of hours to cover the 23 miles from end to end – roughly the distance across the English Channel from where Hitler eyed up the White Cliffs of Dover.

The walls of the Loch plunge down hundreds of feet, in places steeper than a World Cup ski run. This is a bad place to get into difficulties while swimming: just a pool's length out from Horseshoe Crag on the eastern side, there will be nearly 200 feet of water beneath your toes. Two long basins in the centre of the Loch reach 600–700 feet in depth, with a maximum of 754 feet a couple of miles north of Invermoriston. This is deeper than the North Sea between Scotland and Norway and would comfortably swallow up the

British Telecom Tower which dominates the West End of London. Apart from the basins, much of the Loch's floor is 'as flat as a bowling green'. The valley has the classic U-shaped glaciated profile portrayed in geography books but its base has filled with clay and sediment carried in by rivers.[3, 4]

An early attempt (1699) to measure the depth of Loch Ness concluded that it was bottomless – or at least exceeded the 500 fathoms (3,000 feet) contained in a barrel of line. This was reported in a breathless letter to the Royal Society by the Reverend James Fraser of Kirkhill, near Inverness.[5] Fraser also calculated that the mountains around the Loch reached 10,000 feet, further suggesting that mathematics was not his strongest suit.

Much more accurate was the famous Bathymetrical Survey which plumbed the depths of 562 Scottish fresh-water lakes between 1899 and 1906.[3] Massive though this was, the Survey was small fry for its leader, the 58-year-old Sir John Murray FRS, who had cut his teeth on HMS *Challenger*'s 75,000-mile oceanic expedition of 1872–6. *Challenger* had turned up 4,000 new marine species, founded the science of oceanography and was, as Murray put it, 'the greatest advance in the knowledge of our planet since the celebrated discoveries of the 15th and 16th centuries'.[6]

The Bathymetrical Survey began modestly on small lakes with Murray, 23-year-old Frederick P. Pullar and the 'F.P. Pullar Sounding System' for measuring water depth. Pullar's invention involved a bicycle wheel, a revolution-counter and a 14-pound lead weight attached to 1,000 feet of piano wire. It looked makeshift but performed better than the broadly similar instrument patented 25 years earlier by Lord Kelvin. The Survey owed its success to Pullar's invention – and to his death in February 1901 'while attempting to save others from drowning'. He met his end in a pathetically shallow ornamental lake near Stirling, where ice had given way under dozens of skaters. Frederick was the only son of Murray's friend Laurence Pullar, a dry-cleaning magnate. Pullar Senior promptly joined Murray as co-director of the Survey and put down £10,000 to launch it in style.[7] The Survey's 1,800-page final report was dedicated to the memory of his son.

The survey of Loch Ness, which required 1,700 soundings, began in April 1903 and took until the summer of the following year. The results have stood the test of the intervening century, and today's Admiralty charts are remarkably close to those produced by the Survey.

However, all technology has its limitations. In 1970, detailed underwater mapping using a sophisticated sonar scanner picked up features which the Bathymetrical Survey had missed. The floor of Urquhart Bay was not the gently sloping shelf that Sir John Murray believed it to be, but an astonishing underwater landscape of ridges, canyons (over 400 feet deep in places) and caverns – some of which were big enough to conceal an animal up to 30 feet in length.[8]

Through a glass darkly

Even without uncharted depths or underwater caverns, the sheer volume of Loch Ness would make it easy for an elusive creature to hide itself away from those determined to find it. And the Loch has another trick up its sleeve to confound the search: its water.

The rivers which feed Loch Ness drain over 300 square miles of mountain, moorland and forest. Much of the high ground is covered in peat, representing tens of thousands of years' accumulated remains of mosses and other vegetation. Peat is so poor in nutrients that plants such as sundew have to resort to eating insects to make ends meet. Arguably, it supports death better than life. The bodies of those who met their end in the bogs are preserved, beautifully and with an eerie timelessness, right down to the stitching on the sheepskin hat of a Bronze Age man and the tranquil expression on his face when he died. His skin is tanned to a rich teak, having been steeped for 2,500 years in the tea-coloured water which runs out when you squeeze a handful of wet peat in your fingers.

Peat also stains the rivers running into the Loch. When in spate after heavy rain on the uplands, the water is bitter and black, with swirls of off-white froth like the head on a Guinness. Fill a glass from the Loch and hold it up against a sheet of white paper. The water carries a hint of tawny, like a crystal of the palest smoky quartz. Walk out along a jetty in bright sunshine and drop a silver coin into three feet of water; by the time it hits the bottom, it has faded into dull bronze. Drop another coin into six feet of water and it will disappear from view. Fifty feet down, it is pitch-black – and below that there may still be another 600 feet of water.

The hand of man

A glance at the map shows that Loch Ness and the Great Glen have been shaped by human interference as well as the forces of nature. Paradoxically, the largest man-made features are absences: the 'forests' which are no more but whose grand names – Balmacaan, Portclair, Glendoe – still appear on maps. Originally, the whole region was clothed in trees which were relentlessly cleared by Bronze Age farmers, Picts, Romans and industrial revolutionaries. The Forestry Commission tore down the last of the ancient woodlands and then, as if shocked by the nudity of the landscape, replanted hundreds of thousands of acres along the Glen with regiments of conifers destined to become telegraph poles or patio decking.

The most obvious signs of man's presence around Loch Ness are the roads which run down each side of the lake from Inverness to Fort Augustus at its southern end. On the western side, the A82 hugs the shoreline for most of its length, picking up the villages of Drumnadrochit, overlooking Urquhart Bay, and Invermoriston along the way. The road that serves the eastern shore appears more flamboyant. After passing through the village of Dores on the shore of the Loch, it loops inland and climbs over the hills to a height of 1,200 feet before dropping down into Fort Augustus. This road, the B862, is more evocatively known as 'General Wade's Military Road' to commemorate the English commander-in-chief in Scotland during the 1730s. The near-vertical eastern shore of the Loch was impassable, even with gunpowder and forced labour – hence the meandering route inland. As well as creating a network of roads linking the Crown's garrisons, General George Wade sorted out the Scots without being distracted by humanitarian trivia. Although a thug, he was nicer than his comrade-in-arms, William Augustus, Duke of Cumberland, who gave his name to Fort Augustus. Cumberland is remembered as 'The Butcher of Culloden' from the ferocity of his cleaning-up operation in 1746 after the battle of the same name.

Maps show a thin blue line that runs north from the head of Loch Ness, tracking the meanders of the River Ness for four miles to the outskirts of Inverness before swinging east to end in the Beauly Firth. This is the Caledonian Canal, one of the flawed jewels of nineteenth-century British engineering. In 1803, it seemed a clever idea to connect the North Sea with the Atlantic through the Great Glen by linking up Lochs Ness, Oich, Lochy and Linnhe. This would cut several days off the traditional passage around the top of Scotland, which was even more hazardous than usual because Britain was at war with France. Thomas Telford won the contract for the Canal and began work in 1810. When he finished in 1822, five years late and twice over budget, a key plank of the rationale for the Canal had been pulled away; in the meantime, Napoleon had been defeated at Waterloo.

The Canal brought more than ocean-going ships to Loch Ness. The Loch lies over 50 feet above sea level and boats from the Beauly Firth are carried up through a series of locks. Putting these into operation in 1822 raised the water level in the Loch and submerged half of its islands. This was not as cataclysmic as it might sound, as the water rose by only six feet and there were only two small islands to begin with. Both lay close to the shore off Fort Augustus. Dog Island, just 20 yards across, was doomed because it stood just three feet above the waterline. Its larger neighbour, Cherry Island, was nearly circular and originally 170 feet in diameter. After the waters rose, it shrank to 60 feet across, crowded with trees like an over-grown bonsai planting.

An in-depth exploration of Cherry Island was conducted in 1908 by the Reverend Odo Blundell, a feisty Benedictine monk from the Abbey at Fort Augustus.[9] Blundell hired a 'diving dress', survived the incompetence of 'the amateurs at the air-pump' and found that the island was a 'crannog', an artificial lake dwelling from the Bronze Age. Unaware of its intriguing past, Sir John Murray was unimpressed by Cherry Island. In the Bathymetrical Survey's final report, he noted 'the absence of islands in Loch Ness is a striking characteristic, and gives a touch of monotony to the grand and sombre scene'.[10]

Away with the fairies

According to tradition, there have always been monsters in Loch Ness, but this was by no means exceptional. The lake just inland of Gruinard Bay on the north-western coast of Scotland, which the Bathymetrical Survey mapped on 11 August 1902, is called Loch na Beìste, meaning 'the lake of the beast'. As explained in the Survey, a landowner during the 1840s tried to capture the creature by emptying the lake with horse-powered pumps. He failed; the horses did their best for two years, but only managed to lower the water level by six inches.[11]

This was not the only Loch na Beìste. Had Sir John Murray asked around, he would probably have been told that many of the other 561 Scottish fresh-water lochs in his Survey were similarly inhabited. The beast in question did not figure in the Survey's list of lake animals, or in any zoology book. It was a mercurial and sinister entity known as the 'kelpie'.[12]

Kelpies were just one species of the supernatural fauna which inhabited inland lakes across the Celtic diaspora. The name may come from the Gaelic *cailpeach*, meaning 'heifer', although kelpies mostly appeared as horses – hence their alternative name *each uisge,* or 'water horse', which is also the translation of the Irish *capall uisge* and the Welsh *ceffyl dŵr*. Unlike the Loch Ness Monster, kelpies were never an endangered species. They frequented any substantial body of fresh water and some surprisingly small ones. The Loch na Beìste mentioned above (ranked 492nd out of 562 for size in the Bathymetrical Survey) is 700 yards long and only 35 feet deep.

Kelpies often appeared playful and docile but it was foolish to ignore rather obvious warning signs such as dampness and the tendency to materialise out of thin air. Anyone dim enough to jump on a kelpie's back was in for a rough, brief ride. Its hide instantly became adhesive and the water horse would live up to its name, plunging back into the lake and taking its superglued victim with it. The kelpie was part of an international brotherhood of water spirits, many with sociopathic tendencies. Some were quadrupeds, such as the Bulgarian water buffalo and the dragon-like 'knucker' which inhabited 'knuckerholes' in riverbeds around Brighton. Others took human form, including Jenny Greenteeth, a river-dwelling hag in northern England, and the hairy, big-breasted *Rusalka* from Ukraine, who caused swimming to be banned during the week in June when they were at their most predatory.[13]

Apart from spicing up story-telling during long winter nights, the water horse probably served a useful health and safety function by discouraging children from playing near the water – which in the case of Loch Ness could be fatally deep within a few yards of the shore. Here, the deterrent effect of the water horse was bolstered by the grim legend that the Loch never gave up its dead. Alex Campbell, a native of Fort Augustus who is soon to begin playing a complicated role in the story of the Monster, remembered being warned away from the Loch as a child during the 1920s, with the threat that the water horse would get him.[14] Forty years later, former employees of the Caledonian Canal, questioned while on a pensioners' outing, had similar recollections.[15]

These are all fairy tales, but the legend of the water horse has inevitably moulded the saga of the Loch Ness Monster. Tradition predicts that a large lake like Loch Ness would be well stocked with kelpies, which would instinctively be held responsible for sightings of anything unusual. Also, according to Alex Campbell and Constance Whyte, author of *More than a Legend*, the power of superstition was so pervasive that local people refused to talk about peculiar events in the Loch.[16]

A Great Beast

Loch Ness is often a calm and reflective place, and was a natural choice for the Benedictine monks who founded their Abbey at Fort Augustus.

However, some believe that the Loch has a sinister alter ego, and this has attracted devotees of the mystical and paranormal.

An early pilgrim of the occult was Aleister Crowley, who revelled in the *Daily Mail*'s title 'The Wickedest Man in the World' but preferred to call himself 'The Great Beast'. In 1899, Crowley bought Boleskine House, overlooking the eastern shore of Loch Ness to the north of Foyers. He was in his mid-twenties and had done time at Cambridge, where he picked up gonorrhoea and not quite enough Moral Philosophy to be awarded a degree.

Crowley's interests included mountaineering, writing, religion and debauchery. Mountaineering had its ups and downs, including a Mexican volcano which began erupting as he neared the top. His writing ranged from poems such as 'A Place to Bury Strangers' to pornography, some of it unpublishable (British censors judged 'White Stains' to be as distasteful as its title suggests). Crowley saw a healthy overlap between religion and debauchery, and invented a cult which he called 'Thelema' and based around occult ceremonies that made the *Rite of Spring* look tame. Its sole commandment was 'Do what thou wilt' and Crowley obeyed his own orders to the letter. Decked out in the leopard-skin garb of the 'Hermetic Order of the Golden Dawn', he did as he wilt with opium, heroin, women, men and so on.

During his 15 years in residence beside Loch Ness, Crowley went native to the extent of calling himself the 'Laird of Boleskine' and wearing a kilt. However, full integration into the community was never likely, even before he wrote to the local Vigilance Society to complain about the absence of prostitutes in the area. Crowley also put much effort into shocking his neighbours. Perhaps his greatest success was to erect signs warning about the 'Koloo Mavlick'. Even though they had no idea what this was, the locals were so terrified that they would take a lengthy detour to avoid crossing his land.

Many years later, Crowley wondered if the Koloo Mavlick had helped to perpetuate the legend of the Loch Ness Monster. Indeed, he even claimed that he had inadvertently invented the creature.[17] This is, of course, preposterous, because the Monster had a pedigree that went back to over 1,300 years before Crowley arrived at the Loch.

2

Enter the Monster

By the mid-1970s it was estimated that over 10,000 people had reported seeing the Monster. Skimming off the obviously mad and bad left a hard core of roughly 3,000 sightings that seemed plausible. However, even experts could not agree about which of these should be taken seriously, or when the Monster had first reared its head in the Loch.

Father Aloysius Carruth, a monk in the Benedictine Abbey at Fort Augustus and author of *Loch Ness and Its Monster* (1945), had no doubts. The history of the Monster began in AD 565 with a saint and a miracle.[1]

In the beginning

St Columba left his native Ireland in AD 563 to convert the heathen Picts of Scotland to Christianity. For a saint, Columba had occasional humanitarian lapses, such as spiriting up a storm to drown a boatload of ruffians who had made fun of him. He also showed little mercy for various large and terrifying animals which crossed his path as he pressed on towards his primary target: Brude, King of the Picts, in an impregnable fortress just north of Loch Ness.

Columba's most spectacular wildlife encounter came as he and his entourage reached the River Ness. Several Picts were recovering the corpse of a man who had 'been seized while swimming a little earlier by a beast that lived in the water, and bitten very severely'.[2] The crime scene had not been made safe, and the predictable happened when one of Columba's followers, Lugne Mocumin, started swimming across the river to collect a boat. 'The monster, whose hunger had not been satisfied earlier, was lurking in the depths of the river, keen for more prey . . . And with a mighty roar from its gaping mouth, it sped towards the man . . .'

Building on his reputation for improvising his way out of tricky situations, Columba made 'the saving sign of the cross' and commanded the creature to 'go back with all speed'. The monster immediately complied 'as if dragged away by ropes'; Lugne Mocumin was saved and the heathens were converted on the spot. Word of this miracle spread fast and laid the ground for Columba's offensive against King Brude, who succumbed to Christianity shortly afterwards.

This event was recorded faithfully in *Vita Sancti Columbae* (*Life of Columba*)[2] by Adomnán, who succeeded Columba as Abbot of the monastery which he founded on the Isle of Iona. Father Aloysius Carruth believed that this was a genuine sighting and not just an allegory on the banks of the Ness.[1] However, Carruth's general faith in saints might have raised concerns about his objectivity; a rigorous appraisal would need the critical eye of a professional scientist.

One such was Professor Roy Mackal of the Faculty of Science at the University of Chicago. In 1976, Mackal ruthlessly hacked back the 3,000 documented sightings to just 269 'valid' reports of the Monster. Observation No. 1 carried the date of AD 565 (time uncertain) and listed the observers as 'St. Columba, Lugne Mocumin, others.'[3]

Large gaps in the historical record suggest that, after the flurry of excitement with St Columba, the Monster kept a generally low profile for the next 1,300 years.

In around 1520, the 'Monster of Loch Ness' was mentioned as something that nobody had yet managed to kill. The precise year was not recorded, but the swashbuckling Fraser of Glenvachie had just extinguished the last 'fire-spouting' dragon in Scotland.[4]

During the 1730s, workmen blasting out the lochside sections of General Wade's Military Road reportedly saw two 'Leviathans', which they thought might be whales. Describing the event almost 40 years later, a commentator speculated that 'some huge unknown species had made their way in through some subterranean passage and grown too large to return.'[4]

By the turn of the twentieth century, various people had seen animals that differed significantly from the bog-standard water horse. In 1880, a white-faced diver named Duncan MacDonald was pulled hurriedly to the surface while inspecting a ship that had sunk off Fort Augustus. It was several days before MacDonald could describe what had terrified him: 'a very odd-looking beastie' like a huge frog, squatting on rocks near the wreck. He refused to dive in the Loch again.[4]

By contrast, the 'strange creature' which Alexander MacDonald watched 'disporting itself' in the Loch on several occasions in 1888 looked like a giant salamander and evoked not terror but 'a state of subdued excitement'. Also in the 1880s, the 'biggest eel I ever saw' was spotted by Roderick

Matheson from the *Bessie*, a schooner which regularly plied the Caledonian Canal. This enormous 'eel' had other unusual features, with a neck 'like a horse and a mane somewhat similar'.[4]

Although dramatic, these sightings were not common knowledge until 1933, when heightened awareness of the Monster coaxed them into print. Nonetheless, word of the Monster had somehow spread, and far afield. In November 1896, John Keele spotted a piece about the Loch Ness Monster in the *Atlanta Constitution*, a major newspaper on the East Coast of the USA. The story was illustrated with an artist's impression that looked uncannily like the prehistoric reptile which took shape nearly 40 years later, following the frenzy of sightings at the Loch in 1933–34.[4]

In real time

Not surprisingly, the developing story of the Monster was covered in detail in the local press. Since 1817, the *Inverness Courier* has been the main newspaper in the Great Glen, keeping its finger on the pulse of life across much of northern Scotland. The *Courier* has always had a keen eye for prodigies of the animal kingdom. The rare snowy owl which dropped in from Iceland, only to suffer the fate meted out to tens of thousands of grouse each summer; the vast basking shark, too big to measure, shot near Skye; the masses of mysteriously paralysed trout carpeting the surface of a loch in the Orkneys. Even English freaks were reported if they were big enough, such as the seven-foot sturgeon which broke a harpoon while struggling to escape from a flood drain near Doncaster.[5]

The *Courier*'s catalogue of odd events in Loch Ness began on the blazingly hot 1 July 1852. The inhabitants of Lochend were 'suddenly thrown into a state of excitement' by the sight of two 'large bodies' swimming across from Aldourie, a good mile away. A reception party gathered with pitchforks, scythes and an ancient rifle, and variously identified the creatures as a sea serpent, or a pair of whales, seals or deer. As they approached, a 'venerable patriarch' took aim with his rifle, then threw the gun down because only fools shot at kelpies. When the animals finally emerged, they turned out to be just ponies, presumably trying to cool off.[6]

On 8 October 1868, the *Courier* laid down an important landmark with the first contemporary reference to a 'monster' in Loch Ness. A bizarre six-foot carcass washed ashore near Abriachan pier had attracted large crowds. Nobody could identify it – or even agree whether it had come from land or water – until 'an individual better skilled in the science of icthyology' made a spot diagnosis. It was a bottle-nosed whale, minus skin and blubber, presumably dumped by the 'waggish crew' of a sea-going boat to 'surprise the primitive inhabitants of Abriachan'. As the *Courier*'s correspondent remarked, 'The ruse was eminently successful'.[7]

Four years later, a genuine water-monster was reported from Inverness-shire. A 50-foot sea serpent with several humps was seen on two successive days in August 1872, at a range of 200 yards and through binoculars. There were several witnesses, notably a vicar on whose 'intelligent observation and accuracy' the editor of the *Courier* placed 'perfect reliance'. However, this monster was not in Loch Ness, but 60 miles away in the Sound of Sleat, on the county's western seaboard.[8]

Years later, a large and unidentifiable animal was seen moving up the centre of the Loch. The 'fish – or whatever it was' was dark and 'like an upturned angling boat, and quite as big'. Even though the date, place and name of the man who saw it were not disclosed, he was experienced and trustworthy: 'a keeper who dwells on the shore of the Loch . . . of unimpeachable veracity and a first-rate observer'.[9]

The incident passed unnoticed at the time, but surfaced again when another great 'fish – or whatever it was' made its appearance on the Loch, 62 years after the sea serpent had ploughed across the Sound of Sleat.

A night to remember

Entitled 'What was it? A strange experience on Loch Ness', the report in the *Northern Chronicle* of 27 August 1930 was brief but intriguing.[9] Three young men from Inverness, identified only as the sons of well-known businessmen, had been out in a boat fishing for trout off Tor Point on the western side of Dores Bay. It was a fine, calm evening. At about 8.15 p.m., they heard a loud splashing noise and saw 'a great commotion' with spray being thrown up about 600 yards away.

Then the fish – or whatever it was – started coming towards us and when it was about 300 yards away it turned to the right . . . During its rush it caused a wave about two and a half feet high, and we could see a wriggling motion, but that was all, the wash hiding it from view. The wash was, however, enough to cause our boat to rock violently. We have no idea what it was, but we are quite positive it could not have been a salmon.

On making further inquiries, the anonymous reporter unearthed the story of the keeper who had seen the 'fish – or whatever it was' that resembled an upturned angling boat, 'some years ago'. The piece ended with an appeal to the readers. What was it, and had anyone had a similar experience?

Responses appeared in the next issue of the *Chronicle*.[10] 'Not an Angler' suggested a seal, porpoise or giant conger eel. 'Invernessian' quoted an un-named steamer skipper who, some 40 years previously, had seen 'a monster animal or fish' with legs and a furry body, swimming on its back. Most space went to 'Camper', who had seen with his own eyes something like a huge seal with a flat head and a fish in its mouth, 'splashing and snorting' in the River Ness.

Further correspondence was more sceptical. In the *Chronicle*, 'R.A.M.' plumped for a seal, and 'Another Angler' for an otter. In a jocular letter to the *Courier*, 'Piscator' (Fisherman) saw little to take seriously. A fellow angler had seen a 'monster fish' some years earlier – when he and Piscator had enjoyed a drink or two and ended up 'far spent'. Piscator wondered whether the Gaelic Revival had reawakened the water-kelpie to 'take its place once more in the folklore and literature of the Highlands'.[11]

With that, the story faded away and calm returned to the Loch. This lasted until the spring of 1933, when, as Tim Dinsdale later put it, 'The ex-citement begins'. The long silence was about to end, not with a whimper but with one bang after another.

Strange spectacle

The Prime Minister had stated the obvious when he warned that 'this was going to be a very hard winter'. Britain was in the grip of the Depression and there were omens from Europe which pessimists now found frightening. Like everywhere in Scotland, Inverness began 1933 unhappy and insecure. Six per cent of the population were unemployed and crucial revenue had been lost because sportsmen and hunters had stayed down south. There was little to lift the gloom until May, when the Great Jungle Review would hit town with Liberty horses, elephants and zebras – not forgetting Trevor and Romeo, the famous seesawing lions.

According to the *Inverness Courier*, the Germans were guilty of terrible crimes. They had been making whisky and passing it off as the genuine thing; hopefully, they would now be thwarted by having to print the word 'German' before 'whisky' on their labels.[12] Otherwise, Herr Hitler in-sisted that Germany's intentions were entirely peaceful. This comforting

information came in a news item that was as long as the report on the Invergarry Foresters' Annual Dance. Further reassurance was soon provided
by one Herr Zulch, a German Nazi medical student training at Aberdeen,
who received a 'hearty vote of thanks' from the Inverness Rotary Club for
his 'interesting address' about his Fatherland.[13]

Meanwhile, the final sections of the new road down the western side
of Loch Ness were being completed, with much blasting of rock and the
clearing of trees and undergrowth. For the first time ever, much of the Loch
could be seen from a car, and generally from a height that gave far-reaching
views across the water.

By the end of April, the year that had begun so badly was getting worse.
The fishing on Loch Ness was 'lamentably poor'. Further afield, there were
now clear signs that Germany could not be trusted and intended to re-arm
itself.

Then, on 2 May, a piece appeared in the *Courier*, tucked away on page
five. Even though it was shorter than Part II of 'Bird Life in Inverness and
District', it would make history. From 'A Correspondent', it carried the
cryptic title, 'Strange spectacle in Loch Ness. What was it?'[14]

The report began dramatically: 'Loch Ness has for generations been
credited with being the home of a fearsome-looking monster' – and this
was a real beast, not a 'water-kelpie'. It had been seen by a couple driving
along the western shore near Abriachan Pier on the afternoon of Friday,
28 April. Like the Correspondent, both witnesses were anonymous but
their credentials were solid; he was a 'well-known local businessman' and
his wife 'a University graduate'. She spotted it first: a 'tremendous upheaval'
about three-quarters of a mile offshore, shattering the mill-pond-like calm
of the Loch. Her husband stopped the car and they watched the spectacle
unfold:

The creature disported itself, rolling and plunging for fully a minute, its
body resembling that of a whale, and the water cascading and churning
like a simmering cauldron. Soon, however, it disappeared in a boiling
mass of foam.

It was 'many feet long' and its final plunge sent out 'waves that were big
enough to have been caused by a passing steamer'. The couple stayed for another half-hour but the 'beast' did not reappear. Both confessed that there
was something uncanny about what they had seen; this was 'no ordinary
denizen of the depths'.

The Correspondent reminded his readers that, three years earlier, three anglers in Dores Bay had met 'an unknown creature whose bulk, movements and the amount of water displaced at once suggested that it was either a very large seal, a porpoise or indeed the monster itself!' At the time, the anglers had been ridiculed, but now they were vindicated.

The Correspondent tackled the crucial question, 'What was it?' Porpoises were common out at sea but were excluded from the Loch by the weirs on the River Ness and the locks on the Caledonian Canal. Theoretically, seals could swim up either waterway but had never been seen in Loch Ness. This left a large, unknown creature, which previous generations had explained away as the kelpie.

Initially, perhaps because readers of the *Courier* were preoccupied by the grimmer realities of life, the story failed to ignite interest. Indeed, cold water was promptly thrown on the theory by Captain John Macdonald, skipper of the famous MacBrayne steamer *Gondolier*, who had clocked up 20,000 sorties during his 50-year career on Loch Ness. He wrote a long letter which the *Courier* printed in full on 12 May with the comment that 'no one is more fitted than Captain Macdonald to express a sound opinion on the subject'.

The captain knew the Loch intimately 'in all its varying moods, winter and summer – calm, angry, tempest, and perilous'. He had never observed anything like the 'monster' reported by the anonymous Correspondent, but on hundreds of occasions had seen a 'tremendous upheaval' in the Loch. From afar, these disturbances could look 'strange and perhaps fearsome'; close up, they were just shoals of salmon playing at the surface.

Captain Macdonald was surprised by the Correspondent's claim that the Loch had long harboured a water monster, because this was news to him. He had no desire to 'question the credibility of those who witnessed the spectacle' or to 'kill a good yarn that adds to the romance of a beautiful Loch', but he had to set the record straight.[15]

Scepticism still hung in the air ten days later, when 'An Old Angler' who had known the Loch for 60 years, speculated in the *Courier* that the size of the monster might be 'regulated by the strength of a certain beverage'. At this point, the story might have slipped into oblivion, like the tale of the three men in a boat out in Dores Bay. However, that issue of the *Courier* carried some serious suggestions about what the 'Elusive Monster' might be: a big otter, a huge eel or even shock waves from an underwater earthquake.[16] The first viable seeds of the mystery had been sown.

Summertime

Excitement over the Monster steadily spread around the Loch. The first attempt to solve the mystery by catching a specimen was recorded by the *Courier* on 30 May. The brainchild of Mr A. Gray, a bus driver from Foyers, this was float-fishing on a grand scale. A sealed barrel at the end of 60 yards of strong wire was armed with several hooks 'of appropriate strength' baited with chunks of dogfish and skate. The barrel drifted up and down the Loch for some hours, but remained unmolested. Undeterred, Mr Gray told the *Courier* that he would try again.[17]

Also undeterred was the Monster, which put in a brief but spectacular appearance for the passengers on a bus on 30 May. No details about it or the passengers were reported, but the *Courier*'s anonymous Correspondent was more expansive about the sighting the following day. A group of workmen blasting on a hillside near Abriachan watched a massive creature with an enormous head tracking up the middle of the Loch. Surprisingly, it was just yards behind a drifter (trawler), which it followed 'for some distance'.[18]

Temperatures soared during June and eventually reached the combustion point of the car belonging to the Reverend Dr J. Stuart Holden. The Reverend had left it at the roadside to go fishing; his chauffeur was unable to put the fire out because he had gone fishing, too. Neither would have had much luck, because the angling was dire that summer. In a normal year, this would have meant that visitors wanting to spend their way out of the Depression would also be in short supply.

Against expectation, though, visitors came in droves. Many of them piled on board Messrs MacBrayne's steamship, *Gondolier*, to explore the Loch. Throughout the summer, the cruises proved 'very popular, both with visitors and residents'; the four-hour evening cruise from Muirtown Wharf to Urquhart Castle and back, price 2/6d, was full on most days. Towards the end of the season, Messrs MacBrayne had to lay on another boat.[19]

They had all come to look for one thing, and many followed the *Courier*'s advice to the amateur photographer – 'When cruising, take a camera' – just in case.

Although the Monster kept its head down through June and July, it raised spirits through the summer and was a welcome antidote to the woes of the world. The region was still struggling to repair the damage inflicted by the Depression. In Europe, everything was beginning to unravel. As well as the opening night of the Salzburg Festival (Richard Strauss conducting *Fidelio*), the radio now brought rumours of German Jews being treated with great cruelty. Within a few weeks, Hitler had mutated from the ludicrous 'Charlie Chaplin of politics' into a ruthless dictator with terrifying ambitions.

On 21 July, the *Courier* announced the publication of H. G. Wells's *The Shape of Things to Come*. By a chilling coincidence, the following issue reported that Sir Murdoch Macdonald, the MP for Inverness-shire, was concerned that British cities might be bombed from the air, and the one after that ran a feature on England's 'City of the Three Spires' – Coventry.

Terra firma

The Loch's fortunes improved smartly at the start of August. At last, there was 'extraordinary' angling on the River Ness, and although the boatloads of those chasing the 'Elusive Monster' did not know it at the time, some astonishing information was making its way back to Inverness on the mail train from London.

The influx of English tourists had included Mr and Mrs George Spicer, from Golders Green in north London, who had driven up to John O'Groats in their Austin convertible. Mr Spicer, described as quiet and retiring, was Director of Todhouse Reynard & Co., an upper-crust gentlemen's tailors in the West End. Heading south for home on the hot afternoon of 22 July, the Spicers left Inverness and took General Wade's Military Road towards Fort Augustus. Roughly halfway along the 11-mile stretch between Dores and Foyers, they saw something cross the road about 50 yards ahead. It was there for only a few seconds but left them both profoundly disturbed. Mr Spicer reported his 'somewhat exciting experience' in a letter which the *Inverness Courier* published on 4 August.[20]

It was the nearest approach to a dragon or prehistoric monster that I have ever seen in my life . . . [it] appeared to be carrying a small lamb or animal of some kind . . . It seemed to have a long neck, which moved up and down in the manner of a scenic railway, and the body was fairly big with a high back . . . Length from 6 to 8 feet and very ugly.

Other details – feet, tail, head – were lost in the heat of the moment. By the time the Spicers reached the spot, the creature had disappeared, possibly down a tree-clad slope and into the Loch. The usually placid Mr Spicer was overcome with revulsion. 'Whatever it is, and it may be a land and water animal, I think it should be destroyed.'

Spicer apologised for his imperfect description – 'it moved so swiftly and the whole thing was so sudden' – but insisted, 'there is no doubt that it exists'. He appealed for information to satisfy his curiosity about the nature of the animal, and enclosed a stamped addressed envelope for that purpose.

The Spicer sighting soon became famous, as the first suggestion that the Monster had a long neck and could venture out on land. However, the *Courier* seemed reluctant to sensationalise Spicer's story. His letter was printed below the verdict of 'one who knows the habits of otters'. The anonymous expert explained that otters were low-slung, moved quickly with an undulating motion and were perfectly at home on land. He was certain that the Spicers had seen a large otter, possibly carrying a cub in its mouth.

As if in defiance, the Monster made three separate appearances in as many days, having 'moved its beat' (as the *Northern Chronicle* noted) to the southern end of the Loch. All three sightings were reported in the *Courier* by its unnamed Fort Augustus Correspondent on 8 August. The piece was subtitled, 'eye-witnesses' vivid descriptions' – and vivid they were.[21]

The previous Thursday evening, the now-familiar upturned boat was spotted half a mile off Glendoe by Nellie Smith, a maid at the Abbey in Fort Augustus. It was moving in large circles, with 'huge legs which could be seen working quite distinctly'. Two days later, there were two separate sightings half an hour apart in Inchnacardoch Bay. Prudence Keyes, also a maid at the Abbey, saw it first, careering around in great circles and apparently splashing water over its back with legs or flippers. She took the news to Commander R. A. Meiklem, Royal Navy (Retired), a respected local figure whose house overlooked the Bay and who was thoroughly familiar with the Loch. Below them, they could see an animal like a black carthorse lying in the shallows, showing a humped back and with its head apparently underwater. The Commander and his wife watched it for several minutes through binoculars before it disappeared. The animal's back was sharply ridged and its skin grey-black and warty, like a toad's.

The Monster was back again in the next issue of the *Courier* on 11 August, in three separate items.[22] First was Mr Spicer's blunt response to their otter expert: the creature he had seen was far too big to be an otter, and it should definitely be killed. Next was another sighting, by a couple driving down the western side of the Loch, who assumed they had seen a big black rock until they described its appearance to a shopkeeper in Fort Augustus.

The third, the longest piece yet about the Monster, carried the curious title, 'True story of the Monster's life, as told to a Mere Woman'. The anonymous female correspondent had written a witty spoof, but spiked with gentle irony. Because the descriptions varied so much, her Monster answered to the name, 'Otterserpentdragonplesodaurus'. Born in the Ice Age and intensely loyal to Bonnie Prince Charlie, he was kitted out in a backless bathing suit. He had his eye on a career in the talkies in America and was grateful to the 'gentleman who writes to [the *Courier*] from Fort Augustus' for all the publicity.[22]

The piece undoubtedly raised some cheap laughs but, like the doubting Captain Macdonald, the Mere Woman was quickly forgotten. Much more exciting was a clutch of letters about the possible identity of the Monster and another rash of sightings in early September, all reported by the *Courier*'s Fort August correspondent. Several convincing witnesses had seen a creature up to 30 feet in length, looking like an upturned boat, and on one occasion charging across the water 'in a terrific hurry', perhaps reaching 30 mph.[23]

On 20 September, the *Courier* ran a piece entitled 'The Loch Ness Mystery'. This was no longer buried among the minor news items, but was the leading editorial. The existence of the Monster was now fact, not speculation: 'that there exists in Loch Ness some abnormally large creature, there seems now no room for doubt'. Many 'trustworthy witnesses' had seen it, including several 'reliable people' who had watched it at several points on the Loch over the previous few days. Reports of an animal 'like a huge caterpillar', seen simultaneously by several people, had been given first-hand to 'a *Courier* reporter'.[24]

Most striking was the sighting from the Halfway House tea room at Alltsigh, which had turned into a monster hunter's dream. A party of visitors arrived to hear Miss Janet Fraser, the proprietress, calling down, 'Come upstairs, we're watching the Monster'. The afternoon was bright and clear, and the Loch flat calm. Miss Fraser's balcony gave a 'splendid view' of a 30-foot creature resembling a caterpillar, travelling with a worm-like motion along the surface not far behind a trawler that was moving up the centre of the Loch.[25]

The *Courier*'s editorial undoubtedly helped to extend that summer's tourist season – already a record – well into the autumn. Many excursionists headed to Inverness for the holiday in early October. Twice as many buses as usual had to be laid on, but, even with these, the roads around the Loch were thronged with cars and motorbikes. This usually staid community – where, thanks to the energy of the Lord's Day Observance Society, the Sabbath was a fun-free zone – became an all-week hotbed of excitement.

It was also a refuge from the graver news brought by the *Courier*. Germany was ignoring all demands not to re-arm, while Britain's own military activities had claimed an early victim down on the south coast of England: a girl, whose striped beach ball bore a tragic resemblance to the target used for live firing practice by an RAF fighter. Verdict: death by misadventure.

The nature of the beast

Now that the Monster's existence was accepted, the burning question was its identity. The *Courier* conceded that the various eyewitnesses' reports did not obviously add up, but reminded its readers of the old joke about the Irishman who, on seeing a giraffe for the first time, denied that it could possibly exist.

New hypotheses about the Monster's identity were advanced. It might be a giant conger eel, somehow able to thrive in fresh water. Or a bearded seal, like the one shot in Beauly Firth in 1911, whose mounted head now gazed forlornly at the customers of Messrs Watson & Co., gentlemen's outfitters, of Inverness. Or a giant salamander, a species discovered in the Far East a century earlier. This notion came from a local pillar of society, Colonel W. H. Lane, who had met one of the brutes while he was fishing in the Burmese mountains and shot it with his rifle. More imaginatively, Mr A. Russell-Smith from deepest Sussex believed that the Monster was descended from a prehistoric reptile which had evaded extinction and 'over the years' adapted to life in the Loch. A possible ancestor was the massive, lizard-like *Enaliosaurus*.[26]

Mr Martin Walker also wrote in from Cumberland. He had no specific theory about the Monster, but helped to reinforce the growing aura of mystery. Loch Ness, he wrote, was almost bottomless and never gave up the bodies of its dead. He went on, 'I was once told by a most learned man that Scottish Lochs are full of weird and wonderful fish and animals, which live undisturbed in their depths for years'.[26]

Now, at last, those weird and wonderful creatures were revealing themselves.

Scoop

Elsewhere, other strange life forms were stirring. Journalists were being mobilised, not to probe Germany's sudden withdrawal from the League of Nations, but to find out what on earth was going on in Loch Ness.

The *Scotsman* pulled their 'Special Correspondent' Philip Stalker off an assignment on a Royal Navy minesweeper and dispatched him to Inverness. Stalker was gripped by what he unearthed. He broadcast a piece on Scottish Regional Radio that praised the reliability and probity of the eyewitnesses, and wrote two long articles for the *Scotsman*.[27]

'A puzzled Highland community' set the scene. Stalker had spent five hours interviewing eyewitnesses and was now convinced that a creature 'of monstrous proportions' inhabited the Loch. Those who had seen it were drawn from 'all classes of the community . . . people in country houses, tenants round the Loch's shores, inhabitants of Fort Augustus and Loch-side villages, and holiday-makers'. Many in the well-to-do sector refused to admit publicly to what they had witnessed because they feared ridicule. It was absurd to suggest that 'liars and leg-pullers have banded themselves together round Loch Ness'.

What was it? A bus driver who had watched it with three of his passengers at the end of August or beginning of September thought it was a colossal eel. The weight of opinion, however, fell on a creature 'resembling in form the prehistoric plesiosaurus . . . however incredible it may sound'.

Stalker developed that theme in his second piece, 'The plesiosaurus theory'. The star witness was a man who, up to then, had refused to believe that the Loch hid anything unusual. Then, 'one afternoon a short time ago', he saw an animal 'fully thirty feet in length' in the water. He watched it raise its long neck, ending in a small head, then pause while moving its head rapidly from side to side as though 'listening to the sound of two drifters [trawlers] coming from the Caledonian Canal'. Finally, it took fright and dived.

This account was more convincing than Mr Spicer's fleeting glimpse of the 'prehistoric' creature from his car. And when the witness was shown a sketch of a reconstructed plesiosaurus, he said that 'it was very like the animal he had seen'.

Stalker's articles, published on 16 and 17 October 1933,[27] caused an instant sensation. Suddenly, the Monster had enough backbone to be newsworthy – and it was too good a story for Fleet Street to leave to provincial Scottish rags. The big London dailies ran with the Monster, and the *Daily Mail* sent one of its reporters up to investigate. The *Courier* was unimpressed and poked fun at the southerners and especially the *Mail*'s pathetic attempt to trap the creature with a leg of mutton.[28]

The Monster continued to reveal itself to the fortunate few and to deny that privilege to many more. Sightings now came every couple of days, mostly of the 'upturned-boat' and sometimes showing two humps or more. Urquhart Bay and the waters off Fort Augustus were establishing themselves as particular hot spots. Those disappointed included the first organised Monster-seeking expedition, a group of ramblers from Glasgow. They retreated home, damp and unrewarded, after a couple of days of trudging around the shoreline in the appalling weather that heralded the start of November.

Meanwhile, the Monster was arousing conflicting passions. R. M. Green wanted to entice the Monster to the surface with six bullocks impaled on appropriately sized hooks, when a waiting warship could then solve the mystery conclusively with its guns and/or torpedoes. Mr Green hoped that his proposal would be 'well received in the right quarters'. More pacifist in tone was Sir Murdoch Macdonald, who told the Annual Dinner of the Queen's Own Cameron Highlanders that everyone in London was talking about his monstrous constituent.[29]

On 31 October, 'A Correspondent' wrote in the *Courier* that the Monsters 'have proved their existence beyond a shadow of a doubt'. For sceptics, however, verbal reports were not hard enough proof, no matter how honest and reliable the witnesses appeared to be. What was urgently needed was a live specimen, or a body or a significant part thereof – or failing those, a decent photograph.[30]

Within two weeks, photographic evidence was obtained, but, for various reasons, the dramatic image was not revealed for another three weeks. And the *Courier*, the Monster's greatest promoter from the start, missed the scoop.

Image capture

The Monster was apparently not in the mind of Hugh Gray on 12 November when he picked up his camera and set off for his usual Sunday morning walk along the lakeside by the mouth of the River Foyers. Now in his early thirties, Gray had grown up in Foyers and had spent half his life working at the British Aluminium Company Plant which overshadowed the village.

From a promontory about 30 feet above the flat-calm surface of the Loch, Gray noticed a 'considerable disturbance' about 100 yards offshore. The back and tail of a snake-like animal that he estimated to be about 40 feet long were thrashing about on the surface. He watched it for some minutes and took five photographs. On returning home, Gray put the camera in a drawer, where it remained until 1 December. He later explained that he assumed he had failed to photograph the animal. Indeed, when the film was eventually developed by a pharmacist in Inverness, four of the five exposures were blank.[31]

However, Gray's fifth photograph made headlines that were picked up far beyond the Great Glen. It showed a sinuous dark object lying on the surface of the water; the appearance of spray being thrown up suggested that it was moving at speed (Plate 1). This photograph was what everyone – believers and cynics alike – had been waiting for. Gray quickly realised the value of what he had left in the drawer for three weeks. Suddenly, the *Inverness Courier* was too small and too parochial. Instead, Gray sold the photograph to the *Daily Record* in Glasgow and the *Daily Sketch* in London. In Glasgow, a hastily convened group of photographic experts pored over the negative and agreed that it had not been tampered with, and therefore concluded that the picture showed precisely what Gray had seen.[31]

The *Courier* caught up with the breakthrough on 8 December, but only through a letter from a reader in Norfolk who expressed great excitement at the 'keen controversy' which was raging over the Monster's identity. The *Courier* and its Correspondent had missed the boat, and arguably the most important piece of evidence to date.

Ticked off

Two days after Hugh Gray's sensational but undeveloped negative began its sojourn in the drawer, a man who seemed ideally qualified to crack the mystery of the Monster arrived in Inverness on the London train. This was Commander Rupert Gould, formerly of the Royal Navy and hailed by the *Courier* as an international expert in unknown sea monsters.

Commander Gould (Plate 2) cut an imposing figure – 6' 4", heavily built and widely admired as an intellectual whose writings could excite the man on the Clapham omnibus. It all started in 1916, when Gould found some priceless jewels of British history lying under 120 years of dust in the National Maritime Museum at Greenwich. These were the original chronometers which John 'Longitude' Harrison had constructed in the mid-eighteenth century to solve the greatest scientific problem of the age, how to pinpoint the east–west position of a ship anywhere on the oceans. Unpaid, Gould restored all the instruments to working order – a Herculean task which took 12 years. Gould made his reputation with this magnificent achievement and his book, *Marine Chronometers* (1923), which remained the definitive work for over 75 years. Unfortunately, his labours of love left his wife cold; she divorced him, messily, in 1927.[32]

Gould was an oddball, and most at home either alone or in the company of like-minded souls, such as the all-male bibliophile club known as 'Ye Sette of Odd Volumes'. He indulged his love of the 'odd and peculiar' in *Oddities, a Book of Unexplained Facts* (1928), which unpicked mysteries such as the Devil's footprints and the Indian rope trick for the general reader. The book made money and led to *Enigmas, another Book of Unexplained Facts* (1929) and then to his greatest obsession after Harrison's chronometers: *The Case for the Sea-Serpent* (1930). His erudite but lucid dissection of the evidence persuaded many that sea serpents existed and prowled the oceans at large.

This interest made it inevitable that he would end up at Loch Ness looking for the Monster that some had described as 'serpentine'. His expedition was financed by a wealthy friend, Alexander Keiller, who was rich because Keiller's Dundee Marmalade had long graced millions of breakfast tables across the British Empire. Keiller was fascinated by strange things, having funded and supervised the excavation of the stone circles at Avebury in Wiltshire. He was also fond of sex, sometimes on a near-industrial scale and overlapping with the preferences of his acquaintance, Aleister Crowley.

Keiller's generosity, however, stopped short of car hire, and Gould had to make do with a tiny second-hand motorcycle which he christened Cynthia. For just 2/6d in petrol, he circumnavigated the Loch twice, interviewing more than 50 witnesses and accumulating evidence that 'the same or a very similar creature [had] been observed in the Loch as long ago as 1827'. He narrowly missed seeing it himself, when the Monster appeared in Urquhart Bay barely an hour after Cynthia had whisked him off to Inverness.[33]

During his inquiries, Gould managed to trace witnesses who until then had been veiled in anonymity – the 'local businessman' and his 'University graduate' wife who started the whole thing by seeing the 'Strange Spectacle'; the Correspondent from Fort Augustus who had reported it; and even the three young Inverness anglers whose boat had been rocked by a massive wake one fine evening in the summer of 1930. For now, though, those details remained under wraps.

Gould returned to London in early December, just in time to see Gray's photograph of the serpent-like creature splashed across the papers. He sent a press release about his research to the Press Association, and received the brusque reply that they did not want anything to do with the Monster because they did not believe in it. Luckily, *The Times* took his piece, in which he concluded:

The evidence indicates clearly that the monster is a large creature of anomalous type, agreeing closely and in detail with the majority of the observations collected in my book, *The Case for the Sea-Serpent* (1930).[34]

And those who wanted to know more would have to wait until the publication of his new book, due out in spring 1934.

Big game

Perhaps it was the publication of Hugh Gray's photograph on 6 December that galvanised the *Daily Mail* into organising the definitive expedition to find the Monster. This was a bold initiative for a newspaper in late 1933. There was plenty of other news to go after, but the public would welcome a break from the grimness of world events, especially now that Hitler was occupying the Rhineland. The jaunt would produce an entertaining travelogue, and there was always a chance that it could pull the scientific discovery of the century out of the Loch's notoriously opaque waters.

The *Mail* duly assembled its dream team. Veteran *Mail* reporter W. F. Memory would file daily reports about the expedition's progress, while the authoritative lens of experienced wildlife photographer Gustave Pauli would be ready to capture the conclusive photograph of the Monster. That left only the man who would lead the enterprise. The *Daily Mail's* choice was interesting, and would haunt the search for the Monster for decades to come.

The path that brought Marmaduke Wetherell (Plate 3) to Loch Ness was even less conventional than Gould's. After no obvious early career, he abandoned his native Leeds for Africa in 1909, at the age of 25. Following several years as a farmer in Zambia, he gravitated into the world of cinema, where he found that his face fitted. He went on to act in 25 films, beginning with a minor (and silent) part in *The Splendid Waster* (1916) and graduating to leading roles in big features such as *Livingstone* (1925), in which he starred as the eponymous doctor and wrote the script.

Along the way, Wetherell built up a reputation as the archetypal English adventurer: a bold explorer and an accomplished big-game hunter. Somehow, he also managed to collect two significant badges of recognition from the scientific establishment – Fellowship of both the Royal Geographical Society and the Zoological Society.[35]

This unique bundle of skills made Wetherell the ideal man to solve the mystery of Loch Ness – at least in his eyes. The *Daily Mail* agreed and sent him up to Inverness with Memory and Pauli on 18 December, apparently with the aim of nailing the Monster before Christmas.

Other newspapers followed Wetherell's exploits from a safe distance, wary that the whole monstrous affair could implode into farce. Hoaxers had recently taken several British papers for implausible rides and left them looking ridiculous. A couple of years earlier, the press had chased after a mongoose that supposedly lived behind panelling in a farmhouse on the Isle of Man. This was no run-of-the-mill mongoose. It spoke English, Manx and a smattering of Russian, claimed to be called Gef, and for a mongoose was surprisingly au fait with world affairs. Some papers had given the story credibility for an embarrassingly long time.[36] Now, even though such figures of authority as Rupert Gould were injecting some respectability into the Monster, there remained the possibility that the *Mail* could still be hoisted by a mongoose-style petard.

The odds were weighted heavily against Wetherell, as they had been against the tens of thousands who had made the pilgrimage to Loch Ness and returned home disappointed. Fortunately, according to Wetherell, the balance was redressed by his immense skill in tracking and hunting down big animals in the African bush.

The first two days of the expedition revealed nothing of note. On the third morning, 20 December, Wetherell's self-confidence was spectacularly rewarded when he discovered a short train of massive footprints on the wild stretch of the eastern shore between Foyers and Fort Augustus. The prints were eight inches across, showed four short-clawed toes and reminded Wetherell vaguely of the spoor of a hippopotamus. Nearby was another series of prints, also large but differing in shape. Finding them was a massive stroke of luck, as most of the shoreline of the Loch consists of bare rock or pebbles. The Monster had considerately crossed a narrow stretch of soft clay behind the shingle beach. The *Mail's* own F. W. Memory, who had endured many cold hours in a small boat with Wetherell and his ego, expressed surprise at their good fortune. Wetherell explained that this was not luck, but the triumph of his years of experience.

The *Mail* was predictably ecstatic.[37] Block headlines on the front page – MONSTER OF LOCH NESS IS NOT A LEGEND BUT A FACT – laid the ground for Wetherell's sensational account of finding the spoor 'only a few hours old . . . and where I expected to find it', and his extrapolation of the footprints into 'a very powerful soft-footed animal about 20 feet long'. For definitive identification, Wetherell had to defer to a higher authority. Plaster casts of the footprints were made and shipped down to the Natural

History Museum in Kensington, in London, where they would be examined by some of the world's greatest zoological experts.

The initial statement by Dr W. T. Calman, Keeper of Zoology at the Museum, laid the ground for something truly extraordinary:

The cast does not represent the track of any known fossil reptile such as the dinosaur, neither is it the impression of any known aquatic mammal in the British Isles or the rest of the world.[38]

However, the necessarily painstaking investigation was deferred, as Christmas was now looming. The world would have to wait until the New Year to discover what really lay behind the mystery that had turned 1933 into a thrilling cliffhanger of a year.

Happy Hogmanay

Christmas Day 1933 in Inverness was 'extraordinarily mild', with no hint of frost or snow. Boxing Day was similarly benign, and provided splendid views across Loch Ness for all those stuck in the hundreds of motor cars which formed almost unbroken queues along the roads between Inverness and Fort Augustus. The hotels were full and the visitors were spending liberally. Thanks to the Monster, the place was booming and, temporarily at least, the Depression was someone else's problem.

The New Year was seen in quietly across northern Scotland, but not in Inverness. The floodlit Town Hall and the newly installed illuminations on the Greig Street Bridge gave the town 'quite a gay appearance' and attracted a 'surprisingly large gathering' at the Exchange in good time for midnight, when some of the 'more enthusiastic spirits' fired rockets across the river.

As the *Courier* put it in its last editorial of 1933, it had been 'A Monster Year'.[39]

Naturally enough, the BBC's 1933 Year Round-Up featured the Christmas Day Speech by King George V, the second King's Speech to be broadcast. But according to presenter Lawrence Wager, 'there was only one topic of conversation this year . . . everyone seemed to know someone who might actually have seen it'.[40]

The microphone was then passed to a softly spoken man with a lilting Highland accent, identified only as the Fort Augustus Correspondent for an Inverness newspaper. Towards the end of May, he had been told about a large creature in Loch Ness while visiting a couple whom he knew: Mr and Mrs Mackay, the proprietors of the Drumnadrochit Hotel. The Correspondent went on:

I knew it was a good story, something quite out of the ordinary. On getting home, I puzzled my brains only on one point: in what word could I refer to the creature? At last, 'monster' suggested itself – and that is how I introduced the Loch Ness Monster to the newspaper world.

This answered two of the many questions surrounding the Monster that nobody had heard of 12, or even eight months earlier: the identities of the unnamed witnesses of the 'Strange Spectacle', and how the Monster had acquired its name.

But who was the softly spoken Correspondent from Fort Augustus?

3

Revelations

By early 1934, the outbreak of Monster fever was claiming victims across a surprisingly wide area. Down in England, the Monster had been debated in Parliament; no particularly useful conclusion had been reached. Across the Channel, the French press – perhaps swayed by six centuries of the *Vieille Alliance* with Scotland – identified the Monster as the only good news to emerge from the misery that had been 1933. Less charitably, a government spokesman in Austria suggested that all this Monster business was just a cynical ploy to entice tourists away from the much more seductive charms of Vienna and the Tyrol.[1]

From the Americas came reminders that they, too, could claim large, unknown aquatic animals as their own. Could the Monster of Loch Ness be related to those in lakes in Patagonia and Chile, or to Ogopogo, the serpent-like creature that inhabited Okanagan Lake in British Columbia? On the far side of the Pacific, the Japanese were well and truly hooked. Just as the *Glasgow Herald* had reassigned Philip Stalker to the Monster story, the great Tokyo daily *Mainichi* diverted its ace reporter Ken'ichi Ishikawa away from the 'rosy seventh Heaven' of Cambridge to the 'abyss of sordid vulgarity' south-west of Inverness. Ishikawa pulled no punches in describing his fruitless trek around the Loch – but still managed to raise, rather than lower, the temperature of Monster believers in the Land of the Rising Sun.[2]

Back in the Great Glen, the gigantic prints of Wetherell's 'very powerful, soft-footed' animal were reawakening memories that had lain dormant (or had been actively suppressed) for months, years and as far back as 1914. Emboldened, several people now owned up to having seen large animals, generally 10–20 feet long and variously described as 'strange', 'most peculiar' or 'horrifying', lying on a beach, in the swampy river mouth in Urquhart Bay, or even in bracken beside the road to Fort Augustus.[3]

The excitement gathered such momentum that it was only temporarily set back by the definitive verdict of the Natural History Museum on the plaster casts of Wetherell's spoor. On 3 January 1934, Dr W. T. Calman, the Keeper of Zoology, announced: 'We are unable to find any significant difference between these impressions and the foot of a hippopotamus.' There was worse to come. All the prints had been made by the same foot (right hind), which had a shrunken appearance, suggesting that the owner of the limb was long dead.[4] It was not clear who had been hoaxing whom, but the footprints' provenance was tentatively linked to a hunting trophy that had been converted into an umbrella stand, the property of a local resident.[5]

The *Daily Mail* did its best to limit the reputational damage to its expedition and its glorious leader, 'Mr M.A. Wetherell, the big-game hunter'. Under the headline, 'MONSTER MYSTERY DEEPENS', Dr Calman's report was printed alongside the provocative question, 'How did it get there?' and the evidence found by a shooting-party suggesting that the Monster 'may exist on a diet of deer'.[4] Furthermore, five arts students from Edinburgh University ('who have given up their holidays to hunt the Loch Ness Monster') had found another footprint. This one was nine inches across, had four clawed toes and looked like that of a giant amphibian. 'In all probability,' the *Mail* promised, 'plaster casts will be taken.'[6] However, tantalised readers would have been disappointed, as the 'MORE FOOT-PRINTS' story died at that point.

In fact, a plaster cast was taken and duly inspected by Dr Calman, who later reported the outcome in a memo to the Museum's Trustees:

A second cast of another footprint was examined some days later. Although more obscure it strongly suggested the footprint of a Rhinoceros. The 'Daily Mail' did not publish the report which was sent to them of the second footprint.[7]

Marmaduke Wetherell, Fellow of the Zoological Society, African adventurer and big-game tracker extraordinaire, survived the hoax, but only just. His credibility bobbed up briefly two days later when his photograph appeared again in the *Daily Mail*, this time beside a serious, bespectacled young man who was intently examining a tattered strip of animal pelt held in his hands (Plate 4)[8]. Arthur Grant, a 21-year-old veterinary student at Edinburgh University, had returned for the Christmas holidays to the family home outside Drumnadrochit. Thanks to his training, Grant had inside knowledge of a wide range of animals. He also had an astonishing story to tell.

At around 1 a.m. on 4 January, he had left a party in Inverness – 'perfectly sober', according to witnesses – and headed south along the main Fort Augustus road on his motorbike. The night was clear, with moonlight bright enough to read newsprint by. Just before the turning to Abriachan, he spotted a 'huge mass' at the roadside. This quickly materialised into a large animal that cleared the road in 'two great bounds' and charged down into the lake. Grant had a 'splendid view'; in fact, he nearly rode into it. He estimated it to be about 20 feet long, with a 'very hefty' body, powerful flippers front and back, a small eel-like head on a long neck, and a five-foot tail with a rounded tip (Figure 1). The animal had disappeared by the time Grant clambered down the slope below the road to the water's edge.

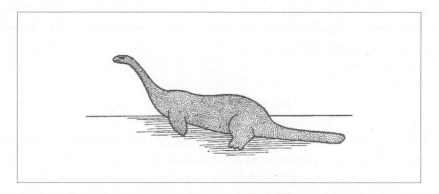

Figure 1 Drawing of the Monster by Arthur Grant, prepared for Rupert Gould. Grant maintained that this was 'more accurate' than his original sketch, reproduced in the *Daily Mail*. From Gould RT *The Loch Ness Monster and Others* (1934), p. 89.

Grant and his brother returned to the scene at first light. No tracks were visible but the account was vivid enough for the brother to send word to Wetherell, who arrived with Pauli and Memory later that morning. They were not the only ones to have picked up the news. By the time they met, dramatic new evidence had appeared on the shore in the form of bizarre three-toed footprints, a pile of bones and a partially skinned sheep.

This photo opportunity turned out to be Wetherell's last stand. Stung by the footprint hoax, the *Daily Mail* decided that the expedition was unlikely to deliver the Monster. The big-talking star of *Livingstone* filed a valedictory report which identified the Monster as a large seal (he claimed to have seen one in the Loch, but everyone else was looking the other way).[9] Then, with an uncharacteristic lack of pomp, Marmaduke Wetherell took the train to London and announced that he was returning to Johannesburg. The *Courier* celebrated his departure with a gleeful editorial entitled 'Mud', which mocked the gullibility of the *Daily Mail* and London's gutter press. The piece eclipsed coverage of the potentially catastrophic news that the

Royal Navy was threatening to pull out of its base in the Moray Firth.[10]

Arthur Grant, the earnest and sober vet student, proved to be a solid witness. He gave sworn statements to the police and the Edinburgh Veterinary College, and Rupert Gould was happy to 'entirely accept' his account of the Monster.[11] Unlike Wetherell, Grant had nothing to gain from his tale and seemed shy under the gaze of the press. His credibility grew as Wetherell's reputation shrivelled away.

By now, Wetherell was out of sight and out of mind, and heading into an African sunset of his own making. However, he had not written himself out of the script. In the best tradition of big movie stars, Marmaduke Wetherell was plotting to ride out again.

Unnatural selection

Hoaxers might have taken the Monster's name in vain, but they had little impact on the rapidly increasing numbers of believers around the world. A steady stream of sightings, mostly reported to the *Courier*'s unnamed Correspondent, raised the total to nearly 50 by the end of February. Following the Wetherell debacle, many witnesses now 'asked for their name not to be divulged' to avoid ridicule or being pestered by the press. The anonymous local who wrote in about the 12-foot water animal (short-necked and with no obvious tail) which he/she had seen in July 1914 signed off as 'MORE THAN ONE', because a friend had seen 'a small similar one', just three feet long.[12]

The new wave of sightings was dominated by the 'upturned boat', sometimes with two or three humps and often churning up a massive wash. One hump, watched simultaneously by 'several reliable witnesses' at Drumnadrochit and Invermoriston (four miles apart), seemed to show the ridged back which Commander Meiklem had described eight weeks earlier.[13] What lay before and aft of the humps remained uncertain, although a 'well-known Ross-shire lady who does not desire her name to be published' watched a long, serpentine neck with a small swan-like head emerge from the Loch on 27 February.[14] But what kind of animal was it? The assorted body parts seen by different witnesses were like pieces of a jigsaw puzzle that could not yet be pushed into place to make a coherent picture.

Seals were back in the frame, even though many doubted that they could find their way up seven miles and the 50-foot drop of the River Ness to enter the Loch. Readers of the *Courier* were treated to a photograph (exceedingly rare in the newspaper at that time) of the head of a bearded seal, specifically the one mounted on the wall of Messrs Watsons' emporium in Inverness. Mr William Berry, 'one of the keenest punt-shooters' ever to prowl the Beauly Firth, had bagged this remarkable beast in January 1911. It was nine feet long and sported a magnificent Kaiser Wilhelm moustache. Pink fish flesh poured out when its belly was cut open, but the species was too rare to threaten salmon stocks. Berry's punt gun had taken care of the entire population ever recorded from Scottish waters.[15]

There were more imaginative suggestions. The descriptions by the Spicers, Arthur Grant and the unnamed Ross-shire lady pointed to something that was both fabulous and unbelievable – a plesiosaur. This was the prime candidate for a correspondent from Patagonia (where a plesiosaur-like animal had been reported from Lake Nahuel Huapi), and for several others closer to home. Ken'ichi Ishikawa of the Tokyo *Mainichi* disagreed: whatever was there could not be prehistoric. So did Dr Richard Elmhirst of the Scottish Marine Biological Association, who somehow calculated that the odds against the Monster being a plesiosaur were 17 million to one. Professor William Beebe, Director of the Tropical Research Department at the New York Zoological Society, also argued that the Monster could not be 'some survivor from a prehistoric age'. Beebe favoured a giant squid, and offered to go down in his legendary deep-sea submersible, a steel and plate-glass capsule capable of diving to 3,000 feet, to prove the case.[16]

Strong support for Professor Beebe came from Mr W. U. Goodbody, who praised this as 'the most feasible theory yet . . . coming from a scientist of Dr Beebe's attainments, it has a good deal of authority behind it'. Goodbody also used the opportunity to lament the apathy of British scientists over 'the possibility of a strange creature being in Loch Ness'. Instead of looking at the evidence, they had 'contented themselves with expressions of disbelief'.[17]

Goodbody's views carried authority, as he was a member of the Ness Fishery Board. He had also seen the Monster, on the wintry afternoon of 30 December, about two miles north-east of Fort Augustus. He and his two daughters, all 'accustomed to looking at boats and other objects over water', had watched it through binoculars for 40 minutes. However, Beebe's giant squid did not easily fit with their Monster – which was about 16 feet long and showing between two and nine humps that projected a foot out of the water. Shortly after the New Year, Goodbody's daughters saw it again, together with their mother and two friends. This time, it was a single-humped 'upturned boat' about five feet long, moving fast off Urquhart Bay. Goodbody admitted that it was hard to reconcile these two observations, but believed that a single animal which could vary its shape was more likely than two different unknown species existing in Loch Ness.

As for hypotheses about otters, seals and other known animals, these were 'understandable, but mistaken'.[17]

Also ruled out was the owner of the hefty skull found on the shore near Aldourie on 1 February. This caused great excitement, until the proprietor of the Sporting Stores in Inverness identified it as belonging to a brown bear. This was quickly confirmed by the Royal Scottish Museum in Edinburgh. How the skull found its way to Aldourie was another of those mysteries. The brown bear had roamed the Great Glen, but back in the age when the Great Caledonian Forest still had trees. The last one had been killed in 1057.

The bear's skull laid the ground nicely for a two-page photo spread in the *Berliner Illustrirte Zeitung* exactly two months later.[18] The article claimed that the Monster had been captured with torpedo nets, identified as a descendant of the plesiosaur and displayed in Edinburgh before huge crowds. A significant clue was the publication date (1 April 1934), suggesting that, in spite of everything, the Germans had retained some vestiges of humour.

Meanwhile, British scientists maintained their deafening silence, apart from unhelpful comments about untrained, impressionable observers and mass hallucinations. Their stance seemed increasingly peculiar, especially when eminent Americans like Professor Beebe were throwing their hats into the ring. Surely the identification of a fabulous new species had to be the zoological discovery of a lifetime? Why were no British scientists prepared to take the risk and investigate the Monster properly? By way of reply, the scientists repeated the party line that they could do nothing without stronger evidence.

As if to call their bluff, this was soon provided. Early on the last day of March, a London gynaecologist set off on the long drive up to the Beauly Firth for a few days' duck shooting on the marshes where he had recently bought wildfowling rights. He was a crack shot, but was also interested in photographing birds as well as bringing them down. To that end he had borrowed a fancy quarter-plate camera with a telephoto lens.

That expression 'and the rest is history' is a cliché, and wretchedly abused. In this instance, though, it is entirely justified.

Bound by Oath

In early 1933, Robert K. Wilson MB BChir (Cantab) FRCS (Edinburgh) was aged 34 and, in professional terms, had arrived (Plate 5). After qualifying with a first in medicine from Trinity Hall College, Cambridge, he specialised in gynaecology and surgery and had built up a thriving private practice at 42 Queen Anne Street in the West End of London. The address said it all. Wilson's consulting rooms were just a block from Harley Street, one of the world's centres of excellence in that most delicate and skilled of all medical specialities: the extraction of money from the wallets of the wealthy.

Wilson led a more exciting life than his brass plaque on Queen Anne Street might suggest. His early years could almost have been lifted from one of Marmaduke Wetherell's movies. The sixth son of a medical missionary, Wilson was born in Madagascar and sent back to England to boarding school, with his eldest sister *in loco parentis* until his eighteenth birthday. This was the moment that Wilson had been waiting for – the earliest day on which he could join the army. As this was in October 1917, he was sent straight to the front line in France, exactly as he had hoped. There he picked up wounds, a mention in dispatches for conspicuous bravery and a thirst for action.

After the war, Wilson went up to Cambridge to study medicine. The adrenaline of active service left him perpetually restless and he remained a reservist for the Royal Army Medical Corps, hoping that war would break out again. He also became fascinated with guns, becoming one of the first forensic experts in firearms murder cases and the author of the definitive book on the machine pistol.[19]

Away from the heat of battle, Wilson was thoroughly professional as a doctor. And in sharp contrast to Wetherell and Gould, who wasted no opportunities for self-promotion and self-aggrandisement, Wilson would prove to be ill at ease in the limelight.

The quarter-plate camera which Wilson borrowed for his Scottish jaunt was owned by his friend Maurice Chambers, man about town, owner of a distinctive yellow Rolls-Royce and keen amateur photographer. Chambers worked as an upper-crust insurance salesman in London but used to drive down at weekends to play the organ at a tiny church on the outskirts of Thornbury, near Bristol, where he had grown up.[20]

The camera might be Stone Age compared with the point-and-shoot digital wizardry of today, but it was state-of-the-art in the early 1930s. It was a splendid contraption, with the lens mounted on a black leather bellows and a magazine containing 'quarter plates' (8 x 10 centimetre sheets of glass coated with photographic emulsion) that slotted into the back. Even with a practised hand and a cool head, switching the magazine for the next exposure was a fiddly business. To slow the process down even more, the operator had to judge and set the exposure time, the aperture and the distance from the camera to the object being photographed.

Wilson knew all about the innards of machine pistols and sports cars, but the quarter-plate camera was unknown territory until that day. He later confessed: 'I was inexperienced as a photographer and was, in fact, very much of the amateur.'[21] Even after a practical tutorial from an expert like Chambers, Wilson would need a good half-minute and a generous dose of luck to capture a reasonable image of a duck, or anything else capable of movement.

What happened after Wilson left Thornbury for the Beauly Firth is best explained in his own words:

I had a fast car in those days and after travelling all night arrived at Fort Augustus too early to get breakfast, so decided to go on to Inverness . . . At about 7 or 7.30 a.m. I stopped by the roadside two or three miles on the Inverness side of Invermoriston at a point where the road is some hundred feet above the loch.

I had got over the dyke and was standing a few yards down the slope and looking towards the loch when I noticed a considerable commotion on the surface, perhaps two or three hundred yards out. When I had watched it for perhaps a minute or so, something broke surface and I saw the head of some strange animal rising out of the water.

I hurried to the car for my camera, then went down and along the steep bank for about fifty yards to get a better view and focused on something which was moving through the water. I was too busy managing the camera to make accurate observation but I made four exposures by which time the object had completely disappeared. I had no idea at the time whether I had anything on the plates or not, but thought I might have.[21]

The accuracy of Wilson's intuition was quickly confirmed. He drove straight to Inverness and found Ogston's chemist shop in Union Street. There he asked Robert Morrison to be particularly careful with the magazines. Morrison asked jokingly if he had got the Loch Ness Monster, and Wilson said that he believed so.

Morrison developed the four plates with due care and without delay. Two were blank. The last plate was indistinct and showed a short, dark shape with a rounded end, curving out of the water and reflected in it, and apparently slipping forwards towards the surface (Plate 7).

It was the third plate which proved to be sensational (Plate 6). The surface of the water is rippled longitudinally, as if ruffled by the wind. In the centre of a ring of ripples sits an extraordinary object, curving forward like a tapered neck and ending in a blunt head. Its reflection is partly visible. Behind it, hard to make out against the water, are suggestions of a low back breaking the surface. Immediately in front of the neck is a small, dark rounded shape.

Morrison immediately saw the importance of Wilson's third plate. Convinced that the photograph deserved a better springboard than the local newspapers, he strongly advised Wilson to sell it in Fleet Street. It is probable that neither of them would have guessed that it would become one of the iconic images of the twentieth century.

Much to the chagrin of the *Inverness Courier*, Wilson's photograph was immediately snapped up by the grubby fingers of the *Daily Mail*. The image, heavily cropped to frame the extraordinary shape rearing out of the water, filled the front page of the issue of 21 April, under the bold headline, 'London Surgeon's Photograph of the Loch Ness Monster'.[22] Inside was a piece reporting Wilson's rather telegraphic account of how he had chanced upon his scoop, with a résumé of the story so far and a rather smug editorial pointing out that the *Mail* had been on the right track all along. Inevitably, there was speculation about what the Monster might be. In the light of this astounding new evidence, the seal theory had now sunk without trace, together with various other marine mammals, fish and amphibians. As if to confirm the adage that 'nothing is stranger than the truth', the most likely candidate now was an aquatic reptile of a design that had not been seen on Earth for over 60 million years.

This photograph had an astonishing impact. It seized the imagination of millions of people around the world, because the editors of countless newspapers saw an opportunity that could not be missed. Wilson's snapshot, tagged with the *Mail*'s label 'The Surgeon's Photograph', quickly became front-page copy in newspapers and magazines in every continent – with one notable exception.

For the *Inverness Courier*, the story was warm rather than incandescent. The piece on page four of its 22 April issue was not much longer than the following one, 'Certainly not a Seal', and the headline, 'Loch Ness Monster seen yesterday by London Surgeon', was distinctly unsensational.[23] There was no photo; to see what the surgeon's camera had captured, the *Courier*'s readers had to demean themselves and buy a copy of the *Daily Mail*.

In a brief interview with the *Courier*'s Special Correspondent, Wilson revealed that his camera had been waiting in case he saw the Monster, and he had taken the photo about two miles north of Invermoriston. He thought that the animal's neck rose two or three feet out of the water, but it had all happened too quickly: 'I am not able to describe clearly what I saw, as I was so busy taking the snaps, and when I had finished the object moved a little and submerged.'[23]

Wilson was as important as the image itself in cementing the Photograph's reputation as a true likeness of the Monster. The title 'Surgeon' carried great weight. Here was an astute professional man, bound by the Hippocratic Oath to be honest and upright. His anxiety not to speculate beyond the bare facts provided a refreshing contrast to the posturing of Marmaduke Wetherell. Wilson also had impeccable society credentials. He was known personally to the editor of *The Field*, an obligatory prop in the drawing rooms of upper-crust Britain. The editor made it known that the Surgeon's Photograph was absolutely genuine, and that was good enough for those whose opinions really mattered.[24]

In April 1934, with memories of the Wetherell farce still fresh, the time was right for something extraordinary but believable to emerge from the peaty waters of Loch Ness. The Surgeon's Photograph probably converted more people into true believers than anything else. It carried the Monster safely through the Second World War, nudged some notable sceptics into joining the ranks of the faithful, and, half a century later, was still one of the central planks in the evidence that the Monster exists.

The Photograph also pushed into the foreground a question that now had to be taken seriously. Could this peculiar body of water, which even scientists found difficult to understand, turn out to be a unique habitat – one which somehow had sustained an animal species that, elsewhere on the planet, had died out millions of years earlier?

Rising star

As the Monster's influence spread during the spring of 1934, it provided some much-needed light relief. Incarnated as a multi-humped prehistoric reptile, the Monster had already become a familiar fixture at functions in the Highlands and Islands and on postcards, often with local props such as Castle Urquhart and monks at the Abbey in Fort Augustus. To the obvious disgust of some of the audience, the dragon in the new production of *Siegfried* at London's Covent Garden was redesigned along similar lines.[25]

Another spin-off was the film *The Secret of The Loch*, a thriller which helped to found the tradition of the Hammer horror movies.[26] The director was David Lean, 26 years old and just setting off on the path which would take him to blockbusters such as *Brief Encounter* and *Lawrence of Arabia*. It starred Seymour Hicks as Professor Heggie, an irascible zoologist battling to convince his peers that the Monster exists, and a gigantic iguana that specialised in chewing through divers' breathing equipment. The Monster was seen only in tantalising glimpses until the very end, which helped to build dramatic tension and cover up the awfulness of the special effects.

The Monster also fired up creative minds further afield. For no obvious reason, the 'Loch Ness' reared up in women's fashion houses around Boston, as a green woollen twin suit with dangling trimmings in grey fox fur. Like Siegfried's dragon and *The Secret of The Loch*, the 'Loch Ness' was just a spot of good fun and not to be taken at all seriously.

Back on the Loch, the number of sightings continued to climb steadily. In late May, Brother Richard Horan from the Abbey at Fort Augustus glanced out across the Loch and found an animal with a long, graceful neck and a broad white stripe down its front, gazing at him from barely 30 yards away.[27] Three weeks later, members of the Inverness Scientific Society and Field Club were amazed to see a massive black object moving steadily up the Loch near Abriachan – *against the wind*.[28] The Society's members were well informed and students of nature and were taken particularly seriously.

So was Sir Murdoch Macdonald, MP for Inverness-shire. While driving with his son early on the morning of 8 August, MacDonald was astonished to see his most famous constituent in the loch off Invermoriston. It was clearly visible as two blackish-grey humps that spanned about 15 feet, moving slowly south. The sight was witnessed by a semi-naked Yorkshireman who sprang out of his caravan with binoculars and confirmed that it was neither a tree trunk nor a boat. The MP was on his way to see the Secretary of State for Scotland, who was happy to accept the Monster as a reasonable excuse for starting their meeting late.[29]

Into print

The spring of 1934 saw the publication of the first two books about the Monster. Each shamelessly plugged the author's agenda and established the tradition of pruning evidence to squeeze the Monster into the mould which the author had created for it.

First off the mark was the formidable Colonel W. H. Lane. A veteran of the Indian Army and the pacification campaigns in Burma during the 1890s, Lane had also helped to excavate the ruins of Babylon. Having swapped the road to Mandalay for the one to Fort Augustus, he was now in restless retirement in Tigh na Bruach, a large house set among Scots pines on a bluff just south of Invermoriston. The name means 'the house on the edge', and so it was: perched 30 feet above the shore and looking out east across the Loch to the rugged mountains opposite which had forced General Wade's Military Road to swing inland.

Lane had no doubt that 'some huge beast did actually inhabit Loch Ness' and knew exactly what it was – a giant salamander like the one that he had shot in the Chin Hills of Burma, but bigger. Its picture featured on the cover of Lane's 18-page book, *The Home of the Loch Ness Monster* (Plate 8).[30]

The giant salamander had only been known since the 1820s, when two were captured in Japan and exhibited in the zoo at Leiden. To the dismay of the keepers and 50 per cent of the exhibit, the species turned out to be cannibalistic. The giant salamander can live for over 60 years, grows to nearly two metres and weighs up to 40 kilogrammes. It is an ugly brute, like a huge, morbidly obese newt, with clumsy feet and tiny eyes embedded in its triangular head. Being largely nocturnal, it is rarely seen and is also at home in fast-flowing rivers.

As far as Lane was concerned, the salamander fitted the bill perfectly. For him, the pivotal sighting had been in February 1932, when a Miss K. MacDonald from Inverness had seen a short-necked, crocodile-like monster with 'wicked pig-like eyes' on the top of its head, plodding up the River Ness against a strong current.

Admittedly, reports of long-necked or serpentine animals, sometimes travelling at incredible speed, did not quite fit the picture, but Lane had made his mind up. The Loch Ness Monster was a giant salamander.

Another oddity

The Loch Ness Monster and Others by Lieutenant Commander Rupert T. Gould, which appeared a couple months later, was bigger (over 220 pages) and reached a completely different conclusion.[31]

Gould's approach was laborious, as expected of the obsessive who had taken 12 years to perform curative microsurgery on John Harrison's chronometers. Contemptuous of the 'Stunt Press', he deliberately ignored all newspaper reports and instead interviewed 51 eyewitnesses in great detail. This took three weeks and two complete circuits of Loch Ness on his trusty motorbike, Cynthia. There were some frustrations. Despite on-the-spot discussion with a road contractor and lengthy correspondence with Mr and Mrs Spicer, Gould was never able to pinpoint exactly where the couple had met their 'prehistoric monster'. In addition, Hugh Gray's interesting and undoubtedly genuine photograph languished in its drawer thoughout Gould's visit. And worst of all was missing a personal encounter with the Monster in Urquhart Bay, by just an hour.

The eyewitnesses spanned society and included labourers, farmers, a hotel proprietor, a water bailiff, policemen, assorted businessmen, an AA scout, an engineer and a doctor. Several had seen the Monster more than once; Mrs J. Simpson of Alltsigh held the record with five sightings. All impressed Gould with their clarity and obvious honesty: 'I have never – I am glad to say – been compelled to discard the evidence of any witness because I mistrusted it.'[32]

Gould's intensive questioning teased out some new information. From 1,000 yards away, Miss Janet Fraser of the Halfway House tea room at Alltsigh saw that the creature's head looked like a terrier, with a frill 'like a pair of kippered herrings'.[33] When Gould visited Mr and Mrs Spicer in Golders Green and mentioned the recently released film *King Kong*, Mr Spicer said that his prehistoric beast had looked like the diplodocus but with a longer, more flexible neck.[34] Gould also found the three unnamed men who had been fishing off Tor Point in July 1930 and whose boat had been rocked by the massive wash that provoked the headline 'What was it?' in the *Northern Chronicle*.[35] One of them, Ian Milne, had caught glimpses of a dark back that looked like an enormous conger eel, except that it undulated up and down, not side to side. Milne, who ran a gunsmith's in Inverness, explained that they had not been named because they 'did not wish to be thought self-advertising'.[36]

The Loch Ness Monster and Others, described in the publisher's blurb as a 'masterpiece of careful scientific investigation', was a worthy successor to Gould's earlier books about oddities, enigmas and sea serpents. Each of the 42 Monster sightings was carefully documented and supported by line drawings, maps and explanatory notes about sea-monsters in general and sea serpents in particular.

Gould concluded that the Loch definitely harboured 'a large living creature of an anomalous species' which, for brevity, he called 'X'. From all his reports, Gould synthesised a picture of X: 45 feet long, divided roughly equally into a tail, a rounded body (with flippers or feet) and a long, slender neck bearing a small snake-like head that it could lift clear of the water. The skin was brown and 'granulated', and darker along the ridge of its back. Sometimes seen travelling slowly, undulating up and down like a caterpillar, X was capable of great speed, tearing across the water at up to 30 mph.[37]

Then Gould tackled the problem of X's identity. He swiftly ruled out optical illusions, mass hallucinations, the highly improbable plesiosaur, crocodiles, whales, eels, Professor Beebe's huge squid and Colonel Lane's giant salamander. Seals and sturgeons were just plausible, but did not behave like X. Otters might explain some sightings, such as the family of four swimming in a line in the Clyde Estuary that had nearly fooled Dr Richard Elmhirst, Director of the Scottish Marine Biological Association. Without binoculars, Elmhirst would have been forced to conclude that he had seen a multi-humped marine animal several feet long.[38]

This left just one serious candidate – the sea serpent – although Gould could not decide whether this creature was a long-necked seal (the choice of Dr A. C. Oudemans, the world's greatest sea-serpent expert), 'a vastly enlarged, long-necked marine form of the newt' (Dr Malcolm Burr, London) or a gigantic fish (Dr Petit, Paris). Admitting that it seemed implausible, Gould plumped for the newt.[39]

There were a couple of rum things in Gould's book. He praised the Surgeon's Photograph as 'most satisfactory and characteristic' and gave it pride of place as the frontispiece. However, the sighting did not figure in the catalogue of eyewitness accounts, and the evidence of R .K. Wilson FRCS was dismissed in a few lines. Another enigma was the short-necked 'crocodile' seen by Miss K. MacDonald in the River Ness two years earlier. This was crucial evidence for Colonel Lane's salamander hypothesis, but Gould dismissed it because whatever Miss MacDonald saw was clearly not X (later, he was happy to accept her sighting of a typical long-necked X in May 1934).[40]

He also left some loose ends dangling. Did X inhabit Loch Ness in solitary splendour? (he believed that it did). How had it materialised in Loch Ness? (no idea).

Gould also helped to solve one mystery while creating another. He revealed the identity of the anonymous eyewitness from Fort Augustus who had provided the most dramatic material for Philip Stalker's piece on 'The Plesiosaurus Theory' in the *Scotsman* the previous October.[41] This was Alex Campbell, who was in his early thirties and worked as a water bailiff, patrolling the Loch and its vicinity for the Ness Fishery Board.

Curiously, Campbell came into the story not because he had seen the Monster, but because he denied having seen it. Gould quoted a letter Campbell wrote to his employers on 28 October 1933, explaining that he had been mistaken in reporting the 30-foot, long-necked creature peering nervously around Borlum Bay. Some weeks later, under identical conditions (early morning, hazy sunshine), he had seen the same thing – until the mist melted away to reveal a group of cormorants whose image had been 'magnified out of all proportion' by an optical illusion. Gould included the account to demonstrate that 'natural phenomena' could occasionally mimic a Monster.[42]

Shortly afterwards, Alex Campbell came clean about something else. As well as working for the Fishery Board, he had a part-time job writing occasional short articles on fishing and natural history for the local papers, especially the *Inverness Courier*.

Campbell now admitted that he was the anonymous Correspondent from Fort Augustus who had filed most of the reports in the developing story of the Monster. This began with the 'Strange Spectacle' seen in April 1933 by his friends, the unnamed local businessman and his university graduate wife who were now known to be John and Aldie Mackay, proprietors of the Drumnadrochit Hotel.

Most interesting and successful

Gould's book was a significant landmark in the history of the Monster. It continued to gain stature even though Gould himself never returned to Loch Ness, because he became one of the most charismatic radio personalities of the late 1930s. As the BBC's 'Stargazer', he was the infinitely wise uncle-like figure who explained the wonders of the universe to the children of Britain as they sat spellbound beside the family radio. As 'the man who knew (nearly) everything', he was arguably the brainiest member of *The Brains Trust* and, uniquely, had never been contradicted on air. Gould therefore had the Home Service listeners – most of the adult population – in the palm of his hand.[43] If Commander Rupert Gould had ridden twice around Loch Ness on his motorbike and found enough of a Monster to write a book about, then it must be there.

The Loch Ness Monster and Others also acted as a catalyst for the first systematic search for the Monster, during the summer of 1934. Soon after publication, the book came to the attention of Sir Edward Mountain, the fabulously wealthy Chairman of the Eagle Star Insurance Company and the man who once declined to insure a passenger liner called *Titanic*.[44] Gould's account inspired Mountain to find the Monster. The newspapers announced his plans with a fanfare of excitement: Mountain's trained observers would watch the Loch for several weeks, while he directed operations from his headquarters in Beaufort Castle.[45]

The observers were easy to find. Twenty men, previously ghillies and gamekeepers, were happy to be pulled out of the steadily lengthening dole queue in Inverness and employed as 'Watchers for the Monster' (Plate 3.7). Each was provided with a notebook, binoculars and camera (sealed, to prevent tampering), and instructions in the use thereof. They were also promised a one-guinea bonus for each fully documented sighting.

The expedition began on Thursday, 13 July 1934 and was scheduled to run until the end of that month. Each day, the observers were driven by bus to vantage points at 1–2-mile intervals around the Loch; they began their vigil at 9 a.m. sharp and were relieved at 6 p.m. The operation was coordinated by Captain James Fraser, formerly of the King's Own Cameron Highlanders, who toured the observation posts each day by motorbike. Every evening, Fraser took all new sighting reports (filled in on a detailed proforma) and any cameras that had been used to Sir Edward in Beaufort Castle.

The expedition was quickly rewarded with sightings. Monster-watching proved infectious, and many observers began early and stayed late; this was fortunate, because it soon became clear that the Monster could not be relied upon to work a strict nine-to-six shift. By the end of July, Mountain had collected 20 sightings (mostly through binoculars) and 21 photographs, guaranteed by Kodak to be unmolested. He extended the expedition for another two weeks, after which his intrepid field commander continued the watch for several weeks. Living in a tent, Fraser and an assistant prowled the shoreline day and night, armed with a cine-camera and a six-inch tele-photo lens.

Mountain published his interim findings on 22 September in *The Field*.[46] He was 'quite positive that there is something huge living in the Loch'. The eyewitness reports yielded 'a very good idea of the creature's appearance'. It was dark, at least 15 feet long, and often showed two or three humps; on four occasions, a long neck was visible. The observers had mostly spotted it travelling across the Loch, often throwing up a heavy wash, and one lucky man had watched it for several minutes basking in the shallows. Unfortunately, the photographs all showed impressive wakes but no detail of whatever had caused them.

What could it be? Mountain believed that the Monster was the same as the large marine animals sometimes reported from Norwegian fjords, but conceded that high-quality, close-up photographs or cine-footage were needed. He had toyed with the idea of stunning the Monster with high-voltage electric shocks to bring it to the surface, but reluctantly abandoned the plan because of the risk to the Loch's fish stocks.[46]

While *The Field* article was in press, Mountain triumphantly unveiled his *pièce de résistance* – ten feet of cine-film taken by Captain Fraser at 7.15 on the misty morning of 15 September. This was projected before 'a privileged gathering of scientists and others interested in the subject', on 30 September in Kodak House on Kingsway, London. The film showed a dark object, estimated by Kodak's optical experts to be eight feet long. However, as the footage had been shot through Scotch mist at a range of three-quarters of a mile, nothing diagnostic could be made out.[47]

Mountain threw the gauntlet down to the sceptics by organising the first ever scientific symposium about the Monster, on 8 November 1934.[48] The venue was Burlington House in Piccadilly, the home of the world's oldest biological association, the Linnean Society. The scientific establishment was well represented by assorted professors of zoology and big names from the nation's museums. On display beside the trophies of Mountain's expedition were the Surgeon's Photograph and Commander Rupert Gould, the celebrated author, broadcaster and expert in the unexplained.

The evidence was reviewed and discussed in detail, but, to the frustration of Sir Edward, the scientists were unexcited by the verbal reports and the photographs of wakes and distant objects. Dr Stanley Kemp FRS, the distinguished Director of the Marine Biological Laboratory at Plymouth, confessed that he had no idea what the photographs might show. In summing up, Sir Edward ventured to suggest that Dr Kemp's opinion was wiser than those advanced by others, who had suggested seals or otters.

The exasperation felt by Mountain and his supporters quickly boiled over in the Letters section of *The Field*. Those who knew and hunted seals and otters wrote in to express incredulity at the stupidity and arrogance of city-dwelling scientists. Typical was C. E. Redclyffe (Major) of Thurso. Whatever had been observed in Loch Ness, he thundered, was 'absolutely unlike any of the thousands of seals of various kinds which I have seen'. Another infuriated reader complained that if a quarter of the evidence for the Monster had been accrued for an unknown animal in a lake in Central Africa, the same doubters would already have sent expeditions there 'in the sacred cause of science'. He suggested that 'some eminent authority, suitably disguised' should set off to investigate Loch Ness the following year.[49]

As well as failing to nail the Monster's identity, Sir Edward Mountain's expedition further polarised thinking. Mountain's campaign and its tantalising findings converted many to the cause, and not just among the followers of *The Field*. Entrenched in the opposite corner were the scientific establishment, and they had now declared their hand. There was nothing in

Loch Ness that could not be explained by the usual fauna of Scotland, tricks of the light and a fertile imagination.

National treasure

Intellectual snobbery has always been bad for clarity of vision. The clever scientists who closed ranks in Burlington House may not have noticed (or cared) that the tide of public opinion had turned against them.

The Monster had already made it into *Hansard*, the red-bound volumes that record proceedings in the Houses of Parliament. The first exchange took place on 12 December 1933, when Sir Murdoch MacDonald and William Anstruther-Gray, Members of Parliament for Inverness-shire and Lanark-shire, asked how the government intended to investigate and resolve the matter of the Monster. There was a back story, as MacDonald had recently written to the Secretary of State for Scotland, expressing concern that the Monster was at risk and needed to be protected by a Bill of Parliament. The Secretary of State agreed that the Monster must not be molested. On the day, however, the government had nothing useful to suggest and knocked the ball straight back to them. MacDonald talked about trawling the Loch and Anstruther-Gray about air patrols, and the debate fizzled out.[50]

Their concerns were not groundless. The Monster had become hot property, and those who wanted a slice (or, ideally, all) of it included local wide boys, canny businessmen, impresarios and people who should have known better. Mr Spicer's original call to kill the beast, followed up by the suggestion that a warship could do the job with guns and/or torpedoes, had stirred the Inverness Constabulary into action. *The Times* printed a photograph of a stern-faced policeman in peaked cap and full-length oilskin coat, order papers in hand, standing on Invermoriston pier and warning a group of men wearing ambiguous expressions not to harm the Monster.[51]

Rupert Gould, desperate to protect the Monster, was deeply troubled. A maverick could kill the creature and 'no police-officer could lay a finger on him until after the event'.[52] Gould's fears were shared by the police at the time, although not publicly. Previously classified correspondence, only released in 2011, shows that William Fraser, Chief Constable of Invernessshire in 1934, doubted that his officers could protect the Monster from harm. The threat did not go away; in 1938, police apprehended a hunting party armed with an industrial harpoon gun, en route to take the Monster.[53]

Further afield, others now wanted the Monster, alive or otherwise. Those who preferred the latter included the sea serpent's greatest ally, Dr A. C. Oudemans of Leiden. Oudemans had a Victorian approach to biodiversity, arguing that the best place for this rare and enigmatic species was inside a glass case at the Royal Scottish Museum, where everyone could marvel at it. Gould, usually a fervent admirer of Oudemans, was shaken to his pacifist core by this 'cruel and unnecessary' suggestion.[52]

Being taken alive had obvious advantages for the Monster, but would also have been far from ideal. Trapping strategies ranged from the Alloa trawlermen who believed that they would only need 'serviceable ropes', to the bungalow-sized metal cage built in London by the Steel Scaffolding Company of Regent Street.[54] Various career options were laid out for the Monster in captivity. Bertram Mills, a flamboyant circus owner, put up the eye-watering prize of £20,000 for its safe delivery into his hands. This was a huge sum – the 2015 equivalent exceeds £1.5 million – and Mills took out an insurance policy with Lloyd's of London just in case he had to pay out.[53] A more appropriate residence – albeit far from its habitat and heritage – could have been the New York Zoological Park, but they only offered £5,000.

Missing, presumed alive

Many things become blindingly obvious in retrospect. As far back as 1933, all the omens of war were there in the pages of the *Inverness Courier*, interwoven with updates on the Monster: the Nazis' execution of Van der Lubbe, whom they had framed for burning down the Reichstag; their brazen stockpiling of arms and explosives in Austria; and Sir Murdoch Macdonald's premonitions of death and destruction raining down on British cities from German bombers.

In the event, the Great Glen was insulated from much of the business of war, and Inverness was never bombed. At Invermoriston, Colonel Lane took command of the Local Defence Volunteers but failed in his efforts to have the local population armed with machine guns. Loch Ness was routinely scanned by the Royal Observer Corps for enemy activity, but with less zeal than Sir Edward Mountain's expedition – perhaps because the chances of hostile craft passing undetected up the Caledonian Canal were not that high.

What did the Monster do during the war? Predictably, sightings fell dramatically once hostilities began; people had better things to do than wait beside a lake for an extremely rare event. Nonetheless, reports of the Monster continued to trickle in. Shortly after sunrise on 25 May 1943, Mr B. C. Farrell of the ROC spotted a creature swimming about 250 yards from shore. It was dark brown and had a four-foot-long neck and prominent eyes. Farrell watched it for several minutes, when it sank vertically without a ripple.[56] The following April, the wife of Commander Meiklem (who had picked out the animal's ridged back through binoculars back in 1933) was astonished to see a large black submarine in Borlum Bay. Just as she realised that it was 'an animal', bobbing gently on the surface, it began to move and she and her husband watched it accelerate away fast enough to raise a wash that could have come from a heavy steamer.[57] A year later, to prove that infinite patience can be rewarded, Colonel Lane and three others watched a 'huge black object' out in the Loch below Tigh na Bruach. Through his binoculars, Lane watched it submerge, leaving an obvious, slightly curved wake like a torpedo trail.[58]

The Monster also played a wider propaganda role in the war, for both sides. In early 1941, it was announced that the Monster had been seen lying dead in the water after a German bombing raid. This catastrophic news might have brought the British people to their knees if it had been published in English and in a newspaper with a wider circulation than *Il Popolo d'Italia*, Mussolini's in-house rag. Instead, the British fought back, lifting the spirits of prisoners of war across Europe by mailing them the report of Commander and Mrs Meiklem to prove that the Monster still enjoyed rude good health.[58]

The advent of war profoundly changed the lives of some of those caught up in the story of the Monster, and in dramatically different ways.

Commander Rupert Gould looked big and confident, but he carried invisible scars from his time in the navy. He had been left with a morbid fear of violence which relentlessly closed in on him while Europe spiralled down into chaos. Gould saw out the war from deep in the heart of Surrey, at a safe distance from the sea and action. He used the opportunity to develop his broadcasting career, as a foundation member of the BBC's *Brains Trust*. With his legendary calm and 'coruscating displays of omniscience', he probably did as much to maintain the nation's morale as if he had been in active service.

By contrast, the call to battle stiffened the sinews of Dr Robert K. Wilson, just as it had done a war earlier when he had been only 18 years old. As soon as hostilities began, Wilson closed down his prosperous practice in Queen Anne Street and abandoned medicine to sign up as a front-line soldier. His first posting (to Northumberland) was a frustrating anti-climax for him and a sorry waste of his talents. In 1942, however, the adrenaline surged again when he was promoted to major and sent to train with the Special Operations Executive. This elite group were selected for their prowess as both soldiers and spies. Their trademark was being parachuted behind enemy lines from a low-flying aircraft on a moonless night. Wilson carried out several missions in Occupied France, Holland and Borneo.

When the war ended, he evidently felt that his medical qualifications had been eclipsed by his military achievements. It was several years before he returned to medicine.[19]

The war destroyed Dr Stanley Kemp, Director of the Marine Biological Laboratory in Plymouth, and the thoughtful scientist who admitted that he could not explain Sir Edward Mountain's sightings. On 20 March 1941, four months after the centre of Coventry was levelled by the Luftwaffe, Plymouth was blitzed. The Laboratory and the adjacent Director's house took direct hits.

It would have been difficult to design a more perfect hell for Kemp, who had to choose between trying to save his possessions and the Laboratory's records. He went for the latter, fighting through flames and a shower of molten lead from the blazing roof while the specimens from the shattered aquariums flapped themselves to death around him. From his house, he managed to salvage only a grandfather clock and his parrot.

Kemp never recovered from the experience and died in May 1945, just as the war in Europe was ending. His obituary in the *Journal of the Marine Biological Association* filled 16 pages.[59] Naturally, it made no mention of the Loch Ness Monster.

4

From legend to fact

The Monster did not really belong in the grey austerity of post-war Britain. The everyday grind of rebuilding a victorious but shattered country left little room for freaks of nature, no matter how compelling they might have seemed ten years earlier.

Against that grim landscape, even the greatest zoological discoveries of the century had lost their excitement. Top of the list was a strapping 160-pound fish with a spear-shaped tail and chunky fins that were halfway to becoming legs. The coelacanth looked as though it might have swum with the plesiosaurs, and it had. Originally, it was known from a fossil lineage that began 360 million years ago but apparently stopped dead at the mass extinction which wiped out the dinosaurs and most animals with a backbone some 65 million years ago. However, rumours of the coelacanth's demise turned out to be premature. Just before Christmas 1938, a trawlerman fishing within sight of the South African coast pulled one up, bright blue and very much alive.

Its alien appearance brought it to the attention of Marjorie Courtenay-Latimer, curator of the tiny museum in East London, who in turn alerted Professor J. L. B. Smith at Rhodes University.[1] By then, it was dead and partly decomposed but this single 'living fossil' created a perfect storm. It proved that whatever was written in tablets of stone was not necessarily gospel. A species written off by the geological record could still be clinging to existence in some privileged habitat – and potentially under the nose of civilisation.

As the 1950s dawned, the coelacanth still inflamed the passions of evolutionary biologists, but most normal people could not have cared less. The great British public, worn down by rationing, the descent of the Iron Curtain and the nuclear arms race, had bigger fish to fry. Even for those who had heard of it and understood its significance, it was as though the coelacanth had never come back to life.

Media and messages

During the early 1950s, television was still a luxury in Britain, and radio and books remained the staples which nourished the mind, fired the imagination and lifted the eyes towards brighter horizons.

As well as news and entertainment, the radio brought knowledge. Wildlife became a favourite theme, with programmes such as *The Naturalist* (1953) which made their reputation as much through the personality of the presenter as their content. A rising star on *The Naturalist* was Peter Scott (Plate 10), a well-known ornithologist and bird artist who had been sidetracked during the war into designing camouflage for battleships. Scott was also turning heads with a new concept for conserving wild animals and birds, while making them accessible to the public. His Severn Wildlife Trust at Slimbridge in Gloucestershire quickly became a safe haven for migrating wildfowl and a mecca for those who wanted to see them at close hand.

Television, for so long the poor relation of radio, began to invade the living rooms of Britain during the mid-1950s. The tiny screens delivered a fuzzy grey picture which left colour and excitement to be filled in by the presenter's enthusiasm and the viewer's imagination. Cued in by the success of *The Naturalist* on radio, the BBC began to roll out television programmes about the living world to ever bigger and more receptive audiences. The flagship series were Peter Scott's *Look* and David Attenborough's *Zoo Quest*.

Look started as a one-off in 1953, but, thanks to Scott's command of the small screen, soon grew into a regular fixture that eventually ran until 1970. The first programme was an outside broadcast which showed Scott in his element at Slimbridge and proved that he had found an even better medium than paint. He came across as dedicated, visionary and someone who could make big things happen. This was a man to watch, and one who would go far. And he did: from Scotland to New Guinea, the Galapagos Islands to Antarctica, and the Steppes to the Great Barrier Reef.

David Attenborough's *Zoo Quest* instantly hit the spot when the first programme was broadcast in 1954. Attenborough had been signed up to film London Zoo's expeditions to catch rare and exotic animals and put them behind bars in Regent's Park. However, Attenborough was an irrepressible Young Turk whose personality blazed through from the start; in no time, this was his show. The result was a series of 25-minute masterpieces which were well ahead of their era and thrilled and inspired the first generation of regular television-watchers.

Attenborough's boldest adventure was the six-part *Zoo Quest for a*

Dragon, broadcast to an expectant nation in October 1956. The dragon in question was a massive but secretive monitor lizard, confined to Komodo and a handful of other Indonesian islands. A fully grown Komodo dragon can stretch over ten feet from snout to tail and weigh up to 360 pounds. Despite its bulk, it can move at nightmarish speed – as has been discovered too late by deer, goats and occasional humans. Although huge, the dragon remained invisible to naturalists until 1910. It covered its tracks well in its habitat of dense tropical forest, and, more importantly, local stories of the dragon were dismissed as folklore. It was only in 1912 that a proper expedition was dispatched to Komodo and the first specimens were captured for scientific study.[2]

Zoo Quest for a Dragon was much more than a trek to fill cages; it turned into a magical travelogue, spiced with geography, anthropology and zoology. The adventure built to a thrilling climax via an orphaned orang-utan in Borneo, sulphur miners dicing with death on the lip of a volcanic vent in Java and the electrifying cacophony of the Balinese Monkey Chant. And who could forget the fresh-faced Attenborough, laughing off the moment when he realised that their boat was skippered by a gun-runner who would think nothing of murdering his passengers to steal their equipment?

The dragons, when two of them finally appeared, did not disappoint: vast, prehistoric in appearance and making horrifyingly quick work of the goat's carcass which had tempted them out in front of the cameras. Confronted with Attenborough's footage of these gigantic, other-worldly reptiles, viewers might have wondered how animals that huge had been so hard to find. And why had scientists been so slow in going to find out if they actually existed?

Down below

Meanwhile, an even vaster and more inhospitable wilderness was being opened up to television viewers, thanks to a new generation of explorer-naturalists who went boldly where few had gone before. Man might have deluded himself into believing that he had conquered the oceans; in reality, he had barely ruffled the surface. The two grand masters of the new art form of the underwater documentary were Jacques Cousteau and Hans Hass.

Cousteau was ex-French Navy and exuded copious Gallic charm both above and below the water. In 1956, aged 45, he had carried off the Palme d'Or at the Cannes Film Festival for his epic underwater documentary, *The Silent World*. His expedition boat *Calypso* and the futuristic 'diving-saucer'

Denise were soon familiar on television screens around the world. After a shaky start (dynamiting reefs), Cousteau became committed to saving the oceans, and summarised mankind's collective responsibility in his catch-phrase, 'We are all in the same boat'.

However, that boat might not have had room for Hans Hass, his arch-rival, who made no secret of what he thought about the Frenchman ('For Cousteau, there exists only Cousteau'). Born in Vienna, Hass was a pro-fessional zoologist as well as a film-maker and had proved that academia could march on regardless in Nazi Germany; when Stanley Kemp was stuck in the blazing ruins of the Marine Biological Laboratory in Plymouth, Hass was diving in the Aegean, doing fieldwork for his PhD. Hass also began his career by dynamiting fish but became more famous for balletic diving sequences featuring sharks, 'the most beautiful and maligned ocean crea-tures'. His wife, Lotte, considered by some to be even more beautiful than sharks, often co-starred. One of his early coups was the first underwater film of a whale-shark, which helped to win him an Oscar for *Under the Red Sea*. At 100 feet and 30 tons, the whale-shark is the world's biggest fish – and all the more remarkable because its existence had never been suspected until 1928, when one became entangled in the anchor-chain of an American battleship in Hawaii.[3]

As well as making millions of viewers desperate to swim with the fishes, both men were innovators. Cousteau claimed that he perfected the scuba breathing system; so did Hass. Cousteau also hitched a lift on an invention which revolutionised underwater photography and made him famous. He teamed up with Harold 'Doc' Edgerton, a research engineer at the famed Massachusetts Institute of Technology in Boston, who had pulled off some spectacular photographic feats during the early 1950s. Using his novel electronic shutter that made the blink of eye seem like an age, Edgerton captured the one-thousandth of a millionth of second when a bullet passed through a still-intact balloon, and the fireball of a nuclear explosion was barely 20 feet across.[4] Edgerton also invented the stroboscopic flash unit which enabled Cousteau to take high-definition photographs deep under-water. Cousteau returned the compliment by nicknaming Edgerton 'Papa Flash' – an accolade which the American wore with great pride.

Climate change

Observant readers will have noted that the Monster has not yet been men-tioned in this chapter. However, it had not disappeared. Sightings continued, even when most people had turned their back on the Loch to confront the bigger issues of the war and its aftermath. A thorough researcher would have built up a dossier of over 200 sightings to celebrate the twenty-first anniversary of the 'Strange Spectacle' reported by Alex Campbell in May 1933.

Nonetheless, the Monster had fallen victim to public apathy. Few reports had broken through into the local newspapers, let alone the national and international press. This had always been a tantalising mystery, but the evidence was inconclusive. What was needed more than anything was the official blessing of serious scientists, and that had been conspicuously withheld. Nothing significant had happened since the thrilling days of 1934. In the meantime, the world had moved on, and the Monster could easily have been left behind in its 1930s time capsule.

However, as 1957 got under way, two decades of indifference were about to end. Now that Britons could finally put their post-war privations behind them there was new optimism in the air. This opened the door to the television naturalists who brought astounding creatures from the ends of the world and the depths of the ocean into the nation's living rooms. The can-do zoological expedition, which went into the most inaccessible places on the planet and teased out their secrets, was transforming the way in which people viewed the world. Once again, the public were ready to be excited by the challenges raised by the coelacanth, the whale-shark and the Komodo dragon. These all proved that there must still be unknown animal species out there, waiting to be discovered – and the coelacanth was a game-changer, demonstrating that extinction was not necessarily for ever.

So far, the action had been concentrated in exotic locations and had transported Britons to brighter and better places; the gloomy depths of a Scottish loch were no competition for the Great Barrier Reef or the Coral Sea. The Monster needed an evangelist who would spread the word so persuasively that Loch Ness would at last be explored by the likes of Cousteau, Hass, Scott or Attenborough.

As it happened, an evangelist was already in place, had been hard at work for several years and was determined to convince a sceptical world that the Monster was not just a legend. This turned out to be a winning formula. Before long, converts to the cause would include adventurers, entrepreneurs, card-carrying scientists – and a wildlife expert whose face was instantly recognisable from the television screen.

Surprising information

Purists' eyebrows (and possibly their hackles) might have been raised by a book that appeared on the publisher Hamish Hamilton's Autumn List for 1957 alongside *The Silent World* by Jacques Cousteau. *More Than a Legend* by Constance Whyte had a striking cover by master illustrator Val Biro, showing a stretch of troubled water framed by boulder-strewn slopes (Plate 11). The title gave little away but the subtitle revealed all: *The Story of the Loch Ness Monster.*[5]

By comparison with Cousteau, the 55-year-old Whyte (Plate 12) was unknown and lightweight: previously unpublished, and a non-practising family doctor with no scientific expertise beyond her medical training almost thirty years earlier. However, she had lived beside the Loch for over 20 years as the wife of the manager of the Caledonian Canal, and had done her homework. Rupert Gould completed his research in two weeks; Constance Whyte took almost ten years. As well as a 'great deal of surprising information' obtained from almost 200 'witnesses of integrity', she delved into the archives from Sir Edward Mountain's 1934 expedition, and the diaries and papers of the late Dom Cyril Dieckhoff, formerly Abbott at the Benedictine Abbey in Fort Augustus. A fluent Gaelic speaker, Dieckhoff had followed closely the evolving story of the Monster and convinced himself that it existed.

Whyte's 'surprising information' had not been won easily. Even as a respected local figure and with a doctor's listening skills, she had struggled to win the confidence of the 'ordinary people' of the Great Glen. The barrier, she explained, was the Highlander's reluctance to talk about the Monster, due both to fear of ridicule and lingering superstition about the malevolent kelpie. To complicate things further, many older people apparently saw no need to distinguish flesh and blood animals from the creatures of folklore. But she persevered, and the more she dug, the more she found.[6]

More Than a Legend was built around 80 sightings, many reported by multiple witnesses and some corroborated by others watching simultaneously from other sites. Twenty-five reports were 'genuine, first-hand [and] contributed by persons well known to me'.[7]

Particularly compelling was the new evidence which Whyte gathered after Gould left off in May 1934. Early on a Sunday morning that June, a housemaid who worked for a schoolmaster at the Abbey spotted 'the biggest animal she had ever seen in her life', lying half out of the water on the shore of Borlum Bay. She watched it through binoculars for over 25 minutes, while it turned itself in the sunshine, arched its back into one or more humps and finally slipped quietly into the water. It had a massive, elephant-grey body, two short forelegs (or possibly flippers) and a long, giraffe-like neck with 'an absurdly small' head.

Even more dramatic was the afternoon in July 1947, when a 30-foot creature with five or six dark brown coils charged towards the shore near Inverfarigaig, coming within 30 yards of the water's edge and throwing up a massive wash that sent picnickers running up the beach in panic. The spectacle lasted some minutes and was watched simultaneously by several others at various vantage points. The afternoon was rounded off nicely when three people in a car some miles away were amazed to see the Monster with its head held high out of the water, hurling up spray as it sped down the Loch.

Those who wondered what the Monster looked like at close quarters could have asked Mrs Greta Finlay about what she and her son saw in the shallows near Aldourie on 20 August 1952. Barely 20 yards from their caravan was a 15-foot animal with two or three humps, an erect neck about two feet high and a 'hideous' head carrying two projections that each ended in a blob. The creature was black and shiny, and made Mrs Finlay think of a gigantic snail. It made off as she ran for her camera, then submerged 'in a great commotion which set waves breaking on the shore'.

On 8 October 1954, the bus from Drumnadrochit to Fort Augustus braked suddenly on the road beyond Strone; the driver and all 27 passengers piled out and spent ten minutes watching a 25-foot brown object with three humps 'cruising about' on the Loch. A few months later in April 1955, two boys climbing Mealfuarvonie above Invermoriston saw a two-humped creature moving 'at great speed' across the Loch; it was big enough to be plainly visible to the naked eye at a range of two miles, and they watched it for nearly half an hour.[7]

Whyte catalogued over 60 additional sightings, involving almost 100 witnesses (identified only by their initials), in a 15-page appendix at the end of the book. Some matched reports in Gould's book, such as the humped creature with a frilled neck observed from the Halfway House tea room at Alltsigh,[8] and the 30-foot, long-necked animal which 'A.C.' watched in Borlum Bay for several minutes, turning 'its head and neck about as quick as a hen' before it sank as two drifters appeared from the Caledonian Canal.[9]

Many sightings were clustered around hot spots, especially where rivers ran into the Loch. Temple Pier, on the north side of Urquhart Bay, held the record with hundreds of reports; Alexander Ross, formerly the piermaster there, had spotted the Monster 15 times, the last occasion being on his seventy-eighth birthday. Not far behind were Fort Augustus, where water bailiff and *Courier* Correspondent Alex Campbell had seen it six times, and Foyers where a Mr Grant was 'a mine of information'. Like Gould, Whyte identified flat, calm water and hot days as the most favourable conditions for seeing the Monster.[10]

Those fortunate enough to have seen the Monster spanned all strata of society. Many were 'ordinary people [who] generally tell the truth' and, Whyte was sure, had nothing to gain from embellishing or fabricating their stories.[11] Others were dignitaries and professional people. On 7 August 1944, a blackish animal with two humps and a lifted head put on 'a fine display' for J. W. MacKillop CBE, County Clerk of Inverness-shire, which he judged enough to convince 'even the most rabid scoffer that the Loch held something abnormal'. Five years later, Lady Maud Baillie, who had commanded the local Auxiliary Territorial Service during the war, watched a large black object leaving a foaming wake off Inverfarigaig. With her was the equally no-nonsense Lady Florence Spring Rice, who declared that she had never had a more exciting afternoon; as the wife of a high-flying diplomat, she would have had plenty to choose from. In all, company director S. Hunter Gordon saw the monster three times, once more than Dr Kirton of Fort Augustus.[12]

Picture power

More Than a Legend contained new information from the very start. The frontispiece was the classical Surgeon's Photograph, together with the previously unpublished picture of the famous neck and head, apparently slipping forward and about to submerge, which the Surgeon had 'taken immediately afterwards' (see Plates 6 and 7). The second image, although 'less successful', showed 'how the animal had moved' and was therefore alive.[13]

Pride of place was given to a 'good picture' of the Monster, taken by Lachlan Stuart on 14 July 1951 (Plate 13). The thirty-year-old Stuart was one of a group of woodsmen employed to clear trees from the lochside south of Dores. He lived with his wife and three children in a croft near Whitefield, which looked across the water from 100 feet above General Wade's Military Road. On his way to milk the cow at 6.30 a.m., he spotted a two-humped object moving fast up the Loch, and sprinted back to the croft for his box camera. He and a friend ran down through the trees to the shore, where a creature lying just 30 yards out was showing three angular blackish humps, each about five feet long and three feet out of the water. Stuart managed to take a photograph but failed to capture what happened next: a long neck with a sheep's-sized head appeared in front of the furthest hump as the animal approached them. Then it veered away with much splashing and submerged about 300 yards offshore.[14]

By late afternoon, the *Daily Express* had acquired the film and took a hot-off-the-press proof of the photograph to Dr Whyte at the Canal Office, Clachnaharry, for her opinion as the local authority on the Monster. Her verdict: 'I could not have put forward this photograph with more confidence than if I had taken it myself.' Stuart's snapshot duly made the front page of the *Scottish Daily Express* the following day, Sunday, 15 July 1951. While visiting Stuart's croft a couple of days later, Whyte discovered just how much the Monster was taken for granted by these people; one of the other woodsmen told her that he and others had also seen it during the previous weeks.[14]

At the end of 1954, new technology was brought to bear on the Monster by another group of 'ordinary' people, the exclusively teetotal crew of the fishing boat *Rival III* from Petershead.[15] *Rival* entered the Loch through the Caledonian Canal on the wintry morning of 2 December, with the crew drinking tea below deck while the mate, Peter Anderson, manned the bridge. Shortly before midday, as they passed Urquhart Castle, the echo sounder began to trace out an astonishing object. Its irregular shape was completely alien to Anderson, who was well used to the appearance of shoals of fish on the machine.

Within hours, the chart recording had fallen into the clutches of the press. Before publication, it was scrutinised by the manufacturers of the echo-sounding equipment. The trace corresponded to a large object, perhaps 50 feet long, that had followed the course of the *Rival* for almost three minutes and half a mile – while hanging 480 feet beneath its keel and 120 feet above the bottom of the Loch. Like the skipper, the experts had no idea what it was, but were confident that fish, a waterlogged tree-trunk suspended in mid-water and fakery could all be ruled out.

Identikit

Whyte acknowledged that large aquatic animals were rumoured to inhabit various freshwater lakes beyond the Great Glen, such as the legendarily deep Loch Morar on the west coast of Scotland and Okanagan Lake in Canada. However, she was unconvinced that these 'half-fabulous' beasts actually existed. The only one for which there was incontrovertible evidence was the Monster of Loch Ness, which Whyte graced with a capital 'M' to distinguish it from the 'ordinary run-of-the-mill monsters'. Its presence had long been taken for granted by the locals; now, there could be no doubt whatsoever that 'a very strange creature had its home in Loch Ness'.[16]

Sceptics had suggested seals, mass hallucinations, hoaxes, decomposing tree trunks propelled by gas and boulders shot up from the Loch floor by earthquakes.[17] Whyte regarded all these ideas as ludicrous. There had been numerous hoaxes, but these had been identified and excluded. The scams had included metal cut-outs nailed to tar barrels discarded from the road works; an echo-sounder tracing, just days after *Rival III* made headlines, but so clumsily faked that it fooled nobody who knew anything about echo sounders; and of course the hunting-trophy buffoonery which had disgraced Marmaduke Wetherell. By coincidence, Whyte's own brother had been working in the Natural History Museum when Dr Calman announced the verdict on Wetherell's infamous spoor. This might explain why Whyte took such delight in mocking Wetherell and his sidekick, photographer Gustave Pauli. She nicknamed them 'Pilt' and 'Down', respectively, to remind readers that the famous 'Piltdown skull' had recently been revealed as a fake.[18]

Whyte went on to build up a picture of the Monster from dozens of verbatim accounts, numerous sketches and a handful of photographs. Twenty to fifty feet long, it had a bulky body that showed up to seven humps. Projecting from an egg-shaped swelling at the front was a long, snake-like neck ending in a small flattened head which carried two horn-like structures. The tail was long and blunt-ended. Two pairs of powerful flippers enabled the animal to dive, swim or dash across the surface at up to 30 mph. The skin was dark and almost black when wet, with many observers likening it to an elephant's hide.[19]

The result looked like nothing on Earth – or at least nothing that had been seen on Earth for over 60 million years. Incredibly, this picture fitted the evidence. In 1933, George Spicer had described the 'loathsome' animal that crossed his path as 'the nearest approach to a dragon or a prehistoric monster that I have ever seen in my life'. Lachlan Stuart, the woodsman turned photographer, agreed: 'like a prehistoric monster in a school book'. Sir David Hunter Blair, the scholar and writer who later dropped his title to become Abbott Oswald of Dunfermline, distilled all the wisdom from 'a veritable cloud of eye-witnesses' into the startling conclusion that 'this animal . . . belongs to no existing species but to a period dating back some hundreds of millions of years'.[20]

Even though supposedly long extinct, this species was familiar to millions, young and old. In the early 1930s, children in Drumnadrochit School were asked by their teacher to describe the 'horrifying' animal they had spotted lying in the swampy shallows of Urquhart Bay; without hesitation, they pointed to the picture of the plesiosaur displayed on their classroom wall.[21] They might have been young and impressionable, but what about Dom Cyril Dieckhoff, Gaelic scholar and seasoned observer of the Loch and its phenomena? Dieckhoff had seen the Monster twice, and always referred to it as 'the plesiosaurus'.[22] Constance Whyte was never lucky enough to meet the Monster herself, but took her cue from Dieckhoff when coining her own name for it: *Lochnessaurus*.[23]

Some of the Monster's reported features were hard to reconcile with a plesiosaur. The 'coils' described by some witnesses sounded more like Rupert Gould's beloved sea serpent, but could still represent contortions of the plesiosaur's back and tail. More problematic were the many humps (the record stood at nine) which many had claimed to see. Like all vertebrates, the plesiosaur's spinal column could flex from side to side, but could not realistically bend itself into vertical humps or undulate like a caterpillar. Whyte speculated that the 'humps' might be inflatable air sacs, as found in the African lungfish. Such sacs could also enable the Monster to remain at depth for long periods without coming up to breathe – which could also explain why sightings were so infrequent.[24]

Cul-de-sac

Naturally, the plesiosaur hypothesis was untenable unless there was a plausible explanation for how this archaic species had sidestepped extinction alongside the dinosaurs. Whyte supposed that its original habitat – 'the abyssal depths of unfrequented regions' – had shielded it from whatever cataclysm had engulfed its peers. No fossils of plesiosaur-like animals postdating the dinosaurs' demise had ever been discovered, and no recent Monster remains had been found in or around the Loch. However, the geological record was full of holes – as proved by the coelacanth and other 'living fossils' that had come back from the undead – while the steep sides of the Loch 'that never gave up its dead' would funnel deceased Monsters straight to the bottom.

Whyte admitted that she struggled to explain how the Monster ended up in the Loch.[25] While the Loch's basin was being gouged out of solid rock, the Great Glen lay under thousands of feet of ice. The Loch filled with water only after the ice had melted, by which time the Highland landmass had risen, leaving Loch Ness sitting proudly with its surface 50 feet above sea level. It followed that the animals could only have come in from the sea, where a sustainable population must have survived since the Age of the Dinosaurs, presumably protected in their own 'abyssal depths'.

Their portal of entry into the Loch remained wide open for speculation. Some authorities, including the former Director of London Zoo Aquarium, believed that the Loch and the North Sea were connected by a network of underground tunnels. Whyte dismissed this notion because the Loch's surface would have fallen to sea level if the two were in free communication.

That left the Caledonian Canal and the River Ness, both entering Beauly Firth several miles beyond the northern end of the Loch. Lampreys had smuggled themselves into Loch Ness, presumably stuck to the hulls of boats trekking up the Canal, but this was a no-go zone for sizeable marine animals which would have to run the gauntlet of seven locks and their keepers. The River Ness, where St Columba had met his Pict-eating monster, was a potential way in, especially when in flood. However, Whyte argued that not even seals, smaller and more agile than the Monster, had managed to negotiate the river. One seal had made it to Holm Mills, seven miles inland, but none had been recorded from the Loch itself.

Whyte had a more radical explanation. At some point during the thousands of years during which the ice was melting and the land rose, the forming Loch had opened directly into the sea. This was the window of opportunity through which the Monster's ancestors had entered their dead-end breeding ground and resting place.[25]

The Monster's life cycle was also shrouded in mystery. They presumably ate fish, and the Loch held plenty of salmon, trout and the deep-swimming Arctic char (itself a relic from the Ice Age). In 1943, three boys from the Abbey School had watched three 'baby Monsters', or at least sleek grey lizard-like animals with powerful fore-flippers, slicing through the clear water under their boat.[26] Beyond that, theories of how Monsters might reproduce were anybody's guess.

Anti-establishment

Constance Whyte had great respect for the 'ordinary people' of the Great Glen but was scathing about the stuffed shirts of the scientific establishment.

1. Photograph taken by Hugh Gray at Foyers on 12 November 1933, published in Glasgow *Daily Record* and *Daily Sketch* on 6 December 1933.

2. Commander Rupert Thomas Gould (1890–1948), restorer of John Harrison's famous marine chronometers and the 'man who knew (almost) anything' on the BBC's *Brains Trust*. Gould was fascinated by 'oddities and enigmas' and wrote *The Case for the Sea-Serpent* (1930) and *The Loch Ness Monster and Others* (1934).

3. Marmaduke Arundel Wetherell (1884–1939), film star, director, big-game hunter and leader of the *Daily Mail*'s expedition in December 1933 to find the Loch Ness Monster. (*Stage and Cinema*)

Mr. A. Wetherell (smoking) and Mr. A. Grant examining remains of a sheep found during search for evidence of the Loch Ness monster.

4. Marmaduke Wetherell with Arthur Grant (right) and Grant's younger brother Peter, on the shore near Abriachan, 4 January 1933. (*Daily Mail*)

5. Robert K. Wilson (1899–1969), soldier, doctor and expert on the machine-pistol. His 'Surgeon's Photograph' became an iconic image of the Monster and one of the most famous photographs of the twentieth century. (David Wilson)

6. The 'Surgeon's Photograph', taken by R.K. Wilson on 19 April 1934, approximately two miles north of Invermoriston. (*Daily Mail,* 21 April 1934)

7. Second photograph taken by R.K. Wilson, immediately after the 'Surgeon's Photograph'. This image was not published at the time but was later sold as a postcard of the Monster by Ogston's Chemist Shop in Union Street, Inverness.

8. The cover of Colonel W.H. Lane's *The Home of the Loch Monster*. Published in spring 1934, this 18-page pamphlet was the first book on the topic. (Roland Watson)

9. The team of local men assembled by Sir Edward Mountain to watch for and photograph the Monster in July 1934. Mountain, holding a hat and wearing a light suit, is standing between the men kneeling in the front row. (IMAGNO, Austrian Archives and TopFoto)

10. Peter Scott (1909–89), at Slimbridge in the early 1960s. Photograph by Philippa Scott. (Dafila Scott)

11. The cover of *More Than a Legend* by Constance Whyte, first published in 1957. (Val Biró)

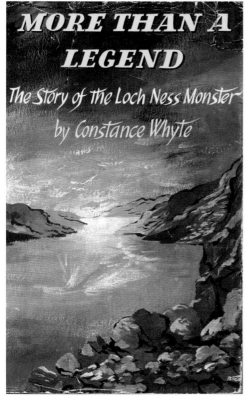

MORE THAN A LEGEND
The Story of the Loch Ness Monster
by Constance Whyte

12. Dr Constance Whyte MB, BS (1902–82), lapsed general practitioner, wife of the Manager of the Caledonian Canal and author of *More Than a Legend* (1957). (Nicholas Witchell)

13. Photograph taken by Lachlan Stuart, below Whitefield on 14 July 1951. (Fortean Picture Library, Ruthin)

14. Photograph taken by Peter Macnab, bank manager and Ayrshire County Councillor, on 29 July 1955. This is the copy which he sent to Peter Scott on 21 April 1961. (Peter Scott Archives by kind permission of Dafila Scott and the Syndics of the University of Cambridge)

15. Tim Dinsdale (1924–87), aeronautical engineer and 'monster-hunter extraordinaire' who caused a sensation with his four-minute cine-film, taken on the morning of 23 April 1960 (see below). (Reproduced by kind permission of Dawn Dinsdale and the Dinsdale family)

16. Frames from the cine-film taken by Tim Dinsdale on 23 April 1960. TOP: a 15-foot open fishing boat with a single occupant, powered by a British Seagull outboard motor, following the same course as the humped object filmed earlier. BOTTOM: the humped object. Note the differences in appearance and in the patterns of the wakes. (Reproduced by kind permission of Dawn Dinsdale and the Dinsdale family)

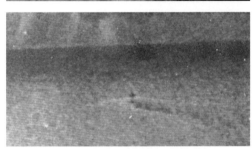

17. Philippa and Peter Scott with Eamonn Andrews on *This Is Your Life*, broadcast in February 1956. (Dafia Scott)

18. Maurice Burton (1898–1992), Curator of Sponges and Deputy Keeper of Zoology at the Natural History Museum, who changed his mind several times about the existence and nature of the Loch Ness Monster. (The Trustees of the Natural History Museum, London)

19. Dr Denys Tucker (left), Curator of Fishes at the Natural History Museum in London, with Professor J.L.B.Smith (right) and Dr Ethelwy Trewavas, outside the Museum in April 1958. Smith described the first coelacanth in 1938. (The Trustees of the Natural History Museum, London)

20. Flashlit photograph taken by Peter O'Connor, reportedly at 6.30 a.m. on 27 May 1960. The image was described as 'astonishing' by Tim Dinsdale, who was allowed to reproduce it in his book *Loch Ness Monster* (1961) on condition that he acknowledged that the animal had been given the scientific name *Nessiesaurus O'Connori* by the Northern Naturalists Organisation. (Fortean Picture Library, Ruthin)

Arrogant and tunnel-visioned, they could not even be bothered to get out of their armchairs and visit Loch Ness to review the evidence for themselves. Of course, the Monster appeared to flout some basic rules of zoology, and pursuing it could turn out to be a waste of time. However, rules were made to be broken, and open-mindedness was supposed to be the hallmark of a good scientist.

Yet professional zoologists had been hostile from the start, rubbishing some of the best evidence. Hugh Gray's seminal photograph of the massive creature ploughing across the Loch's surface was dismissed as 'unconvincing as a portrayal of any living species' (Sir Graham Kerr, Regius Professor of Zoology at Glasgow University), or 'a bottle-nosed whale . . . or a shark . . . or mere wreckage' (Mr John R. Norman, Curator of the Piscatorial Department at the British Museum). In 1934, a roomful of alleged experts at the Linnean Society diagnosed the Surgeon's Photograph as a seal or the dorsal fin of an elderly killer whale. And Dr E. G. Boulenger, Curator of Lower Vertebrates at London Zoo, wrote off the entire Monster business as 'a stunt, foisted on a credulous public, and only excused by a certain element of low comedy'.[27]

All the prejudices and inflexibility of the scientific establishment had been neatly summed up by one of the leading biologists of the 1930s, Sir Arthur Keith FRS. Speaking from the ivory tower of the Royal College of Surgeons, he proclaimed: 'Give me a piece of hair, skin or fur and I will tell you what this animal is.' In the meantime, the failure of the Monster to reach the dissection table – together with all the discrepancies in the eyewitnesses' testimonies – had convinced Keith and his peers that this was 'not a thing of flesh and blood'. The Monster belonged to 'the world of spirits' and was nothing more than a mass hallucination. Keith dismissed it as a problem for psychologists, not zoologists.[28]

Accepting that the Monster 'challenges beliefs which are integral to the mental make-up of most established zoologists', Whyte knew that *More Than a Legend* would set her on a collision course with the scientific mafia.[29] In some ways, though, the landscape now appeared less hostile. Most of the original naysayers were dead or otherwise out of action. The most formidable, Sir Arthur Keith, had died a couple of years earlier, with egg (if not an entire omelette) on his face after the infamous affair of the Piltdown skull. From a detailed analysis of the skull, Keith had championed the proposal that 'Piltdown Man' was a missing link between man and higher apes. Unfortunately, the skull turned out to be a cleverly doctored and chemically aged fake, knocked together from a human cranium and the jawbone of an orang-utan.

But was the new generation ready to take the Monster seriously? Whyte was pessimistic: 'Official research is so controlled that it is extremely difficult for a young scientist with a hunch to go out and follow it up'.[30] So she threw down the gauntlet. Here was an unparalleled opportunity for someone with curiosity and an open mind to make their name by cracking one of the greatest unsolved mysteries in modern biology.

What if no serious scientists rose to the challenge? Then the operation would proceed without them, because the Monster was too important to let slip into oblivion. 'This is a zoological riddle, a really super whodunit,' wrote Whyte, and 'the perfect mystery for amateur zoological sleuths.'[31] This was not mere fantasy. J. L. B. Smith, who recognised the coelacanth for what it was and chalked up the zoological discovery of the century, was a professor of chemistry, not marine biology. His knowledge of fishes was vast, but entirely self-taught; Smith was the epitome of the gifted amateur. Following his example, the prize of the Monster could be carried off by enthusiasts with passion rather than a degree in zoology.

Lasting impressions

More Than a Legend created a sensation when it was published in late 1957. It pulled the reader in, whereas Gould had put many off with his obsessive cataloguing of each sighting and his determination to turn the Monster into a sea serpent.

Whyte's book sold briskly, buoyed up by the enthusiastic reviews quoted on the dust jacket: 'evidence impressively marshalled against the hosts of incredulity' (*Evening Standard*) and 'a most readable and valuable book' (*Sunday Times*). It also won the approval of a professional zoologist who was brave enough to stick his head up from behind the establishment's barricade. This visionary was Dr Maurice Burton, Deputy Keeper of Zoology at the British Museum, Fellow of the Zoological Society and author of the best-selling *Living Fossils* (1953). Burton wrote in the *Illustrated London News* that this was 'the most complete survey ... as convincing as we could hope for in a phenomenon which has never really been investigated'.

Critics might have questioned some fine details. Why did Whyte refer to the person who took the Surgeon's Photograph only as Dr ——, when the identity of Robert K. Wilson, gynaecologist of Queen Anne Street, London W1, had never been a secret?[32] And occasionally readers were left tantalised. For example, when did three anglers watch a 'dark, massive shape' slipping away under the water after it nearly capsized their boat off Tor Point?[33]

Overall, though, the publication of *More Than a Legend* was both a landmark and a turning point in the Monster's fortunes. Since before the war, the legend had been dying a natural death. As a measure of that terminal decline, Rupert Gould's contribution had already been forgotten when Whyte began her research in earnest. She only discovered his book (by then out of print) in 1949, too late to talk to the man himself.

To ensure that Commander Gould was not overwritten by history, Whyte devoted an appendix to him and his works. She ended *More Than a Legend* by proposing an epitaph for him, taken from the title of one of his essays:[34]

There were giants in those days.

5

A contagious obsession

Spring 1960, and a new decade and a gripping new chapter in the saga of the Monster were about to get under way. The Sixties were not quite ready to swing. Carnaby Street was just a place name near Soho, The Beatles (John, Paul, George and Stuart) were still in Hamburg, and the contraceptive pill was waiting to be unleashed. On the global stage, the Americans were five years into the war which they knew they would win in Vietnam, and there was a new space to watch up there, criss-crossed by sputniks and missiles capable of carrying nuclear warheads.

Down in Westminster, the Conservatives were in power and Harold Macmillan was in Downing Street. Up in the Great Glen, it was business as usual. Prosperity had crept back during the Fifties, and Inverness was once again a thriving tourist centre which boasted a unique portfolio of diversions: hill-walking, motor-touring, salmon-fishing and Monster-hunting.

The magnetism of the Monster was as powerful as ever, thanks to Constance Whyte's account of the extraordinary creature that might just reveal itself to a lucky tourist. Accordingly, the excitement was sustained by a steady influx of visitors determined to prove for themselves that the legend had become reality.

On the record

Since Whyte's book was published in 1957, dozens of lucky eyewitnesses had been in the right place at the right time. On the afternoon of 2 February 1959, AA patrolman Hamish Mackintosh discovered that he was not alone while making a call from the phone box beside the landmark monkey-puzzle tree at Brackla, south of Abriachan. A few hundred yards out in the Loch, a slender neck and head that rose about eight feet above a broad, humped body were turning from side to side. Mackintosh flagged down two passing cars and another four witnesses joined him to watch the display, which lasted for some minutes.[1]

Better prepared was Torquil MacLeod, who had been hired by an anonymous gentleman (later revealed to be Sir Charles Dixon) to capture the Monster on film. MacLeod knew the terrain well, as he and his wife Elisabeth had given up everything to pursue the Monster full-time in their converted army-surplus radio lorry. For Dixon's expedition, MacLeod covered all eventualities, with a 16mm cine-camera, range-finding 8x binoculars of the sort used by artillery spotters to measure precisely how far their shells fell off target, and (glimpsed by staff at the Foyers Hotel in the boot of his car) a sawn-off shotgun.[2]

On the afternoon of Sunday, 28 February 1960, MacLeod was driving along the Fort Augustus road a couple of miles south of Invermoriston when he caught sight of something dark moving at the base of the arc of grey scree known as The Horseshoe, about a mile across the Loch. Through his binoculars, he saw 'a huge animal', lying partly out of the water, with a long neck that moved like an elephant's trunk. MacLeod watched it for about nine minutes, until the animal flopped into the water and disappeared. He had not bothered with the cine-camera as the range was too great; unfortunately, the creature did not reappear.

Back at the hotel and 'excited as a schoolboy', MacLeod immediately sat down to record his 'first hint of success'. 'I am rather appalled at its size,' he wrote, having calculated the length of its body (minus the submerged tail) to be about 50 feet. Strangely, he addressed his letter not to his employer, Sir Charles Dixon, but to Dr Constance Whyte.[3]

So much for those fortunate enough to have seen the Monster for themselves. Everyone else wanted evidence that they could inspect, and, in the continued absence of a captive specimen or some body parts, photographic emulsion was the next best option. Since the publication of *More Than a Legend*, the Monster had been caught on film again – and in an image that rivalled the Surgeon's Photograph for dramatic impact.

The picture, published in the *Weekly Scotsman* on 23 October 1958, gave credibility to Torquil MacLeod's estimate of the length of the creature that he had watched beneath the Horseshoe Scree. The photographer was Peter Macnab, a man with unquestioned credentials: a bank manager and a County Councillor in Ayrshire. On the afternoon of 21 October, he and his son had been driving along the road above Urquhart Bay and had stopped for Macnab to photograph the ruins of the Castle. As he was lining up his shot, he noticed a disturbance in the water to the left of the Castle and quickly changed the standard lens of his Exacta 127 camera to a six-inch Dallmeyer telephoto. Through this, he saw 'some black or dark enormous creature cruising on the surface' and took a photograph, 'in a great hurry' and without time to screw the camera on to its tripod. He then picked up a simple fixed-focus Kodak Brownie camera and took a second shot. Meanwhile, his son was 'busy under the bonnet of the car', checking the engine; by the time he looked up, the animal had submerged, leaving just ripples on the surface.[4]

When Macnab showed his friends the photographs, their response was discouraging. Fed up with their 'scepticism and leg-pulling', he destroyed the negative from the second camera and 'nearly got rid of the first as well'. Fortunately, he relented, and an astonishing image was saved for posterity and critical inspection.

His snapshot shows the ruined tower of Urquhart Castle on its promontory, strongly lit from the right, and with its outline clearly reflected in the mirror-calm surface of the Loch (Plate 14). To the left are two long, low black humps at the front of a wake that is heading towards the Castle. From the scale of the tower, the object is obviously massive. Later, a 50-foot trawler was photographed in the same position and, from this, Macnab's creature was estimated to be 50–60 feet in length.

The 'Macnab photograph' quickly became one of the classic images of the Monster, alongside Hugh Gray's 'serpent', the Surgeon's Photograph and Lachlan Stuart's three-humped creature lying in the shallows. It was declared by experts – including some previously staunch unbelievers – to be absolutely genuine and one of the most solid pieces of evidence for the creature's existence.

But there was better to come. Eighteen months later, Macnab's photograph was eclipsed by the image of a humped object ploughing across the Loch – captured on 40 feet of cine-film.

Moving pictures

In April 1960, 36-year-old Tim Dinsdale (Plate 15) had a steady but unexciting job at Heathrow Airport. An aeronautical engineer by training, he had been hooked by *More Than a Legend* and had spent several months analysing records of sightings to create his own picture of the Monster. The creature which took shape on his drawing board was 40 feet long, with a small, rounded head on an eight-foot neck, a massive 20-foot body bearing two smooth dorsal humps and two pairs of spade-shaped flippers, and ending in a ten-foot, blunt-ended tail.[5]

Going through this exercise convinced Dinsdale that the Monster was no figment of the imagination and fired him with a passion that he had never known before. At the start of the Easter weekend, he drove off alone to see Loch Ness for himself. With him were a Bolex cine-camera loaded with 16mm black and white film and a sturdy wooden tripod, both borrowed from a friend; he had removed the front passenger seat from his car to make room for the tripod, which stood beside him, ready for action at all times.

Dinsdale's trip began with a 600-mile drive north, and (at risk of sounding clichéd) turned into a journey that changed his life forever.[6] He booked into the Foyers Hotel for the week beginning Monday, 18 April. Each day from dawn to dusk he patrolled the shores of the Loch in his car. The first two days were completely unrewarded and the third even more frustrating, when a fellow guest at the hotel said that he had watched a large V-shaped wake moving briskly up the middle of the Loch. The following day, Dinsdale filmed a disturbance in the Loch, but it turned out to be a shoal of rocks just beneath the surface. The next day – Friday – also drew a blank.

As the sun rose on his last morning, Dinsdale was exploring the southern end of the Loch. He saw nothing and, all but resigned to failure, drove back towards the hotel and breakfast. Then, at about 8.30 a.m., on the road to Upper Foyers, he glimpsed something far out on the water.

He was about 300 feet above the Loch and estimated the object to be 1,300 yards away; later, expert analysis lengthened the range to just over 1,600 yards. Through his 7x binoculars, Dinsdale saw a rounded, mahogany-coloured object stationary in the water. He had just decided that it was not a boat, when ripples broke out of the far end and the object began to move away, showing a large patch of dark grey on its flank. The sight electrified him: 'I knew at once that I was looking at the extraordinary humped back of some large living creature.' Dinsdale quickly swapped the binoculars for the Bolex, waiting on its tripod, and began to film the object 'in long bursts, like a machine-gun', pausing periodically to rewind the camera's clockwork drive mechanism.

For about four minutes he tracked the object as it carved a zigzag course across the Loch, leaving a distinct V-shaped wake behind it. Close to the far side, and now over a mile away, the object turned south and travelled about a quarter of a mile parallel to the shoreline, gradually becoming less distinct. Finally, it disappeared in a band of dark water where the steep mountainside plunged into the Loch.

Dinsdale estimated that he had used about 40 feet of film, leaving about 15 feet – barely 90 seconds' worth – unexposed. He hurled the car down the road to Lower Foyers and drove straight across a field to the shore, hoping to catch sight of the animal again. Unfortunately, as he later wrote, 'there was nothing to be seen for miles in each direction – just the glassy surface. The Monster had dived back into the depths.'

After breakfast, Dinsdale returned to the spot where he had tracked his Monster and finished off the film on another object which followed a similar zigzag course across the water. This was a 15-foot open fishing boat, standard issue for the Loch, fitted with an outboard motor. At the helm was the proprietor of the Foyers Hotel, who had been briefed by Dinsdale. Through his binoculars, Dinsdale could see clearly that this was a man in a boat, and the wake – a broad 'V' bow-wave with a central streak from the outboard – was distinctive. Dinsdale hoped that this 'control' sequence, filmed from the same viewpoint with the same camera, would be enough to prove that the earlier object was not a boat but an animal.[6]

Dinsdale headed home in high excitement and secure in the knowledge that he had the Monster in the can. The film was developed with due care by Kodak and the results were everything that Dinsdale had hoped for. The humped object looked nothing like a boat, even when single frames were blown up several-fold, and its wake looked quite different from that left by the boat (Plate 16).

Dinsdale's film was quickly recognised as powerful, even sensational evidence for the Monster. He had intended to use it wisely, to convince the scientific community that the creature existed and urgently merited a full-scale investigation. The film was shown in closed session to various scientists and had the desired effect of converting some noted sceptics into fervent believers. Unfortunately, hopes for a dignified perusal and a considered verdict were undermined on 13 June 1960, when the *Daily Mail* published stills from a copy of the film.

From that moment, Dinsdale was catapulted into the limelight and would never escape from it. That same evening, he found himself on *Panorama*, the BBC's flagship news and current affairs programme. Ten million people watched the humped object cutting its V-shaped wake across Loch Ness, listened to softly spoken Alex Campbell describing his dozen sightings of the Monster, and saw Dinsdale being interviewed by a clearly fascinated Richard Dimbleby. Afterwards, the BBC switchboard was jammed with calls from people desperate to find out more about Dinsdale and his Monster.[7]

As Nicholas Witchell later wrote, 'Nessie was firmly back in the news and this was a comeback which was going to last.'[8]

*

Tim Dinsdale soon became a household name in Britain and around the world, due as much to his obvious sincerity and unshakeable belief in the Monster as to the footage captured by the borrowed Bolex. Drawn back by the tantalising prospect of seeing the Monster again, he made repeated pilgrimages to the Loch and slipped into the role for which he would go down in history – as Monster hunter and guru.

This is the extraordinary tale of how a stroke of luck transformed an intrigued amateur into one of the most commanding figures in the saga of the Monster, as told by Dinsdale himself. However, there is a back story, with a plot rich in other ingredients – pride, prejudice and persuasion. It also brings in Her Majesty the Queen, and, more significantly, a player who had been waiting in the wings (and looking the wrong way) but would now begin to move centre stage.

Enter Peter Scott.

Superhero

Even if he had made nothing of the opportunities presented to him, Peter Scott would always have been famous. Towards the end of March 1912, Robert Falcon Scott, leader of the *Terra Nova* expedition to the Antarctic, wrote to his wife and urged her to get Peter (then aged two) interested in natural history. Scott was just 11 miles from salvation when he sealed the envelope, but he knew that his wife would be widowed by the time she opened it.[9]

Peter Scott could have cruised comfortably through life as his father's son. Instead, he took pride in carving his own course through things that took his fancy, such as painting, shooting, sport, the war, broadcasting and wildlife conservation. He was spectacularly successful in all of these; in fact, there was just one area in which he never quite excelled.

Scott's obsession with wildlife began in his late teens, although with a different emphasis. As an undergraduate at Trinity College, Cambridge, he found the Natural Sciences Tripos less exciting than lying in a punt behind a muzzle-loading gun that could (and often did) blast half a dozen ducks out of the sky in one go. It was many years before Scott switched from massacring birds to saving them for posterity, but his passion for committing them to canvas also went back to his student days at Cambridge. His paintings of ducks and geese, often gliding in against a fiery sunset or the steely grey of dawn, were instantly recognisable and just as quickly snapped up.[10]

Before the war, it would have been difficult to predict where Scott's career would take him. As well as painting, sport figured large in his life. His ice-skating was not quite up to Olympic level but his dinghy sailing was; in 1936 he brought home a bronze medal from the Berlin Olympics. By then, he was also dabbling in radio broadcasting and was rated as 'a very strong personality, a good voice and very keen to do work of this sort'.[11]

After the war, Scott picked up broadcasting again and rapidly built a reputation as 'a spell-binder'.[11] He even passed the acid test of ensnaring the 'lost tribe' of war-weary ex-servicemen, perhaps because he had credibility with them. Scott's wartime career had gone much further than designing camouflage for battleships. As well as his Distinguished Service Cross for conspicuous bravery during a skirmish with German E-boats in the English Channel, he had also seen action as an observer with the RAF. One memorable sortie – 'an adventure that I would not have missed for the world' – was a heavily resisted night-bombing raid on Kiel, from which seven aircraft did not return. This 'adventure' closed a grim circle for Scott, as the target was the city where dinghies had battled for the Olympic medals just five years earlier. In action, Scott was renowned for being 'cool, calm and collected'. Emotion and excitement broke through only rarely, such as during the home run to Portsmouth after escorting a convoy through the Channel, when a flight of geese passed low overhead.[12]

Scott also had the Midas touch. Items that he turned into gold included his natural history broadcasts, the conservation initiatives that crystallised in 1961 into the World Wildlife Fund, and Philippa Talbot-Ponsonby (Plate 17). His first marriage, in 1942 to the novelist Elizabeth Jane Howard, foundered after a few years. In 1947, he appointed Miss Talbot-Ponsonby as his personal assistant and soon discovered that her impressive curriculum vitae – Bletchley Park, followed by the British Embassy in Belgrade – had failed to capture her greatest qualities. They married in Reykjavik in August 1951 while on a trip to the Icelandic breeding grounds of the pink-footed goose. The following year saw the birth of their first daughter whom they christened Dafila, the Latin name for Scott's favourite bird, the elegant and personable pintail. Under Scott's wing, Philippa flourished as a photographer in ways that she might not have discovered without him. Soon, she was a regular fixture on the *Look* natural history programmes, not as Mrs Scott but as a wildlife expert in her own right.

And the only box that Scott never managed to tick? As an undergraduate at Cambridge, following yet another reprimand for missing lectures while out wildfowling, Scott confessed to an uncle: 'Anyone can learn the names of fossils and the classification of animals, but I don't want to do things that anyone can do.' Later he wrote, 'as my academic career in Natural Sciences held no great promise I decided to change horses in mid-stream. Instead of a scientist, I would become an artist.'[13]

To the relief of his tutors in Natural Sciences, he abandoned zoology and transferred to the course in History of Art and Architecture. His Ordinary Degree (Class II/III), awarded in 1930, was later described as 'the most undistinguished way in which it has been possible to obtain a Cambridge degree, at least during the current century'.[14] However, Scott's other skills and his pioneering work in wildlife conservation eventually erased that black mark and gave him unique status as a conduit for bringing serious zoological science to the great British public. As a result, Scott was a man with whom the top zoologists in the land – including Fellows of the Royal Society – were happy to talk as an equal.

By the time *More Than a Legend* appeared in 1957, *Look* had gone global as *Faraway Look*, featuring Scott and Philippa on location in astounding places. They showed nature, and man's relationship with it, in the raw. One graphic example was the epic, 52-hour copulation of the praying mantis, the male's performance being all the more remarkable because the female had chewed his head off after just two hours. From New Guinea came close-ups of the magnificent plumage of some of the rarest and most beautiful birds on the planet – not in situ, but adorning the headdresses of tribal dancers.

By the spring of 1960, Scott had become the face of natural history in Great Britain, so when Tim Dinsdale wanted to do his best for the Monster that he had just filmed, he went right to the top. Of course he dropped a line to the Queen at Buckingham Palace, but his principal target was Peter Scott at the Wildfowl Trust in Slimbridge, Gloucestershire.

Hook . . .

According to Scott's biography, the first time he heard of the Loch Ness Monster was in June 1960, in a conversation with Richard Fitter, the scientific correspondent of the *Observer*. They were on a tram in Warsaw, attending the General Assembly of the International Union for the Conservation of Nature. However, when interviewed in 1977, Scott told a different story: his interest in the Monster went back to 1957, when he read *More Than a Legend*.[15] In fact, neither version is true.

At Slimbridge in early 1960, the hyperactive Scott needed a dedicated team of minders to keep him and his diary under some sort of control. The front line consisted of Philippa, Michael Garside, who had succeeded her in the formal role of Scott's personal assistant, and Geoff Matthews, the Trust's scientific director. Matthews was complementary to Scott: thoroughly versed in academic ornithology, with an outstanding PhD from Cambridge on how migrating birds found their way around the world with unerring accuracy. Matthews, Philippa and Garside scanned Scott's correspondence, incoming and outgoing, often adding their own comments. They also grappled with Scott's perfectionism and his fondness for redrafting documents and letters, sometimes several times (a process Garside nicknamed 'scotting').

Naturally, Scott had pondered the vexed question of the Monster – enigmatic, exotic and perhaps clinging to an existence that should have been snuffed out aeons ago. Until now, he had observed proceedings from a safe distance and remained silent. But within a few weeks he would find himself becoming an apologist for the Monster – and in front of the nation's top zoologists and royalty.

This unlikely sequence of events was triggered by two foolscap sheets of dense typescript, addressed to 'Mr Peter Scott, Naturalist, c/o BBC Television, London W1', which reached Slimbridge on 14 March 1960.[16] The letter's tone was respectful but forceful, and the aim was clearly stated up front: to get Scott excited about 'an animal of truly incredible interest and importance'.

The writer introduced himself as 'a qualified engineer [with] no pretensions in the field of zoology', but 'a fair knowledge of animals and birds' and 'an insatiably curious mind which by virtue of my training is both rational and analytical'. He had become 'totally absorbed' by the Monster after reading a magazine article, which inspired him to hunt down *More Than a Legend*. As a result, he spent several months subjecting 100 well-documented eyewitness reports to intensive mathematical analysis. Then he scrutinised the Surgeon's Photograph in minute detail and found conclusive new evidence that this showed a 'truly extraordinary' animal which inhabited Loch Ness.

He had not worked in isolation, but with 'a very learned doctor of science of international repute . . . who has studied the subject for many years and probably knows more about it than anyone else alive today'. Their joint research had convinced them that the Monster existed. Furthermore, the writer added, thanks to 'the knowledge and technical genius of my associate, we have some idea of what it may be – in all probability'.

The distinguished associate was not named, but he was exceptional among scientists. Inexplicably, 'professional zoologists and biologists and the like' had done nothing active apart from refuting the evidence – an attitude that the writer found 'both unforgivable and utterly stupid'. It was time to 'put the Monster squarely in the textbooks … as one of the most astonishing discoveries in history'. Hence this direct appeal to Scott, universally respected as an open-minded man who was passionate about zoology and wildlife, in the hope of winning his 'serious support' which might be 'best done on TV'.

The writer ended with the offer to meet Scott at his convenience 'almost anywhere in the United Kingdom', and a promise: Scott would not be bothered again if, after listening for just 30 minutes, he was left unconvinced that 'this weird, marvellous creature, perhaps akin to the Plesiosaur' really existed.

The letter was sent from 17 Blewbury Drive, Tilehurst, near Reading and was signed by T. K. Dinsdale.

Dinsdale's letter, arriving out of the blue, must have given Scott pause for thought on several counts. Dinsdale was unknown and (except in his own eyes) unqualified. There were no details about the mathematical wizardry that he believed gave the Monster new credibility. And why would Dinsdale write without the co-signature (and maybe without the knowledge) of his highly esteemed scientific associate, whoever that might be?

In fact, Scott already knew enough to be able to piece together some of the background. Three weeks earlier, an article had appeared in the *Illustrated London News* referring to a novel mathematical analysis of a hundred Monster sightings which had yielded a statistical description of the creature.[17] The same mathematical rigour had apparently been applied to the Surgeon's Photograph and had proved beyond doubt that the image was genuine, that the head and neck reared a full six feet out of the water, and therefore that this was a massive animal unknown to science. Dinsdale was not mentioned by name but the author of the piece was well known on the natural history circuit in London, even if his scientific track record was in a completely different area.

Maurice Burton (Plate 18) was the Deputy Keeper of Zoology at the British Museum (Natural History) in South Kensington. Originally appointed in 1949 as the Museum's Curator of Sponges, he had made his name by classifying the sponge specimens collected during the epic *Challenger* deep-sea expedition of 1872–6. This was a massive, tedious labour of love which demanded thousands of hours at the microscope analysing the structure of the tiny spicules of silica that keep sponges in good shape. 'To preserve my sanity', Burton also wrote about wildlife and natural history for the general reader. His output included popular columns in the *Daily Telegraph* and the *Illustrated London News*, as well as a best-selling book, *Living Fossils* (1955).[18]

The Loch Ness Monster had merited only a fleeting mention in *Living Fossils*, but in February 1960 Burton published a series of articles about it in the *Illustrated London News*. He explained that the topic was no passing fancy; he had studied the Monster and its possible identity for many years and 'probably knew as much about it as anyone now living'. Burton's articles subjected the various theories – everything from waves and optical illusions to fish and prehistoric reptiles – to scientific scrutiny, but had left his readers hanging in suspense. Originally persuaded that an oversized eel was responsible, Burton had then flirted with the plesiosaur. Now, it seemed that the recent mathematical analysis of the Surgeon's Photograph had left him wedded to the latter.[17]

Scott replied to Dinsdale's letter a few days later.[19] He was encouraging but also made it clear, very politely, that Dinsdale's claims would not go unchallenged. Scott began by putting himself firmly on Dinsdale's side:

I am very much interested in what you write, as I have for a long time thought it more than probable that an undescribed animal lives in Loch Ness. I should very much like to hear more of your conclusions and deductions.

He added significantly, 'I have no scientific reputation to lose'.

Then Scott demonstrated that he actually knew a thing or two about the Monster, by blowing the cover of Dinsdale's mysterious associate. He wrote that he had been 'interested to read Maurice Burton's article in which he described your analysis'. Next was a little test for Dinsdale regarding his ground-breaking analysis of the Surgeon's Photograph:

I have always been worried about the scale of the object in this picture. The size of the ripples on the waves suggests to me that the creature is only a couple of feet long. But I know nothing of the circumstances under which the photograph was taken . . .

I find myself wondering why the bottom right-hand corner has been touched up, and what kind of camera it was taken with and what the neighbouring exposures on the film look like, and how I can be certain that a model on the end of a stick was not manipulated to make the necessary ripples.

However, he softened the blow by speculating that Loch Ness could be properly explored with 'Cousteau's latest gadget, the diving saucer' which had gone down to 1,000 feet.

And he ended the letter by repeating the words which Dinsdale must have been longing to read: 'Anyway, I'd like to hear more.'

Which he did, by return of post.

. . . line . . .

T. K. Dinsdale to Mr Peter Scott, 26 March 1960.[20]

A 'marathon letter' (four pages of closely typed foolscap), which began with Dinsdale complimenting Scott on his interest and open mind ('how very refreshing') and for identifying Burton as his unnamed associate. Dinsdale explained that he and Burton were following different but complementary paths which converged on the common goal of bringing the Monster the scientific recognition which it desperately needed.

Then he unveiled his revolutionary mathematical analysis, at great length but with remarkably little clarity. This apparently boiled down to something that, for an aeronautical engineer, fell far short of rocket science. Dinsdale had simply tabulated eyewitnesses' estimates of the Monster's vital statistics – length of neck, body and tail, and whether flippers or legs were thought to be present – and calculated the average of each parameter. He then used the averages to create his 'master blueprint' of the creature: neck 6–8 feet long, body 20–25 feet and a ten-foot tail, with a majority vote in favour of flippers rather than legs.

He went on to apply these data to the Surgeon's Photograph, using a circular argument which ran rings around logic. Assuming the creature's neck to be eight feet long (as in the master blueprint), the animal's body in the photograph was probably 20–25 feet in length, while a small, dark shape breaking the surface beside the root of the neck – which had been missed by all previous observers – must be part of the statistically probable front flipper. All this was entirely consistent with the blueprint, which must therefore be correct. QED.

This led Dinsdale to ask Scott why he thought that the animal's neck in the Surgeon's Photograph was only two feet long. He admitted that Scott was 'an expert on the appearance of water surfaces' – as indeed he was, having spent decades looking at rivers, lakes and oceans through the eyes of a wildfowler, sailor and artist. Surely, though, the 'line of windblown wavelets, marching almost diagonally through the picture' could only suggest that the area of water photographed 'must be a large one'? Dinsdale gave Scott a tip to help him interpret the image correctly: 'Hold it at arm's length for best effect.'

What about Scott's suggestion of skulduggery? Dinsdale could not absolutely rule out 'someone going out in a boat with a model, or jogging it about from the shore' but thought that 'a faker would prefer to keep his activities rather private'. And the bottom right-hand corner, which Scott believed had been touched up? Yes, it looked blurred, but, if deliberate, it had been crudely done and seemed pointless as it was so far from the creature itself. Anyway, the *Daily Mail* would not have published an obvious forgery.

Dinsdale enclosed a carbon copy of the original document he had sent to Burton, entitled 'Analysis of a Photograph of the Loch Ness Monster Portrayed on the Front Page of the Book, <u>More Than A Legend</u>, by Constance Whyte'. This began rather grandly:

OPINION. I have studied this photograph very carefully and, for a number of sound reasons, believe it to be genuine. If it is genuine, it is clearly of great importance, because it portrays a type of animal unknown to Science.

Clearly, Dinsdale argued, it *was* genuine. The London surgeon in question (anonymous, as he was in *More Than a Legend*) was 'reputable, honest . . . and obviously not a hoaxer or publicity seeker'. By his own admission, the surgeon was also 'a pure amateur' at photography, whereas as any photographer knows, 'really good faking is extremely difficult without careful preparation and the right equipment'. If this picture was a fraud, Dinsdale insisted, 'it could only have been done with extraordinary brilliance by a truly master faker . . . a man of remarkable psychological cunning'.

Dinsdale added his analysis of the 'second Surgeon's Photograph', which is revealing in a different way. This picture, also reproduced in Whyte's frontispiece, shows the head and neck tipping forward as though about to submerge. Dinsdale pointed out that the object appears much smaller than in the classic image and suggested that the surgeon had 'removed the telephoto lens to make sure he got the subject on the film'. He did not seem to have considered the possibility that the *Daily Mail* had enlarged and cropped the classic photo to make its front page more dramatic.

Scott's later comments show that he saw straight through Dinsdale's verbal prestidigitation and pseudomathematics. However, his attention was seized by something else that Dinsdale had enclosed: his personal copy of *More Than a Legend*. Scott had not seen the book before, but now found it convincing. It became one of the stock weapons in his battle to persuade sceptics that the Monster must be properly investigated.

It was therefore Constance Whyte, not Tim Dinsdale, who nudged Scott to take his first hesitant step towards the Monster. However, Dinsdale was the man of the moment. For the attention of Michael Garside, Scott wrote on the top of Dinsdale's letter, 'I could see him in London sometime'.

This notion took decisive shape in the letter that Garside sent Dinsdale on 29 March:[21]

Mr Scott has asked me to thank you for your letter and the copy of the book, both of which he was most interested to read . . .

Unfortunately, he will not be in town with time on his hands until early May. I wonder if you would find it convenient to meet him at the Savile Club, W1, at 2.30 p.m. on Wednesday 4th May 1960?

Dinsdale immediately fired back a handwritten note: 'This would suit me admirably . . . In the meantime, I hope very much to be able to dig up some more rather special information.' He signed off with a flourish and his professional qualifications: ARAeS, GIPRODE, AMSLAE. There was a supporting document: a pencil sketch of a plesiosaur on a leash held by a stick-man at a door marked 'Savile Club', saying 'Absolutely insists on meeting you; says she never believed you were real' (Figure 5.6).[22]

This long-awaited meeting never took place, because of 'rather special information' which Dinsdale captured on film beside the road to Upper Foyers, just before breakfast on Saturday, 23 April 1960.

. . . and sinker

T. K. Dinsdale to Mr Peter Scott, received at Slimbridge on 30 April 1960. Handwritten and undated, clearly written by a triumphant man in a hurry.[23]

I spent six days at Loch Ness and shot approximately 50ft of 16mm film in b&w through a 135mm lens on the phenomenon known as the Loch Ness Monster – and believe me that's what it is, both a phenomenon & a monster! . . .

I have written to Her Majesty the Queen asking her for an immediate audience with a view to getting the whole area turned into a national reserve & do not intend publishing my news on the film until I have had her reaction.

I'm going to need you as an ally, as this move is of paramount importance . . . if the monster is to be saved from the onslaught of the multitudes.

Can you possibly see me before the 4th May? & arrange for 16mm cine projection & 35mm still projector & the best glass bead screen you can find. This film must be seen in complete darkness for best effect & through the best equipment.

Until I have heard from the Queen, I must ask for your absolute confidence, because whatever happens the press must be kept out of this until I have had the chance to present my whole case.

I know I will have your support when you have heard my story.

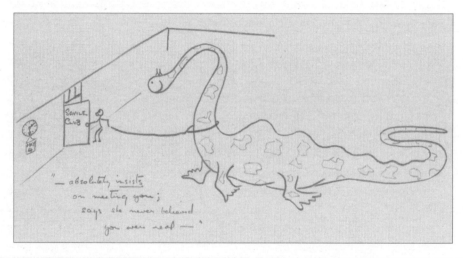

Figure 2 Sketch by Tim Dinsdale, accompanying his letter of 30 March 1960 to Peter Scott. Reproduced by kind permission of Dafila Scott and the Syndics of the University of Cambridge.

The balance of power had shifted and Dinsdale was now calling the shots. On 3 May, accompanied by Philippa and projection equipment, the VIP whose diary had been completely full for weeks ahead made the trek from Slimbridge to Tilehurst, and was duly impressed by what Dinsdale had to show.

A couple of days later, Scott received a brief thank-you note from Dinsdale, with an update and a request.[24] Despite Dinsdale's insistence that the press must not be informed until Her Majesty had replied, the Gaumont news-reel company had somehow got wind of the story and were going to run a brief news item but without showing the film. The request was for Scott to use his influence in high places. As he was still waiting to hear from the Queen, Dinsdale would be 'very much indebted' if Scott would kindly 'pop the right word at the Palace' and reassure the Queen that Dinsdale was 'not just another deluded person'.

Scott did as Dinsdale bid him. He was already well known to the Royal Family, partly through the elevated circle in which his parents had moved. The Queen's first visit to Slimbridge, where she inspected five trumpeter swan cygnets airlifted from Lonesome Lake in British Columbia, had taken place in April 1952, just a few weeks after her accession to the throne.

Now, though, the usually sure-footed Scott blotted his copybook. His letter of 6 May, to 'Dear Martin' (Colonel The Honourable Martin Charteris, Buckingham Palace), went through several drafts.[25] Scott explained the background: 'a chap called Dinsdale . . . a rather serious-minded aeronautical engineer who lives on a housing estate outside Reading and works at London Airport . . . an educated man with a scientifically trained mind and in my view no crank or crack-pot'. Scott had seen the film, which he found 'extremely impressive' even if it was 'not absolutely conclusive'. The object filmed could only be 'a new animal' or 'an extremely costly and elaborate hoax'; if the latter, Dinsdale was not party to it. The film was persuasive enough for Scott to have arranged for it to be shown to a group of top zoologists in two days' time.

The Queen and the Duke of Edinburgh might be interested to see the film now, but Scott felt it would be better to wait for 'more completely irrefutable evidence'. In the meantime, it was apparent that Dinsdale was regretting his impetuosity in writing to Her Majesty, and Scott had a proposal to smooth things over. The animal needed a scientific name. All that could be given at present was a '*nomen nudum*', a purely descriptive label that made no assumptions about whether it was a reptile, amphibian, fish or mammal. Scott's suggestion was *Elizabethia nessiae*.

Charteris replied on the same day he received Scott's letter, which Her Majesty had been 'very interested' to read. He had originally dismissed Dinsdale's 'somewhat over-heated' missive as the work of a crank, but would now reply to him. However, there was no way that the animal, if it existed, could be named after the Queen. It would be 'most regrettable' if the whole thing ultimately turned out to be a hoax. Moreover, it was not very appropriate to attach the Queen's name to something that for so long had been known as 'the Monster'.[26]

Scott was very likely stung by this gentle but unmistakable rebuke, but by then he had an even bigger problem to worry about: Dinsdale.

Meeting of minds

Charteris' letter left the Palace on 11 May, the same day that Scott invited his 'top zoologists' to a confidential screening of Dinsdale's film at London Zoo. The supporting acts were Dinsdale's explanation of his novel 'statistical analysis' and the first-hand testimony of a witness from Fort Augustus who had seen the Monster at close quarters on several occasions.

Scott had managed to assemble nine experts, headed by Sir Arthur Landsborough Thomson (President of the Zoological Society of London), Richard Harrison (Professor of Anatomy at the London Hospital and an international authority on marine mammals) and Dr Francis Fraser, Keeper of Zoology at the Natural History Museum. Not invited, however, was Fraser's Deputy, the well-known zoological writer and Monster expert who 'probably knew as much about it as anyone living', Dr Maurice Burton.[27]

Scott hoped that the event would line the Monster up for serious scientific investigation; instead, it was a disaster. Another letter that went through several drafts (and was vetted, at Scott's request, by Thomson of the Zoo) was the fulsome apology that Scott sent to the members of his expert group a good two weeks after the ill-fated meeting. He was sorry for having wasted their time. Instead, he should have asked them to read *More Than a Legend* beforehand and then shown the film without comment. Dinsdale should not have been allowed to speak. He had been 'hopelessly unimpressive' and his so-called statistical analysis nothing more than 'rather a poor rehash of Constance Whyte'.[28]

Some of the experts had suggested that Dinsdale's film was a hoax, and one speculated that the 'Monster' was a mini-submarine. Scott did not think so, and could only hope that the whole dismal affair had not prejudiced them against the need for a 'large-scale zoological effort' to find out what was there – even if it turned out to be a known species. This exercise was, he argued, a responsibility of British Zoology.

A copy of *More Than a Legend* was enclosed with Scott's letter to each of the experts, with the parting hope that those who felt 'less sceptical' after reading it might be prepared to join a 'small, perhaps unofficial study group', representing the Zoo, Natural History Museum, Nature Conservancy and the Council for Nature.[28]

From the tone of his letter, he was not entirely confident of success.

Correspondence trial

It was clear from the letter that Dinsdale sent to Scott after the fiasco at the Zoo that he knew he had done considerable damage.[29] Even clearer was his determination that nothing must stop his crusade to bring the Monster the top-level attention it so urgently needed.

First, Dinsdale thanked Scott for putting the word in at the Palace. A 'most welcome' letter had arrived from the Queen's Secretary, conveying Her Majesty's thanks for Dinsdale's 'most interesting' communication. Dinsdale was also grateful for Scott's 'stalwart efforts' at the Zoo. Then he dived straight into a post-mortem of the meeting. He was sorry if he had let Scott down by 'becoming irritated and showing it', but the 'lack of enthusiasm and rather patronising stone wall tactics' displayed by the assembled experts 'gave me the pip'.

Scott's reactions had also given him cause for concern. 'I sensed, perhaps incorrectly, that perhaps you may have one or two doubts about the whole thing . . . your reputation, committing yourself _vice_ the Queen.' Maybe Scott thought that he was a crank after all, and that his 'Monster' was indeed a mini-submarine. If so, then Scott should back out, and immediately, before it was too late.

But whatever Scott decided, Dinsdale would march on regardless . . .

. . . because when I saw that huge, miraculous creature, ploughing through the sombre black water of Loch Ness, I reached out with my magic lens across thirteen hundred yards and grasped it by the tail . . .

I may be shaken to bits and half-choked in the sea of stupidity, apathy and doubt that surrounds it [but] I will never let go, for I have seen the truth, just as I have seen it in the words of a hundred ordinary people who describe the creature so clearly through the medium of my simple analysis.

Time was now running out. The Queen had declared herself interested, whereas Scott's experts had been given their chance and had blown it. For three weeks, Dinsdale had kept his secret – the product of '80 hours of unrelenting self-discipline' – but soon he must share it with the world. 'I am prepared to wait one more week', he wrote, 'if that is alright with you.'[29]

Scott's response was placatory, but exasperation broke through. 'Dear Tim', he wrote on 22 May, 'For heaven's sake hold your horses on the film! I can't do a thing to help under an ultimatum that you're going to give it to the wolves in a week.'[30]

Scott explained that he was definitely not pulling out but had to remain open-minded and was not yet convinced that a large, undescribed animal was there. He was 'rather concerned' about Dinsdale's attitude to the experts. 'It is the scientist's duty to be sceptical', especially where hoaxes could be involved; he reminded Dinsdale of the 'brilliantly realistic' model sperm whale in the film *Moby Dick*. Like it or not, the scientists had to be on board if the Monster were ever to be generally accepted. They would need hard evidence – 'the kind of proof that does not exist' – and time. After all, Darwin had waited 20 years before publishing his work.

That was obviously far too long for Dinsdale, and Scott had been pulling more big strings in the meantime. The BBC Natural History Unit were keen to show the film, but 'only at the right moment . . . and exactly as we tell them'. They were also prepared to buy the rights immediately for £60, which Scott regarded as a 'pretty good offer'.

However, it was not good enough for Dinsdale. He wrote back immediately, welcoming Scott's continuing involvement, but pointing out that he had a wife, four children, a mortgage and 'no funds to speak of'. 'Not being a fool', he had sought 'confidential professional advice' which had stuck a price tag of several thousand pounds on the film. For the moment, he would hold back on releasing the footage but this was getting harder as the news had somehow broken out ('by what route I am baffled').[31]

This letter reached Scott on 25 May. A week later, Dinsdale wrote that he would have to 'make use of the film sooner than expected'. An unspecified 'development' had cropped up. Instead, Dinsdale waxed philosophical about the very little world we live in. As 'the custodian of truth', he was fighting prejudice against 'all almost mythical creatures', and even though the Monster had given him 'a vicious shake', he would never let go of its tail.[32]

From there, despite Scott's cajoling and warnings, it was only a matter of days before the Monster was out of the can and on to *Panorama* and the front page of the *Daily Mail*.

Damage limitation

Scott was left to pick up other pieces in the wake of Dinsdale's passage. First was an angry handwritten letter from Dinsdale's esteemed scientific associate, which hit Slimbridge on 3 June. Maurice Burton, having tried for several weeks to restrain Dinsdale's 'exuberance', was appalled to learn that Scott had convened a high-level meeting at the Zoo, and without inviting him. Even worse, Dinsdale was refusing to tell him who had attended.

Burton insisted that he had to know, in case he bumped into one of them and was put on the spot about the object in the film. In fact, Burton had somehow sneaked a preview of the film and had studied it 'with the greatest possible care'. In his opinion, there was nothing to support Dinsdale's claim that this showed a massive unknown animal – except for one tiny, subtle detail, which Burton's expert eye had spotted but which he omitted to share with Scott. Now, would Scott kindly tell him who had been there?[33]

Scott replied politely on 11 June, just two days before the film hit the headlines and *Panorama*. Unfortunately, the meeting had been held *in camera*, and Scott could not break that confidence. Otherwise, Scott levelled with Burton. Dinsdale's film had initially impressed him but repeated viewing had highlighted its 'limitations'. Scott agreed that Dinsdale's exuberance was regrettable and 'greatly lessens the chance of informed support for further efforts'.[34]

Burton's response, on 13 June, was terse. 'Judging by this morning's *Daily Mail*, Dinsdale is more than exuberant'.[35]

Big time

It had been a momentous spring for the Monster. The front page of the *Daily Mail* and *Panorama* had won over millions. For the moment at least, the four minutes of grainy black and white cine-footage had trumped all the exotica that Scott, Attenborough, Cousteau and Hass had brought back from their far-flung locations to entrance their followers. Dinsdale had appeared from nowhere and shot to international prominence, leapfrogging Maurice Burton and Constance Whyte to seize the crown as the greatest living authority on the Monster.

As the summer solstice approached, Peter Scott may well have regretted his new partnership with Dinsdale, who, if not mad or bad, was certainly dangerous to know. A few months earlier, Scott had blithely claimed that he had no scientific reputation to lose. However, he could not afford to estrange the zoological establishment, whose support was crucial to his core business of building up Slimbridge and his other conservation enterprises.

Elsewhere, others were coming to realise that the Monster could be both enticing and treacherous, just like the kelpie. Those preparing to pursue the Monster included two dozen bright young things from the universities of Oxford and Cambridge, who enjoyed the support of senior professors at both institutions. In stark contrast was a respected marine zoologist who was about to cross Scott's path, to the consternation of his minders at Slimbridge. Embittered and isolated, the zoologist did have a scientific reputation (and a job) to lose, and his entanglement with the Monster was threatening to make that happen.

Meanwhile, up in the Great Glen excitement and expectations were riding high. Some, of course, had seen it all before. They included a man who wrote a polite note to Peter Scott, explaining that the Monster subsisted on eels, in which Loch Ness abounded, rather than salmon, as was commonly supposed. The writer had the authority of knowing the Loch like the back of his hand and a job that entailed expert knowledge of the Loch's fish stocks and handling tens of millions of salmon eggs every year.[36]

His letter to Scott began by expressing his thanks for having been brought all the way down from Fort Augustus to London to relate his eyewitness accounts to the 'interesting' meeting at the Zoo. The writer was water bailiff and special correspondent to the *Inverness Courier*, Alex Campbell.

6

Gentlemen and beasts

Lying at the feet of Captain Briggs Gilpatrick on the deck of the Gloucester schooner *Gatherer*, she was not a pretty sight. To be fair, she had just been cut out of a shark's stomach and was partially digested. But even before she met the shark off the Grand Banks of Newfoundland, she had never been much of a beauty, with an ungainly head stuck on the front of a yard-long, eel-like body topped with spines. However, it was not just her ugliness that caught Captain Gilpatrick's eye. In all his years at sea, he had never seen anything like her. Neither had the expert on deep-sea fishes to whom he sent her body, Dr George Brown Goode, chief icthyologist at the Smithsonian Institution in Washington, DC. Shortly after, Goode proudly reported the discovery of this new species, which he named *Notacanthus phasganorus*. That was in 1881. Unlike the coelacanth, Goode's peculiar fish promptly sank back into obscurity and 70 years went by before it surfaced again.

In 1951, a comprehensive bulletin about *Notacanthus phasganorus* was published by the British Museum (Natural History) in London.[1] The author was Dr Denys Tucker, the Museum's Curator of Fishes and one of its rising stars. When the Museum acquired him in 1949, Tucker was only 26 years old but already had a curriculum vitae which some of his senior colleagues would have coveted. Energetic and productive, he quickly made an international reputation for himself and was the obvious choice to edit an English-language version of the icthyologist's bible, *Frischwasser Fische der Welt* (*Freshwater Fishes of the World*). And when Professor J. L. B. 'Coelacanth' Smith visited the Museum in 1958, Tucker headed the welcoming party (Plate 19).

Everything went swimmingly until 22 March 1959, when Tucker visited Loch Ness to see what all the fuss was about. There, just like Captain Gilpatrick of *Gatherer*, he met something outside his experience: a dark hump moving across the water off Glendoe, which matched nothing in his encyclopaedic knowledge of freshwater fish and their predators. He did not recognise his creature until he watched Tim Dinsdale's film just over a year later. By then, Tucker had sifted through all the known facts about the Monster and had been forced to reach an astonishing conclusion. The only animal that fitted the bill was *Cryptoclidus*, one of the long-necked plesiosaurs which, according to all other evidence, had been wiped out at the end of the Cretaceous period, 65 million years ago.[2]

Tucker began giving lectures about his plesiosaur hypothesis to scientific gatherings in London, Cambridge and elsewhere. He was evidently a persuasive speaker, but the Monster remained a force that could divide and conquer. It provoked a particularly sophisticated exchange of views during Tucker's talk on 'Loch Ness – the case for investigation' to a distinguished zoological club in London. For reasons of propriety, expletives were deleted (as were the names of the club and sundry professors) from Tucker's account of this meeting of minds:[3]

There was only one hostile critic, ******* ********, who sneered at Gould and ridiculed almost all the evidence and said it was clearly a seal. To my great delight he was practically lynched by the rest of the audience and howled down; Professor ********* for example, was shouting: 'You shut your ******* mouth! What the ******* hell do you know about it?'

Initially, Tucker seemed unaware that he was storing up trouble for himself. His bosses were less than thrilled about the new line of research which was taking their Curator of Fishes away from the Museum's core business. It was highly likely that Tucker was wasting his time and their money chasing a myth. There was no hard evidence for the Monster – no body to dissect, no bones to piece together, not even a shred of tissue to analyse under the microscope. Also, the Museum had learned the hard way to stay well clear of anything that could be a hoax. A couple of years earlier, they had been presented with a piece of skin covered with coarse hair, purportedly from the scalp of a yeti. Regrettably, the Museum's experts proved beyond doubt that the hairs belonged to a pig. As the Museum's Director later complained, 'this piece of objective identification was written off by the faithful as mere incompetence on our part'.[4]

It was therefore not surprising that the top people in the Museum would not touch the Monster, or anything to do with it, with a barge pole. Now one of their own people had broken ranks and was shouting about a notion even more preposterous than the Abominable Snowman. To make things worse, he was an internationally respected authority on aquatic zoology, which could lead the ignorant to believe that he was on to something – and that the Museum was backing him.

The Museum Trustees had their principles and were determined to stick to them. For the avoidance of doubt, they had issued a memo to all staff:

The Trustees wish it to be known that they do not approve of the spending of official time or official leave on the so-called Loch Ness Phenomenon . . . They take this opportunity of warning all concerned that if as a result of the activities of members of the staff the Museum is involved in undesirable publicity, they will be gravely displeased.[5]

Unfortunately, any hope that Dr Tucker would see sense failed to materialise. Tucker, now transformed into a rabid evangelist by having watched the Monster with his own eyes, did not seem to notice or care that he was on a collision course with his employers.

Human sacrifice

Peter Scott became aware of Tucker and his beliefs while trying to persuade Dr Francis Fraser, Keeper of Zoology at the Natural History Museum, to join his Monster review group. Others, including senior Oxbridge professors, had already been recruited, but Fraser firmly declined to have his arm twisted. Under the heading CONFIDENTIAL, he wrote plaintively to Scott on 24 June 1960:[6]

I am curious about the phenomenon . . . if only there were one photograph that would give me something to go on, but there is not. I said something of this sort to Dinsdale at our meeting.

Fraser's misgivings had brought him 'a lot of criticism . . . some of it not very pleasant'. Nonetheless, he would hold his line until solid evidence came in.

It was not Dinsdale who was making Fraser's life miserable. Fraser confessed to Scott that he was having problems with 'a member of the zoological department' who had 'developed an intense interest in the subject and has pressed me to have the Museum involved'. The whole business had already caused Fraser 'so much worry and consumption of time that I have had just about enough of it'. As a result, Fraser could not get involved, 'either as Keeper or as a private individual'.

Any mystery about the identity of the unnamed member of staff was dispelled a few days later, when Denys Tucker managed to push his way into Scott's diary. He came across forcefully. Scott wrote to Fraser on 27 June: 'Having just met the chap you refer to, I am not surprised that you have "had just about enough of it"'.[7]

Scott tried again to recruit a senior scientist from the Natural History Museum, and failed. He approached the Director, Terence Morrison-Scott, only to find that Tucker had got there first. Like Fraser, Morrison-Scott was steering clear of anyone with 'an emotional desire to prove a pre-conceived and passionately-held belief [in] a relict reptile from the Mesozoic which has survived the last 100 million years'. It was not just a question of bad science. Inevitably, such a person 'would twist the matter into saying that the Museum agrees with that view'.[4]

To the top brass at the Museum Tucker was now a loose cannon – uncritical, dogmatic and unconcerned about damaging the institution's reputation. To Scott, Tucker looked somewhat different. Here was an acknowledged expert on aquatic zoology who had seen the Monster himself and was so convinced by the evidence that he was prepared to risk confrontation with his superiors.

Scott found this a difficult tightrope to walk, as shown by his letter to Tucker after their meeting:[8]

I was enormously interested in what you had to tell me . . . I trust that your personal problems will resolve themselves to your satisfaction. What a pity that they should occupy so much of your time . . . Life is really too short for such red herrings.

Tucker's lengthy reply was vetted by Scott's minders on arrival at Slimbridge. When it reached Scott's desk, it carried a prominent warning, handwritten in pencil:[9]

This sort of letter should not be answered. It is most important that you (Peter) should not become involved in his Museum row. Phil.

Scott added a tick and 'Agreed', and carried on much as before.

Tucker's letter, essentially a rant about 'the Museum battle', was his last letter on the Museum's headed notepaper. The next, dated 5 July, was from his home address in Wimbledon.[10] Enclosed were copies of correspondence showing 'just how severely' the Museum's Trustees had dealt with him, a Curator of 11 years' standing and by far the youngest DSc in the place: suspended with immediate effect, barred from the library and non-public areas of the Museum, sacked as of the end of July.

The Trustees had done more than kill his job and his salary. They had deprived him of the oxygen that scientists need to survive: interaction with peers and access to a library and research material. What had been his crime? In Tucker's view, he had been victimised for exercising academic freedom to pursue a legitimate research topic which the dinosaurs at the Museum happened not to like. He admitted that 19 'disagreements with the local bureaucracy' were on his employment record, but was quick to point out that no complaints against him had been upheld.

Scott's response was diplomatic, expressing regret at the Trustees' decision and his hope for a satisfactory outcome.[11] On the positive side, Scott had been invited to write a feature on the Monster for the *Sunday Times* and asked Tucker if he could refer, 'with proper acknowledgement', to his theories.

Naturally, Tucker leaped at the opportunity to settle his score with the unbelievers, 'academically distinguished and downright bloody-minded'.[12] He also told Scott about dramatic new evidence that strengthened their hand. A photograph showing the Monster's flank and long neck had been taken at a range of just 25 yards by Peter O'Connor, a fireman and amateur naturalist from Gateshead. O'Connor had been courageous as well as lucky, running along the shore towards the 20-foot creature as it lumbered off into the water.

There was an update on the 'Museum row'. Supported by his union, Tucker was fighting to be reinstated. Held in reserve if his appeal failed was his Member of Parliament, followed by the High Court. Tucker knew that justice was on his side. 'The battle will be won by the man with the quietest conscience,' he wrote to Scott, 'and that's me.'[13]

Monster Blues

Denys Tucker was out of a job but his self-sacrifice had not been in vain. On 14 June 1960, Scott received a confidential letter from Peter F. Baker, a PhD student in physiology at Emmanuel College, Cambridge. Under the heading CAMBRIDGE LOCH NESS EXPEDITION, typed in red capitals, Baker put Scott in the picture. Six months earlier, Dr Denys Tucker had delivered a stunning lecture about the Monster in Cambridge, with such conviction and passion that a group of students decided on the spot to go and find it for themselves.[14]

Baker's plans were a far cry from traditional student pursuits such as Rag Week and the Boat Race. Thirty-five students from Cambridge and Oxford had joined forces to drum up an echo sounder from Marconi and enough money from the *Scotsman* to kit themselves out with binoculars, cameras and walkie-talkies. They had also charmed their way to the very top, winning the blessing of both universities' Professors of Zoology. Sir Alister Hardy of Oxford and Carl Pantin of Cambridge, both Fellows of the Royal Society, were well known to Scott, who was trying to recruit them to his Monster review group.

The Joint Oxford and Cambridge Expedition to Loch Ness, led by Dr Tucker, aimed to obtain solid evidence that the Monster existed. To convince sceptics, Baker explained, this would have to be better than Dinsdale's 'very interesting' film. The Loch would be monitored continuously for four weeks from late June, focusing on the known hot spots for sightings. Oxford, divided into three groups, was to cover the southern end around Fort Augustus, with Cambridge split between Urquhart Bay and the eastern shore from Dores to Foyers. Meanwhile, a 15-foot boat carrying the echo sounder would cruise up and down the Loch, scanning the water from side to side. The popular press were being deliberately kept in the dark; any significant discovery would go through proper scientific channels, because 'the place for its publication is *Nature*'.

Scott wrote back warmly, adding his blessing to those of Professors Hardy and Pantin, and asking Baker to keep him fully informed of progress. Baker did so – which meant telling Scott shortly afterwards that a fly had dropped into the ointment at the last moment. Unfortunately, for personal reasons, Dr Tucker was no longer able to join the Expedition.

Scott knew why. Tucker's appeal against his sacking had turned nasty and, for the moment at least, he did not dare to leave London.[12]

Great and good?

The fallout from the Tucker affair must have strengthened Scott's resolve to find serious scientists prepared to look at the evidence for the Monster, rather than dismissing it out of hand. He knew the calibre of the people he needed: Professors from the Golden Triangle of Oxford, Cambridge and London, and Directors of internationally respected institutions such as the Natural History Museum and London Zoo. Unfortunately, some of these had always been distinctly anti-Monster, and a few previously open minds had been closed by Dinsdale's performance at the Zoo a few weeks earlier.

Scott settled down to his charm offensive in late June 1960. First on his hit list (and the first to say no) was Sir Arthur Landsborough Thomson, President of the Zoological Society of London. Thomson could not swallow the notion that 'an inhabitant of a tropical swamp has survived for untold millions of years, through glacial epochs'. Sightings were mostly 'ordinary phenomena seen by unskilled observers', with 'some lying and faking thrown in to make it really difficult'. He concluded, 'it is better for such bodies as the Zoological Society not to be officially represented'.[15]

Also beyond persuasion was Vero Wynne-Edwards, Regius Professor of Natural History at Aberdeen, who could not see how a large animal could materialise in a lake that only came into being with the melting of the glaciers that had filled it throughout the Ice Age. Loch Ness did not harbour residual marine species such as the shrimp, *Mysis relicta*, which proved that it had never opened directly into the sea. The 'very telling points' made by Wynne-Edwards impressed Geoff Matthews, the trusted Scientific Director at Slimbridge, who wrote a note for Scott: 'I must say I like this.'[16]

Recruiting a big name from the Natural History Museum would have been a major coup, but Tucker had comprehensively poisoned those waters. Terence Morrison-Scott, the Director, declined with a hint of regret: 'You sit on the fence as I do – though I have certain biological reasons for perhaps looking over the fence with curiosity rather than sitting on it.'[4]

Scott had more joy with Richard Harrison, Professor of Anatomy at the London Hospital, who was dubious but replied with a reluctant yes. Harrison, a noted expert on marine mammals, thought that the 'Monsters' were inanimate objects and that 'somebody was trying to pull Mr. Dinsdale's leg'. Public money should not be wasted, but an informal committee could look at new evidence and 'might go some way towards protecting the good name of British Zoology'. Wearing his anatomist's hat, Harrison added, 'I would only start to be impressed if there were a body to be dissected!'[17]

Sir Alister Hardy FRS, Lineacre Professor of Zoology at Oxford, also yielded, while making it clear that his support for the Oxbridge students' Expedition should not be confused with belief in the Monster. He was deeply sceptical. The Loch had been filled with ice for thousands of years, since when nothing bigger than a seal could get in from the sea. As no 're-mains, bones, etc.' had ever been found, the 'Monster' was probably caused by wind and waves. Nonetheless, the phenomenon was worth investigating. Hardy would join Scott's group after the summer vacation, which he was spending in the company of plankton at the Marine Biology Station in Millport, on the west coast of Scotland.[18]

Three more recruits quickly signed up. Leo Harrison Matthews of the Zoological Society of London followed up the suggestion by Sir Solly Zuckerman FRS, Secretary of the Society, that this could be 'a good thing for the Society, for publicity reasons'. In turn, Matthews proposed Desmond Morris, who had seen Dinsdale's film and was 'full of ideas'. Morris, a Fellow of the Zoological Society and former Curator of Mammals at the Zoo, was a popular but controversial figure whose research interests ranged from the artistic skills of chimpanzees to the sexual behaviour of humans. He was the easiest catch of the lot, taking just three lines to confirm that he was delighted to accept Scott's invitation. The haul was completed by Carl Pantin, Professor of Zoology at Cambridge, who had encouraged his undergraduates to investigate Loch Ness. Pantin was content to join the panel, as long as it did not get bogged down as a clearing house for everything brought in by 'any casual tourist'.[19]

Scott celebrated Pantin's acceptance with a doodle in royal blue ink on the back of his letter: an apparently exhausted plesiosaur, lying on its back with its eyes shut.

The summer brought drama, with the unexpected conversion of Professor Sir Alister Hardy FRS into a believer. On his way home from Millport, Hardy had diverted to West Kilbride in Ayrshire and the home of Peter Macnab, who had photographed the two-humped, 50-foot creature cruising towards Urquhart Castle. Macnab was grilled at length by the fearsome professor, who later gave his verdict to Scott: 'I am quite satisfied that Mr Macnab is a perfectly trustworthy person. He is a Bank Manager.' The Macnab photograph was, quite simply, 'the strongest bit of evidence yet'.[20]

By the end of July 1960, Scott had signed up his advisory panel. But he had overlooked someone who believed that a panel of Monster experts would be incomplete without him; someone who could potentially turn the dream team into a nightmare.

Tim Dinsdale.

Great expectations

Because of a prior engagement on 20 June, Scott missed Dinsdale's 25 minutes of glory on *Panorama*. He had not yet managed to obtain a copy of the film but had listened to a tape recording. 'It sounded good,' he wrote to Dinsdale, 'but [Alex] Campbell did not sound very convincing.'[21]

Scott's absence from the ten million who had watched the programme must have disappointed Dinsdale, as did the non-arrival of an invitation to join Scott's Monster review panel. On 21 June, Dinsdale wrote:[22]

I would like to be on the committee and naturally would love to be involved in an expedition. I am not a zoologist . . . however, I do know a great deal about the subject and the Loch . . . and am utterly determined to knock spots off all the baloney that has gone on around that amazing creature.

Then he launchèd into fulsome praise of the 'astonishing' new photograph of the Monster in the shallows, taken by the plucky Peter O'Connor from just 25 yards away.

His letter contained a throwaway comment which must have given Scott a feeling of Tuckerian *déjà vu*. Dinsdale explained that he might be tied up for a while with 'personal affairs and this business of changing companies . . . My present MD has turned anti-monster and done me a bit of no good behind the scenes, I fear. Incredible.'

Scott's reply contained no reference to Dinsdale's problems at work and no invitation to join the panel.[21] Instead, he sent his own commentary on O'Connor's 'astonishing' photograph (Plate 20). Neatly embedded in Scott's letter were some tests of Dinsdale's objectivity, and perhaps of his suitability to join the high-powered academics on the Monster panel.

Scott had done his homework on O'Connor and exchanged detailed letters with him. It turned out that O'Connor was 26 years old, belonged to the Northern Naturalists Association in Newcastle and had long been interested in the Monster. The previous year, he had hit the headlines with his plans for a well-equipped expedition of 60 Northern Naturalists to investigate Loch Ness. Unfortunately, the police had taken a narrow-minded view of the investigative equipment (Bren guns mounted on canoes, harpoon guns, underwater spear guns, machetes and possibly a bomb) and had prohibited the trip. Now, older and wiser, O'Connor had come in peace with the noble aim of proving that the Monster existed. He insisted that his photograph clearly showed a plesiosaur-like creature, for which he proposed the formal scientific name, *Nessiesaurus o'connori*.[23]

Everything about O'Connor's photograph left Scott sceptical. 'I am afraid I think it is a fake,' he wrote to Dinsdale.[21] The newspapers could have smartened up the picture for publication, but most of the water had been painted in, the lighting was unconvincing and the scale looked much smaller than the 20-foot beast which O'Connor claimed. 'I think you do yourself an injustice by so easily accepting this kind of thing as evidence,' he told Dinsdale. 'This kind of faking, hoaxing (call it what you may) is one of the greatest difficulties we have.'

This brought Scott back to another picture that troubled him – the Surgeon's Photograph – and the chance to fold in a cautionary tale about someone whose gullibility undermined his credibility.[21]

I nearly lost faith in Denys Tucker's judgement when he spent 20 minutes 'interpreting' the frontispiece of Mrs Whyte's book to me. When he tries to tell me [the neck] is 12 feet out of the water, I know he is lying. I am inclined to lose faith in his critical faculty and to wonder whether all the other interesting things he told me are based on such doubtful evidence.

Having set this little trap, Scott added significantly, 'I hope you see the point of all this'. Surely Dinsdale would remember that he too had 'interpreted' the finer details of the Surgeon's Photograph to Scott?

If Dinsdale saw the point, he did not take the hint. Geoff Matthews added a comment in pencil before passing Dinsdale's three-page reply to Scott. It read simply, 'I think he gets worse'.[24]

Dinsdale had subjected O'Connor's photo to his 'gimlet eye', and, informed by 'the comparative background of reports known to me as a result of the analysis', was certain that it could not be 'a rotten fake'. To accuse someone of faking a photograph was effectively branding that person a cheat and a liar. O'Connor could not be, because he had risked his life going after the Monster. Moreover, Maurice Burton had interviewed O'Connor and was 'very impressed by his manner and personality' – especially when O'Connor left instructions with him in case he died while pursuing the creature.

Anyway, O'Connor's Monster was the same size and shape as the one that Dinsdale had filmed, and the length of its neck (nine feet, according to O'Connor) was close to that which Dinsdale had calculated from the Surgeon's Photograph.

Dinsdale was even more excited about another photograph – an enlarged frame from his own cine-footage, broadcast during the 'really marvellous *Panorama* film', which showed 'an amazing transformation' of his Monster. This had doubled Dinsdale's 'already utter determination never to give in until this marvellous creature is accepted and understood'. When they eventually caught one, he and Scott would have 'a mountain of fun', whereas the doubters would finally have to accept that they had treated the Monster-believers as badly as the establishment had dealt with Galileo in his day.

Now, what about that expert committee with all those professors and zoologists? In a handwritten note that sprawled down the margin of the typed letter, Dinsdale reminded Scott that 'Engineer or not, 6–700 hours of study puts me into the "expert" category on this subject'.[24]

Scott's reply indicated that their minds were not going to meet over O'Connor's photograph.[25] And the stamp of approval that Dinsdale longed for – an invitation to join Scott's panel of Monster experts – did not materialise.

Scott was not Dinsdale's only target. In late July, Scott received a further letter from Colonel The Honourable Martin Charteris, which suggested that patience was wearing thin at the Palace.[26]

Dinsdale had sent the Queen another 'rather emotional letter', asking for an audience to prove to Her Majesty's satisfaction that the Monster existed, so that she could give her blessing to 'official and effective protection of these animals'. Charteris would tell Dinsdale that he must write instead to the Secretary of State for Scotland, but, in the meantime, could Scott let him know 'very briefly, the present state of the investigation'?

Scott did so, and Charteris wrote back to express his thanks, this time from the safe haven of the Royal Yacht *Britannia*, at Lerwick. Charteris would deal 'very gently' with Dinsdale but had a suggestion for Scott's consideration:

I'm sure that you would be right to enlist a psychologist amongst your team, as there is obviously something about the Loch Ness Monster which makes normally sane and balanced people behave in a highly emotional manner. Even if of no use to you, he would have an interesting time examining the causes of the Loch Ness Monster neuroses.[27]

Scott did not reply to that one.

Exposé

With its circulation approaching one million, the *Sunday Times* was the best-selling quality weekend newspaper in Britain throughout the 1960s. It was known for good journalism and eye-catching features by big names. A high point in the summer of 1960 was the series by Ian Fleming on 'Thrilling European Cities', which he later turned into a travelogue that was published hard on the heels of *The Spy Who Loved Me* (1962).

Those leafing through the *Sunday Times Magazine* of 14 August 1960 to find out what thrilled Fleming in Vienna might have been distracted by the piece which filled the front page.[28] Liberally peppered with extraordinary pictures, it was entitled, 'The Loch Ness Monster – fact or fancy?'. The author was an even bigger name than Ian Fleming: Peter Scott.

It had been a busy few weeks for Scott and he was about to fly to Naples with Philippa to preside over the Olympic jury for yacht racing. Nonetheless, the *Sunday Times* article had gone through several drafts and caused him some soul-searching in the process.

Scott began his article by setting the scene for something intriguing, that could turn out to be either zoological or psychological: sightings consistent with a large animal with a humped back and a long neck, up to 50 feet long. He cited the 'excellent' book by Constance Whyte, which built on Rupert Gould's research, the striking photographic evidence and, most recently, Tim Dinsdale's film, broadcast on *Panorama*. All this begged obvious questions. Did the Monster exist? If so, what was it and how had it got into Loch Ness? Did the Loch hold a colony of the creatures?

Dinsdale was given due prominence. The object he had filmed was, in Scott's opinion, either a large aquatic animal or a boat. Dinsdale was convinced that it was the former, and had filmed a boat following the same course to prove the point. A still from the *Panorama* film showed a broad, curving V-wake with a distinct central streak, headed by a light blob. Dinsdale was pictured at his post in what could have passed for a publicity shot from *Look*. Wearing a sou'wester and an alert expression, binoculars slung around his neck, he was seated behind a tripod carrying his trusty long-lensed Bolex.

What could the 'monster' be? Scott pointed out that ideas ranged from long-necked seals and outsized eels to wave formations and psychological phenomena. And, of course, there was the plesiosaur theory. This had been around since the 1930s but had recently been developed by Dr Denys Tucker, whose research had highlighted the 'striking correspondence' between the Monster's reported features and the Cretaceous plesiosaur, *Cryptoclidus*.

Dr Tucker had come up with solutions for some awkward puzzles around this notion. Professor Wynne-Edwards had proposed that a breeding colony of about 25 individuals would be needed to sustain the species, and doubted that the Loch held enough vegetation or fish to feed them. Tucker's answer was eels, migrating up the River Ness from the sea and then getting trapped in this 750-foot deep cul-de-sac. As for the complete absence of mortal Monster remains, Tucker believed that stones could be responsible. Crocodiles were known to swallow stones, perhaps to adjust their buoyancy and stop themselves from being marooned on the surface. Such stones had been found in fossil plesiosaurs, and could take recently deceased Monsters straight to the bottom of the Loch.

The plesiosaur theory would have gained support from the accompanying pictures: the prehistoric profile in the Surgeon's Photograph, Macnab's

50-foot, double-humped creature heading for Urquhart Castle and an artist's reconstruction of the candidate itself (reproduced, with fine irony, by kind permission of the Natural History Museum), complete with lugubrious expression and rounded flippers like canoe paddles.

Scott wound up on a speculative but positive note. The priority now was to investigate Loch Ness properly. This was the logical extension of the Oxford–Cambridge Expedition, inspired by Dr Tucker, which was actively exploring the Loch and had already reported a sighting and unexplained echo-sounding contacts. Eventually, specimens might be caught and studied in captivity. In the meantime, a panel of distinguished scientists had been established 'to examine any evidence which may be forthcoming'.

Casual readers might well have concluded that the Monster existed and was a plesiosaur, and that scientists now needed a kick to set them off on its track. But did Peter Scott believe in it? He had written the article and was backing efforts to find the Monster, but had kept his own cards very close to his chest.

Scott was sitting on the fence – and even with careful reading of his article it was impossible to tell which way he would topple if someone tried to force the issue by giving the fence a really good shake.

As always, the back story was somewhat different. The frequent references to plesiosaurs were partly explained by Tucker's kind offer to check early drafts of the article and to 'correct factual inaccuracies'. He also used the opportunity to insert his own theories, and Scott let them all stand.[29]

The biggest picture in the article was the Macnab photograph, printed above a legend which quoted from a letter that Macnab had written to Scott: '[I saw] some moving disturbance in the water – and of course, the knowledge of something large living in the Loch had been with me for a long time . . . "Undulation" is descriptive of the movement'. Yet, while first drafting the article, Scott had confessed to Jack Lambert, the associate editor of the *Sunday Times*, that he believed this photo to be a fake.[30] It was only the dramatic conversion of the arch-sceptic Professor Alister Hardy FRS, and Hardy's assurance that Macnab and his photograph were both genuine,[20] that had persuaded Scott to change his mind.

What about Tim Dinsdale? Far from being pleased at seeing himself in the *Sunday Times*, he was now writing a caustic letter to Scott about a blunder in the article. His relationship with Scott might have frozen over completely if he had seen a confidential note which Scott sent to a friend at Imperial College a short while later.

'Poor little Dinsdale,' Scott wrote, 'is so certain he's filmed a plesiosaur that it's no good talking to him any more.'[31]

Mailbag

The *Sunday Times* feature stirred up a great deal of interest, at least as measured by the volume of mail received at Slimbridge. 'We are inundated with correspondence,' Scott observed to a friend.

The first rush included what Scott described as a 'sad' letter from Tim Dinsdale, castigating Scott for having printed the wrong still from his movie in the article. It was not the Monster, but the boat sent out afterwards to prove that it and its wake looked completely different. If only Scott had seen *Panorama*, or even the blown-up frames that Dinsdale had tried to show him.[32] In his defence, Scott had been sent only galley proofs of the typescript and had left the illustrations in what he assumed to be the safe hands of the picture editor. 'I am furious that they put in the wrong still', he wrote back to Dinsdale, adding that he hoped to watch *Panorama* after returning from his jury service at the Olympics. Scott also offered some gentle advice to Dinsdale: 'Meanwhile, don't write too often to the Queen, or she will get browned off with the whole thing.'[33]

This provoked a sulky response from Dinsdale, surprised that Scott had heard about his supposedly private letter. His main concern was that, without royal patronage, the creature would be killed. He gathered that some newspapers had put a bounty of a quarter of a million pounds on the Monster's head. Dinsdale was aware that some thought him 'soft in the head', but assured Scott that even if plesiosaurs did not seem plausible south of the border, they certainly were up at Loch Ness.[34]

Many other letters followed, sent in by people from all walks of life and ranging from complete belief in the Monster to extreme scepticism, and from unquestionable sanity to something else.[35]

Several correspondents informed Scott that the Monster had entered Loch Ness through subterranean channels which opened into the North Sea. Scott drafted a generic 'the bath would empty if you pulled the plug out' response for Mike Garside to send out to them, including the gentleman who could prove that the Loch drained through an underwater river into the Jura Mountains in Switzerland.

There were some cracking ideas for forcing the Monster to the surface for proper study and identification, including heat (generated by underwater pipes burning paraffin) or electrical impulses projected into the water.

Numerous other monsters now surfaced. These included one with a cow-like head (Orkney, 1942); a carcass seen by someone too shy to write in himself (Blakeney Point in Norfolk, not sure when exactly); and an 18-foot scaly beast with a long neck, spotted trudging through bracken (Loch Ness, many years ago). The last letter, from a lady in Inkpen, Berkshire, illustrated some of the problems with verifying second-hand accounts. The person who told her about his Monster was of good character ('He did not drink') but would be difficult to contact ('This man is now dead').

Some gems were scattered through the dross. Several people believed that waves, which Scott had rather dismissed in his article, lay behind many sightings. An experienced sailor had watched three distinct grey humps surging up the middle of the Loch, well behind a trawler. To his naked eye, it was the Monster; through 8x binoculars it was obviously the boat's bow-wave, standing up prominently against the glassy calm of the surface.[36] Even more bizarre things could happen when the bow-wave from a boat hit a steep barrier, such as the rocky sides of Loch Ness. On a calm day, the reflected wave front could clash with the incoming wash and throw up a 'standing wave'. This could look big and black, with a variable number of humps, and seemingly could either remain stationary (sometimes for several minutes) or move in any direction. One splendid example that reared up from the Loch turned out to be the wake of a trawler that had passed by over 25 minutes earlier.[37]

The list of inanimate possibilities was completed in a letter from Maurice Burton, who had changed his mind again. His latest article in the *Illustrated London News*, entitled 'Muck monsters', floated the idea that the whole thing could be explained by masses of debris which rotted quietly in the depths until they had trapped enough gas from decomposing vegetation to be lifted to the surface. There, they would appear with the characteristic disturbance, could move across the surface depending on how the gas escaped, and then sink again when their buoyancy had fizzled away. 'I wonder if this is what the bank manager saw,' Burton mused.[38]

The plesiosaur hypothesis might have excited the public but failed to impress someone who knew about the creatures. Dr L. B. Tarlo, an expert on fossil vertebrates at the Natural History Museum, had just returned from presenting a paper to the Palaeontology Section of the International Geological Congress in Copenhagen. Tarlo agreed that the fossil record might not hold all the answers, but he was 'reasonably certain' that plesiosaurs could not have evolved into anything like the multi-humped form proposed for the Monster, even over 100 million years. Evolution simply did not work that way: 'entirely new groups do not suddenly arise'. In his reply, Scott committed himself more clearly than in the *Sunday Times*: 'I do not myself believe in the plesiosaur theory as the most probable explanation (if there is a monster at all) . . . but think that a descendant of an extinct form could be responsible.'[39]

Another warning shot was fired across Scott's bows by Commander J. C. Turnbull, Royal Navy (Retired). After reminding Scott that they had met during 'the 1939–45 unpleasantness', Turnbull got down to business:[40]

I want to warn you about Gould. In your article, it seems that you were prepared to accept Gould as an authority and statements in his book as evidence . . . In fact, he was the most gullible of men, with no sense of humour and quite incapable of distinguishing between 'evidence', hearsay and a leg-pull.

Turnbull's damning verdict was evidence-based. Years ago, Gould had fallen straight into a neat trap laid for him by Dr Richard Elmhirst, the distinguished Director of the Scottish Marine Biology Station at Millport – the same institution with which Sir Alister Hardy enjoyed his planktonic relationship. Elmhirst set Gould up by describing a multi-humped phenomenon that he had watched on the River Clyde, but without revealing that his binoculars had shown this to be a family of otters swimming in a line. What happened?

Gould at once exclaimed 'a sea serpent!' . . . He always heard what he wanted to hear and never waited for a true explanation.

Having proved to his own satisfaction that Gould was dangerously impressionable, Elmhirst let him in on the secret. Gould later cited Elmhirst's account, with a quaint drawing of four otters swimming in procession, in *The Loch Ness Monster and Others* as a fine example of how a banal species could fool the unwary into believing that they had seen the real thing – but failed to explain that he himself had been duped.[41]

In strictest confidence

Correspondence about the *Sunday Times* article had tailed off by 12 September 1960, when a well-composed four-page letter arrived that would give Scott the chance to try out his new panel of Monster experts.

It was from Torquil MacLeod, the naturalist whose undercover Monster investigation was funded by an anonymous gentleman.[42] His credentials: a lifelong passion for Scottish wildlife and the Monster, an experienced sailor and confident in judging size and distance across water. His mission: to film and photograph the Monster. His cover story: working from his converted army radio truck, supposedly making a wildlife film. Progress to date: two sightings since beginning his mission 11 months earlier in November 1959. Until now, on his employer's instructions, he had not told anyone else, except for Constance Whyte, also sworn to secrecy. It was Constance Whyte who had suggested that he should contact Scott.

MacLeod described in detail his first sighting at the end of February. The animal below the Horseshoe Scree, which he watched for nine minutes through 7x50 binoculars, was dark in colour, with a body 45 feet in length, a long neck and a hefty fore-flipper.

The second sighting was more tantalising, as the animal was all but submerged, but more robust because there were nine other eyewitnesses, mostly on a yacht which had chased it. On 7 August, MacLeod and his wife were driving two Australian friends south from Inverness towards Abriachan when he spotted something odd off Tor Point about 2,000 yards away. A yacht, travelling at about six knots down the mirror-still Loch, was being overtaken by a wake 'of no clear origin'. The crew evidently noticed it, as the yacht suddenly changed course towards it. For the next 20 minutes the yacht wove around as it tried to catch up with the object causing the wake. This revealed itself intermittently as a low, sleek hump about ten feet long. Flippers were not visible, but two pairs of splashes, several feet in height, appeared occasionally, as with a four-oared rowing boat.

Finally, the animal submerged and the yacht continued to Fort Augustus, where MacLeod was waiting to interrogate its crew. The couple and their four children on board added to what the MacLeod party had seen from the shore. The animal was about as large as the yacht (48 feet long, 10 feet across the beam) and greenish-black. Their closest approach had been about 140 feet away – too close for comfort, the father admitted, with the children on board. MacLeod wrote it all down – except for the names of the crew and their yacht, which he forgot in the heat of the moment.

MacLeod brought this information to Scott to bolster the case for a proper investigation of the Monster. He argued that too many expeditions had been poorly planned and executed; for example, the Oxford–Cambridge Expedition, which he had watched in action that summer, was worthy but amateurish. The aim of a definitive expedition must be:

The close-range photography of the animal in its natural habitat and the taking alive and unharmed of a specimen for positive identification by a competent authority, and <u>the return, alive and unharmed, of the animal to its Loch</u>, and then the authorisation of a suitably responsible body for the protection, preservation and observation of the species.

MacLeod could almost have been quoting the founding principles of Slimbridge, the International Union for the Conservation of Nature and the nascent World Wildlife Fund. Scott wrote back that he was 'enormously interested and intrigued' by MacLeod's letter and promised to bring his evidence before the scientific committee.[43]

Group therapy

Through the autumn of 1960, Scott accumulated evidence to present to his panel of heavyweight zoologists. A great deal would hang on their initial reactions. Scott could not risk a repetition of the Dinsdale-induced fiasco at the Zoo, which had nearly scuppered Scott's entire Monster strategy. To smooth the way for the experts, Scott decided to convene an informal 'Loch Ness Study Group'.

Richard Fitter was let into the secret in late September, and agreed to be secretary. Scott also intended to invite Constance Whyte, Denys Tucker, Torquil MacLeod ('cloaked in secrecy ... but sounds like a sensible sort of chap') and MacLeod's anonymous sponsor, who turned out to be Sir Charles Dixon. And there was someone else. 'I ought to ask Dinsdale,' Scott explained to Fitter, 'as he did originally excite my interest. I doubt that he can help us much, but it would be a kindness.'[44]

Scott's contacts had obtained a suitably imposing setting – Meeting Room 3 at the House of Commons – and the Group was scheduled to assemble at 3 p.m. on 19 October 1960. Other formalities had to be observed. As a precaution, Scott declared his intentions to Sir Solly Zuckerman, in his role of Chief Scientific Adviser to the Armed Forces. Zuckerman was happy to confirm that no Admiralty interest would be affected by an in-depth investigation of Loch Ness, adding, 'I very much hope that you find the beast'.[45]

Scott wrote to Dinsdale just three days before the meeting – and three weeks after inviting everyone else. Apologising to Dinsdale for the short notice, Scott said that it would be good to have him in the group. Dinsdale confirmed by return of post that he would be there, and promised to 'behave with decorum'.[46]

For once, he did. So did Denys Tucker. The confidential minutes of the Loch Ness Study Group, chaired by Scott, showed that its sole purpose was to select evidence for detailed review by Scott's scientific committee. It was agreed that this should consist of *More Than a Legend*, the Macnab photograph, the Dinsdale film and Torquil MacLeod's written account of the sighting on 7 August during which the yacht gave chase. The latter would be supplemented, if possible, by any photographs taken by the yacht's crew, who were currently being traced. No plans were mentioned for the group to meet again.[47]

Afterwards, Constance Whyte wrote to express her delight and excitement at the meeting and its outcome. Scott's committee of eminent scientists would have plenty to consider, especially as some other recent witnesses had 'seen it on the shore'.[48]

By contrast, Torquil MacLeod suddenly lost momentum. He had worked out, from the times recorded of boats passing through the Caledonian Canal on 27 August, that the yacht in question was called *Finola* – but stopped at that point. He sent a brief letter to Scott, explaining that he had not been well since attending the meeting at the House of Commons. With apologies, he would have to leave further inquiries to Scott and the others. Scott wrote back to confirm that they would continue to chase the family on *Finola*, as their photographs could provide crucial evidence.[49]

Year end

As 1960 approached its close, it was clear that this had been a bumper year for the Monster, and one of mixed fortunes for its followers. The creature had now embedded itself firmly in the public consciousness. Peter Scott had written a big newspaper feature about it and had recruited eminent scientists to look at the evidence. He was careful not to say whether he believed in the Monster or not, but he was sticking his neck out for it. This must mean that he thought it existed. And if Scott believed that it was real, it must be.

Some tantalising new evidence had emerged from the month-long Oxford and Cambridge Expedition during the summer. Several sightings had been recorded, but no convincing photographs or film taken. Bruce Ing had twice watched a humped object about eight to ten feet in length crossing the Loch, with a clear view through powerful 9x binoculars off Achnahannet. The smart 'searchlight' echo-sounding unit lent by Marconi proved its worth. This could detect, lock on to and track an object moving underwater up to 1,200 feet away. Various promising traces were recorded and identified as large fish or inanimate objects – except for one 'dense' object of indeterminate size, which apparently dived from the surface to 60 feet and climbed back up, too fast to be a fish.

Peter Baker and his team managed to fill a page in three successive issues of the *Scotsman*, and promised to return to pick up the mystery.[50] Tim Dinsdale, whose film the bright young things had hoped to better, looked down on their efforts with a degree of condescension: 'They worked hard . . . but their results, although intriguing, were not conclusive.'[51]

Throughout all this, the zoological fraternity had been bitterly divided and had not come across well to the public. Dr Denys Tucker, the international expert at the Natural History Museum, believed in the Monster and had seen it with his own eyes. Other scientists (who had not seen it themselves) clearly thought that Tucker was a heretic, and had stuck their knives into him. *The Times* had reported that the Museum had sacked Tucker because he refused to recant. What on earth was going on with these people?

Tucker would have drawn consolation from an extraordinary event which took place during the summer. Had it been made public, what happened when the Professor of Zoology at Oxford dropped in on Peter Macnab, bank manager of West Kilbride, would also have sent shock waves reverberating through believers and non-believers alike. At last, an academic had deigned to leave the safety of his ivory tower and follow the example of Rupert Gould and Constance Whyte. This notoriously uncompromising sceptic had sat down to listen to an eyewitness and had been converted into a believer on the spot.

Two wives should be given the last words for 1960. The first was Mrs Dinsdale. A couple of weeks before Christmas, Tim Dinsdale added a PS to one of his long letters to Peter Scott:[52]

Wendy, my wife, has just told me that unless someone catches the Loch Ness monster next year, she'll go and catch the creature herself, because she's fed up with it!

For the attention of his minders, Scott scribbled, 'I like the P.S.'

Later that week, Scott received a letter from Mrs Elisabeth MacLeod.[53] She was sorry that Torquil had not been able to chase up the *Finola*'s crew or reply to Scott's most recent letter. Unfortunately, he had not just been off-colour after the House of Commons meeting. He had been diagnosed with Hodgkin's lymphoma, a rapidly progressive blood malignancy, and was now gravely ill and undergoing chemotherapy in hospital in Edinburgh.

She hoped that he would get home again, but, if he did, this would not be until after the New Year.

Phenomenal investigations

The year 1961 will be remembered as one of threatened boundaries. In January, Nikita Khrushchev exhorted all those fighting 'wars of national liberation' to go for it. This was music to the ears of Ho Chi Minh in North Vietnam, but not to John F. Kennedy, recently installed in the White House. Khrushchev and Kennedy were also preoccupied with a much bigger frontier. Russia was still ahead in the Space Race, despite a setback just before Christmas when they had to blow up a wayward Sputnik and its two canine cosmonauts to prevent them from falling into foreign hands. Meanwhile, test pilot Yuri Gagarin was preparing for a flight that would last just 108 minutes, roughly five times as long as an intercontinental nuclear missile needed to fulfil its destiny. Back on Earth, cultural limits were being stretched. The Beatles (John, Paul, George and Pete) were back in Liverpool and moving up the bill at the Cavern Club, but only on condition that they did not play rock 'n' roll.

And, down in London, the search for the Loch Ness Monster was beginning to infiltrate the corridors of power.

The man for the job

Meeting Room 3 at the House of Commons had been the grand setting for the first and only meeting of Peter Scott's informal Loch Ness Study Group on 19 October 1960. Anyone concerned that this was not an appropriate use of the Palace of Westminster would have been directed to the person who pulled strings to make the venue available – David James, Conservative Member of Parliament for Brighton Kemptown (Plate 21).

The Monster might have seemed an unconventional interest for an MP from the South Coast of England, but James was not a conventional man. When he appeared on *This Is Your Life* a couple of years later, Eamonn Andrews introduced him as 'a walking adventure story', writer, Laird of a Scottish castle and one of the wartime heroes from *Great Escape Stories*; the guests that evening included the former Nazi policeman who saw through James's disguise and sent him back to his prisoner-of-war camp.[1]

David James was born on Christmas Day 1919 and put down for Eton shortly thereafter. Home was Torosay Castle on the Isle of Mull, in which the family tree had been rooted since the 1860s and where an upper-class childhood provided useful life-skills in deer-stalking, fishing and sailing. James tolerated Eton until the age of 17, when something better came along: a four-masted Finnish sailing ship and a year-long trip to Australia and back. On return, his father took the prodigal son off to see the Spanish Civil War, where a sniper nearly put paid to young David's further career plans. From there, it was a natural progression to Oxford, but – thanks to the war – only for a year.[2]

Like Scott, James saw action at sea, but his naval career ended unhappily in late February 1943 off the Hook of Holland. He was rescued by the German crew who had sunk his motor-gunboat and was sent to the Marlag prisoner-of war-camp near Bremen. On arrival, he began hatching plans to escape. His first attempt, disguised as Lieutenant Ivan Bagerov of the Bulgarian navy, took him 50 miles to the docks at Lübeck and within spitting distance of safety on a Swedish boat. The giveaway was not the fake uniform (his Royal Navy greatcoat with the gold braid rearranged), or his dreadful Bulgarian accent, or even the phonetic rendition of 'Bagerov'. His papers had been beautifully forged using a 2B pencil, but were not good enough for the eagle eye of Paul Schädrich of the *Wasserschutzpolizei* (Waterways Police). Realistically, there was only one answer to Schädrich's question, 'Which camp did you escape from?', and James was promptly returned there.

He tried again a few months later, playing a traumatised Swedish sailor with gruesome facial burns (an amateur dramatics recipe) that deterred close inspection. This time, he found a Finnish steamer at Lübeck that was bound for Gothenburg with a cargo of oranges, a sympathetic stoker and room for one under the boiler. From there, it was plain sailing to neutral Sweden and a flight home with the RAF. James spent the rest of the war in Antarctica on the clandestine Operation Tabarin, where his roles included intelligence, mapping and dog-handling. This set him up admirably for a job in 1948 as on-location Polar Advisor for the epic film *Scott of the Antarctic*.[3]

After the war, civilian life was enlivened by inheriting Torosay Castle in 1945, by parallel careers in writing and publishing and by a growing interest in politics. His books included *Escaper's Progress*, the tale of his exploits at Marlag, and a well-regarded biography of Lord Roberts of Kandahar.[4] Setting up house in East Sussex lined him up to fight the Brighton seat for the

Conservatives, on a reactionary, anti-Communist, anti-divorce ticket that attracted vigorous heckling from left-wing students at the new University College of Sussex. James won the seat in the 1959 General Election which returned Harold Macmillan as Prime Minister. According to the *Sunday Times*, the 40-year-old James was 'one to watch' – as was the new MP for Finchley, Margaret Thatcher.[5]

Those who watched James's progress included Peter Scott, who had met him briefly during the war. In some ways they were kindred spirits, with a privileged upbringing, an unspectacular university education and passions for adventure, the sea and risk-taking. Scott had even dipped his toe into the hazardous waters of politics, but only for long enough to fail in his bid to become the Conservative candidate for Wembley North in 1945.

By early autumn 1960, Scott had decided that another prong was needed for his offensive to force the scientific establishment to investigate the Loch Ness Monster properly. As well as his heavyweight academics, he needed someone with political muscle to make sure it all happened. David James was the obvious choice, and Scott went to see him. At first, James shied away from the Monster, explaining that he had gone into politics to make his reputation, not to ruin it. But Scott's powers of persuasion, assisted by *More Than a Legend* and the four minutes of Dinsdale's film, won him over.[6]

James booked the room at the House for Scott's Loch Ness Study Group on 19 October 1960 and then broke parliamentary rules by allowing the meeting to go ahead without him, when 'an obscure bug' laid him low. His commitment to the cause was undoubted from the start, as he later explained in a two-page article in *The Field* entitled, 'Time to meet the Monster'. Under the subtitle, 'Loch Ness and its secret: fact or fiction?', David James MP expressed a sentiment that some might regard as coura-geous for a new Member of Parliament: 'I assert my wholehearted belief that there is an unidentified species in Loch Ness.' It was clear from his tone throughout that he intended to prove the case.[7]

Enticing James on board was a major coup for Scott, but the new recruit carried a parasite which can hitch a lift in any strong character who is used to taking command – the risk that he might end up running the show his own way.

Acid test

Scott's expert panel 'to discuss the Loch Ness Phenomena' met on 12 April 1961 in an even more auspicious setting than the Palace of Westminster.

Burlington House on Piccadilly was home not only to the Linnean Society, where Sir Edward Mountain had failed to win over the sceptics in 1934, but also to the Royal Society. This was the most respected scientific body on the planet, and Scott's primary target. If the Royal Society could be persuaded to give its support, then the establishment would follow suit – and so would public opinion and funding. Three of Scott's academic recruits were Fellows – Leo Harrison Matthews of the London Zoological Society and the duo of Oxbridge Zoology Professors, Carl Pantin and Sir Alister Hardy. Significantly, the previously sceptical Hardy had confided to Scott that he believed Peter Macnab's photograph of the two-humped creature to be genuine.

In addition to Desmond Morris (authority on primate sexuality), Scott drafted in H. R. Hewer (seal expert), and two friendly ornithologists, Richard Fitter and Geoff Matthews, the Scientific Director at Slimbridge. The panel was completed by Gwynne Vevers, the Curator of the Aquarium at London Zoo. Vevers was not just knowledgeable about fish. His colourful career had taken him from gannet colonies to the pigmentation of starfish and predicting the route that the *Bismarck* would take on leaving the safety of her Norwegian fjord in May 1941.[8]

The 'Phenomena' that the expert panel would consider had been carefully selected by the informal hanging committee, and Scott's final list comprised just five pieces of evidence. Torquil MacLeod had responded to chemotherapy well enough to make the trip down to London; he was still a sick man, and Scott told Garside to settle his travel expenses 'at once'. MacLeod would describe his two sightings: the 50-foot, long-necked creature at the foot of the Horseshoe Scree, and the object whose wake had overtaken the yacht *Finola*. The *Finola* sighting would be backed up by others. The MacLeods' Australian friends, Mr and Mrs Seddon Smith from Murrumbeena, had received a telegram from Scott –

FOR PURPOSES SCIENTIFIC INVESTIGATION PANEL CAN YOU PLEASE CONFIRM BY CABLE AND AIRMAIL LETTER DETAILS OF SIGHTING OF LOCH NESS MONSTER LAST AUGUST[9]

– and responded 'GLADLY AIRMAIL LETTER FOLLOWING'. The crew's account would be given by Mr Lowrie and his eldest son, who had photographed the wake.

Scott also called in the Reverend Leslie Dobbs from St Michael's Church in Colehill, Dorset. On 23 October 1960, the Dobbs family had watched an arrow-shaped wake moving across the mirror-calm Loch, and, for a few seconds, saw a 'very large, black animal' with two humps break the surface. As reported in his *Vicar's Letter* in the parish magazine (alongside news of the 1st Colehill Brownie Pack), Dobbs was instantly transformed into a true believer by 'the phenomenon which shattered all our sceptical jokes'. He signed his piece, 'Doubting Thomas (Except I See)'.[10]

Next were two key items of photographic evidence, the Macnab photograph and Tim Dinsdale's film. Finally, Peter Baker would present the tantalising sonar tracing recorded during the Oxford–Cambridge Expedition the previous summer, showing the strange 'dense' contact moving up and down between the surface and 60 feet down, too fast to be a fish.

Altogether, an enticing menu to whet the appetite of clever people with curiosity and persuade them that there was enough *prima facie* evidence to explore the Loch thoroughly. But would Scott's carefully chosen experts sit down and tuck in?

On the day, Scott chaired the meeting, but only 'to start the ball rolling' and introduce the witnesses. The day went as well as it could have done. The witnesses were all judged 'quite honest . . . If any fraud had been perpetrated, it was by persons unknown to all concerned.' After the witnesses withdrew, the discussion was lively and positive. All agreed that there was enough evidence to present to the Royal Society, ideally through its Inter-Service Committee, chaired by Sir Solly Zuckerman. With his support, the navy would make available sophisticated equipment such as sonar buoys to locate the creatures. Scott closed the meeting in high spirits, asking everyone to respect 'complete security' over the proceedings.

The next day Scott drafted the minutes and quoted the group's consensus statement: 'We are of the opinion that the evidence of a large animal in Loch Ness is sufficient to justify serious zoological investigation.' At present, there were 'insurmountable difficulties' in allocating the animal to any known species of aquatic vertebrate – which made the case for finding it all the more compelling. It was therefore appropriate to approach the Royal Society for support, through Solly Zuckerman FRS.[11]

Scott circulated the draft to the panel, asking for a quick response. Two weeks later, the minutes had still not been agreed – because, after further thought, Professors Pantin and Hardy had both developed cold feet.[12] They had decided that the evidence was not yet strong enough to take to Zuckerman; Scott would have to water down his conclusion and remove any reference to the Royal Society for the time being.

This was a bitter disappointment to Scott, but there was a further complication. Someone on the panel had leaked the confidential draft minutes to a man who was convinced that he must be at the centre of anything to do with Loch Ness and its Monster: Tim Dinsdale. And to prove the point, Dinsdale had been busy behind the scenes.

Hot copy

The first intimation that Dinsdale had written a book about the Monster reached Scott in a letter from a publisher at Routledge & Kegan Paul. *The Loch Ness Monster* would be launched in a few weeks' time and was destined to be 'helpful and successful'. The publisher was impressed by Dinsdale's professional skill in analysing photographs, but added ominously: 'I have tried to keep him down to earth'. The proofs were enclosed in case Scott had time to write a few words for the cover; Scott politely declined.[13]

An agitated letter from Constance Whyte revealed that she had also been nobbled to provide some cover blurb.[14] She could not understand how Mr Dinsdale had produced a book 'on so short an acquaintance with the subject'. The thing was 'crowded with inaccuracies, large and small' and had clearly been written in haste and in ignorance of zoology and the Loch. Whyte was 'very worried' that it might undermine serious attempts to prove that the Monster existed. What did Scott think, as 'captain of the team' which might be pulled apart when the book was launched? Scott wrote back, saying that he shared her concerns.

The following day brought a letter from Dinsdale himself, mentioning that Scott might be asked to review his forthcoming book. 'It probably isn't perfect,' he admitted, 'and I don't ask or expect any favours – only that you treat it fairly'. Scott again pleaded a lack of time, and waited for Dinsdale's book to arrive.[15]

Loch Ness Monster, by Tim Dinsdale A.R.Ae.S., was dedicated to Wendy, his 'patient' wife, and their four children 'who all *know* that fairy stories are true'.[16] Dinsdale's passion had originally been ignited by an account in *Everybody's Magazine* of an animal spotted on a fine June morning in 1939. It was 'like a monster of pre-historic times, measuring a full 30 feet from tip to tail, with a long sinuous neck and a flat reptilian head' and dived 'in a turmoil of water and setting up a miniature tidal wave'. This was 'the most incredible sight' that the witness had seen during 40 years on the Loch. The witness worked as a water bailiff and his name was Alex Campbell.[17]

At that moment in February 1959, Dinsdale fell victim to 'the germ of curiosity' against which there was no immunity. This led to his analysis of one hundred eyewitness reports (10,000 words and 18 square feet of paper), from which he calculated that the average Monster had a 9-foot neck, 20-foot body, a blunt 10-foot tail and two-point-something humps (rounded down to 2.0 for convenience). Finer details included hefty flippers or feet, eyes like the slit in a darning needle and a neck frill 'like a pair of kippered herrings'. Perfectly at home in the Loch (which it could cross at up to 35 mph), the Monster had also been spotted on land and could clear a road in a single bound. Up close, it was 'a formidable creature of most evil appearance'.[18]

Dinsdale brought the register of sightings up to date. Below the Horseshoe Scree, a 50-foot creature had been watched through binoculars by an 'anonymous gentleman' (Torquil MacLeod was still working under cover when the book went to press). After that came the Reverend Dobbs' moment of epiphany the previous summer, then the creature that stalked the yacht *Finola* and – just before the New Year – a massive humped animal which 50 boys and three canteen staff at the Abbey School in Fort Augustus watched for ten minutes off the mouth of the River Oich. The star witness had seen the Monster many times: 'in his fifties, of slight build and scholarly appearance ... [showing] reserve and absolute sincerity, making no attempt to impress me or dramatize his account ... I knew I had met a man who spoke the truth'. This old hand was none other than Alex Campbell.[19]

Numerous photographs had been taken of the Monster.[20] Dinsdale explained his expert analysis of the Surgeon's Photograph, much as he had done to Scott the previous year. Someone who understood a 'fundamental physical law' (e.g. an engineer) and held the Photograph at arm's length would see 'obscure details' and a 'secondary meaning' which proved beyond all doubt that this was the image of a living animal of huge scientific importance. Assuming its neck to be six feet long, the animal's body would stretch to about 14 feet, as predicted by Dinsdale's calculations. It also matched the sketch drawn by the 'anonymous gentleman' and labelled, 'I think the Loch Ness Monster looks like this' (Figure 3).

A 'very important photograph', snapped in 1955 by Mr P. A. MacNab [sic], showed a two-humped creature gliding towards the ruins of Urquhart Castle. To fully appreciate its significance, Dinsdale suggested that the Monster and its wash should be examined '*closely* and *intelligently*' before measuring them against the 50-foot high tower of the Castle.

Even better was Peter O'Connor's photograph, 'without doubt the closest and most revealing still photo ever taken ... and therefore of unique importance'. Dinsdale had personally grilled O'Connor, 'a perfectly straightforward person', about aspects of the photograph that might be misinterpreted by sceptics. Why was the flash needed two hours after sunrise, and why was the sky black? Answer: O'Connor's cheap Brownie Flash camera underexposed everything. How had the Monster allowed O'Connor to get so close? Answer: thanks to his Marine Commando training, O'Connor could walk silently through water. Dinsdale's verdict: this was not 'the incredibly cunning work of a fraud or hoaxer', but 'unposed and taken in a hurry by someone *who really was* faced with a most unnerving situation'. There was just one slight awkwardness. O'Connor allowed Dinsdale to reproduce the photograph only if he stated that the Northern Naturalists' Organisation had named the creature, 'Nessiesaurus O'Connori'. Dinsdale did so, but buried the reference in the Acknowledgements, where few would see it.

Figure 3 Sketch drawn by Torquil MacLeod, following his sighting of a large animal at the foot of the Horsehoe Scree on the eastern shore of Loch Ness, 28 February 1960. Reproduced by kind permission of Routledge & Kegan Paul.

The evidence headed for its climax when the reader joined Dinsdale in his car, crammed with camping and photographic gear, for his epic 1,000-mile round trip to Loch Ness during Easter week 1960. The pivotal point of the book – and indeed of the entire 1,400-year saga to date – was the four minutes when Dinsdale stopped his car above Foyers and 'through the magic lens of my camera, reached out across 1,000 yards, and more, to grasp the monster by the tail'. Scott would have recognised those words from one of Dinsdale's letters to him a year earlier.[21]

What was the Monster? Dinsdale's statistical analysis led him straight to the answer: 'an evolved species' of the long-necked plesiosaur, *Elasmosaurus*

(Plate 22). This was confirmed by Alex Campbell when Dinsdale showed him a picture book of prehistoric reptiles. Campbell picked out *Elasmosaurus* and 'in a quiet voice . . . with no intentional drama', said, 'That's it – that's it, only with a shorter neck'".[22]

Dinsdale bolstered his argument with a photograph of a fossilised elasmosaur skeleton dug out of a clay pit near Peterborough. The 40-foot *Cryptoclidus* had sported a 19-foot neck with 76 vertebrae and yard-long, paddle-shaped flippers with five digits. Putting flesh on those bones would produce a creature remarkably similar to Dinsdale's reconstruction. 'If it is not,' Dinsdale admitted, 'then I have no idea what it could be.'[23]

The distasteful subject of hoaxes was dismissed in a few lines. Dinsdale had been impressed by the 'undoubted sincerity' of ordinary people such as Lachlan Stewart [sic], Hugh Gray and Alex Campbell. The unnamed London surgeon, being a member of the medical profession, was 'not without ethical standards'; if his Photograph turned out to be a fake, it had been done 'with great cleverness'.[24]

As well as frequent typographical errors – Drummadrochit, MacNab, Otawa, halucination and Jurrasic – *Loch Ness Monster* contained some curious omissions. Dr Maurice Burton, once Dinsdale's esteemed scientific associate, was not mentioned at all. Neither was Denys Tucker, who had already put his money on *Elasmosaurus*. Peter Scott was anonymised as a 'very famous person, a naturalist and artist' who had found time for Dinsdale and his ideas. They had met at the famous person's club, and Dinsdale had sent him a cartoon, drawn in fun, but which 'was to prove prophetic'. This showed a stick-man leading a grinning, long-necked plesiosaur-like monster – exactly as in Dinsdale's letter to Scott, but without the nameplate of the Savile Club.[25]

Loch Ness Monster ended with an upbeat statement of Dinsdale's objectives – PROTECTION, PRESERVATION AND STUDY – and a call to arms:

Who, therefore, will join with me in the search for the truth – about the shores of this strange and beautiful place?[26]

Dinsdale evidently hoped that his book would be a bestseller and that thousands would flock to join him at Loch Ness. He was right on both counts.

Publish and be damned

You wait years for a book about the Loch Ness Monster, and then three come along at once. Of the two books that followed hard on the heels of Dinsdale's *Loch Ness Monster*, the less contentious was the third edition of *More Than a Legend*.[27] Constance Whyte's only update was a three-page 'Note' at the front; everything else was preserved in its original 1957 time capsule. Dinsdale believed that his film had created the greatest splash in three decades, but Whyte dismissed it in one line as further evidence that the beast had flippers. She gave more space to other post-1957 milestones, including Nobel Laureate Richard Synge's testimony (1958), the AA patrolman's sighting at Brackla (1959), and the *Finola*'s close encounter in the summer of 1960.

Dr Denys Tucker got an honourable mention as 'the first professional zoologist to visit and explore the Loch' and for having come out as a believer in the existence of a large, unknown animal. The Oxford–Cambridge Expedition of 1960, inspired by Tucker, was described neutrally as having done 'most valuable work on the fish population of the Loch, a potential food supply for monsters'.

Whyte ended her Note with a sorrowful sideswipe at the inertia of the scientific establishment. The Monster 'challenged beliefs which are integral to the mental make-up of most established zoologists', but that was no excuse. Across the Atlantic, the lake creatures of Canada were being taken seriously and actively studied. British scientists had a golden opportunity in Loch Ness but were letting it slip through their fingers. They now had to look to their laurels.

Problems and solutions

More controversial was *The Elusive Monster*, with a subtitle guaranteed to narrow the eyes of believers: 'A Re-examination of the Loch Ness Problem'.[28] The author was Dr Maurice Burton, now retired but formerly a senior zoologist in the Natural History Museum. Since 1933, he had nursed 'a fairly consistent belief [that there was] something in Loch Ness that merited serious attention', and had written many articles about what this might be.

Over the years, Burton had changed his mind a lot. He went first for a giant eel and then, after reading *More Than a Legend*, 'an animal of the plesiosaur type'. Recently, he had gone off that idea, writing in *New Scientist* (22 September 1960) that 'the plesiosaur theory begins to look a bit tattered'. Now, just a few months later, the plesiosaur was dead in the water. By page 30 of his new book, Burton had declared that there was no large, unknown animal in Loch Ness after all. Sightings could be explained by a combination of natural phenomena and the normal fauna of the Highlands.

To scotch any rumours that he was just another armchair speculator in London, Burton had spent a week at Loch Ness with 'my own team' and scanned a two-mile stretch of shore from 5 a.m. to 10 p.m. each day. They also took a dinghy out into the Loch over measured distances to check whether human visual acuity could match the claims of some eyewitnesses. These experiments demolished some of the evidence which Gould, Whyte and Dinsdale held most dear. For example, Miss Janet Fraser of the Halfway House tea room at Alltsigh could not possibly have seen 'a frill like a pair of kippered herrings', without binoculars, at a range of 1,000 yards and looking towards the sun. Similarly, reports of an eye 'like the slit in a darning needle' (seen from 500 yards), or skin that appeared 'blistered' (from 750 yards) or 'like a tree trunk' (over a mile away, in poor visibility) merely confirmed that the human brain is good at embroidery, and raised doubts about everything that these witnesses claimed to have seen.[29]

Burton's team had tried to tempt Monsters up from the depths by towing a 'rubby-dubby bag', crammed with rotting fish guts and a dash of pilchard oil, behind their dinghy. This was a great hit with sharks down in Cornwall but left the Monster unmoved. Indeed, the week-long expedition yielded just one 'meagre' sighting, of a small humped object that appeared briefly at the front of a chain of ripples. Otherwise, they saw things that might have fooled the casual observer, but which to their expert eyes were just tricks of the light or the wind.[30]

Burton had a new theory to explain 'Monsters' that surfaced in a welter of foam. These were not animals but mats of rotting vegetation, brought to the surface by the gases of decomposition.[31] The phenomenon was well known in deep freshwater lakes in Norway. Up to 60 feet long, some vegetable monsters were reported to have a neck, a horse-like head and even prominent eyes. They surfaced suddenly, occasionally with enough force to crack the ice on frozen lakes, and could speed across the water before submerging again. Significantly, the 'monsters' inhabited lakes that collected the effluent from sawmills and were devoid of fish. Some had been chased and captured – revealing a raft of sawdust, pine needles and sticks, bound together by skeins of stinking algae. Methane produced by rotting organic matter was trapped within the debris and eventually lifted the raft to the surface, where the escaping gas could jet-propel the 'monster' across the water before it lost buoyancy and sank back to the bottom. Burton believed that this also happened in Loch Ness. Most Monster sightings were near the mouths of rivers, where debris was washed into the Loch. Gas could explain violent eruptions of foam and explosive sounds like 'a gale' and the 'express train' which had startled the Eden School Mountaineering Club in August 1960.

A correspondent had told Burton that algal slime, possibly nourished by sewage, collected near Dores, Drumnadrochit and Fort Augustus (all sighting hot spots) during warm weather. This inspired Burton to bubble air under sheets of algae in an aquarium. He managed to create mini-monsters with up to three humps that could scud across the surface before sinking. Burton urged his readers to do their own experiments at home, using a glass of water and a 'medicinal effervescent tablet' (presumably the legendary indigestion cure Alka-Seltzer) to demonstrate how vigorously generated gas could apparently breathe life into inanimate objects.[32]

By now, believers in the Monster might already have reached for the Alka-Seltzer, because Burton had carted some of their holiest cows off to the abattoir. The Spicers were dismissed ('I find it difficult to believe this story'), as was the vet student Arthur Grant ('ridiculous'). Hugh Gray's 20-foot serpent was a 'tree-trunk buoyed by the gases of its own decay'. The Surgeon's Photograph, with its 'complete lack of detail' and the 'incredibly' small head, was probably a branch. Peter O'Connor's story was riddled with too many inconsistencies to be believable. And Tim Dinsdale had simply filmed a boat. The object had followed the course which locals used to cross the Loch from Foyers. Also, Dinsdale had made the throwaway remark (quoted in the *Daily Mail* but never repeated) that, through binoculars, the animal's back had looked 'like an underfed horse'. Burton believed that the knobbled appearance was explained by a boat's crew wearing sou'westers.[33] He made plain his disdain for his erstwhile collaborator in a footnote to the 'List of works referred to in the text': '*The Loch Ness Monster* by Tim Dinsdale . . . appeared after this book had gone to press. It contains nothing to make me alter any of the opinions expressed here.'[34]

Some 'Monster' sightings were apparently of living animals, but Burton could not bring any hope to those who longed for something exotic or pre-historic. Being cold-blooded, reptiles could not survive in water that was so close to freezing point. This ruled out plesiosaurs and their descendants. Instead, this was a species well known around the Loch but easily capable of alarming those unfamiliar with them – the otter. Adults can grow to nearly six feet in length, stretch their necks out surprisingly long, swim like a torpedo, nip up steep wooded slopes and bound across roads. Close up and appearing without warning, otters could be sinister and intimidating. Expanding an otter to 20 feet or more was a challenge, but allowing for observer error and tricks of the light, this was still possible.[35]

There were also hints of a larger, long-necked otter. No specimen had ever been captured, but its description had reached Burton via the aristocratic pen of Count Christopher Vojkffy, who had seen a drawing made by Abbé Henri Breuil. This was not exactly contemporary. Breuil, known as the 'Pope of Prehistory', had sketched it from a painting of 'what was believed to be a large aquatic animal' near Nîmes in Provence – in the famous Stone Age caves at La Baume-Latrone.[36]

Those who had clung on during Burton's switchback exposition could still have been thrown off by his final twist. He confessed that there was still 'the outside chance [of] an unusual animal in Loch Ness'. Suggestive evidence included the Surgeon's Photograph and the 'remarkable' three-humped photograph by Lachlan Stewart [sic], which evidently was neither a plesiosaur nor matted vegetation. If such an animal existed, Burton thought that a good place to look for it would be Cherry Island.[37]

The Elusive Monster was launched with a fanfare of publicity and the certainty (printed boldly on the dust-cover) that Burton, 'the first trained zoologist to have written a book on the subject', would transform the debate about the Monster, which until now had been 'all too apt to inspire ribald disbelief or a faith that borders on religious mania'.

The jacket blurb also carried a health warning. Many would find the book fascinating and original but it might exasperate 'those who have closed minds on the subject'. That category presumably included Tim Dinsdale and Burton's former colleague at the Natural History Museum, Dr Denys Tucker. Burton gave even less space to Tucker than to Dinsdale. He did not mention him at all.

Compare and contrast

The three books provided rich pickings for a review in *Oryx*, the international journal devoted to animal conservation.[38] For good measure, the reviewer also threw in Rupert Gould's *The Loch Ness Monster and Others*, even though it was now long out of print. The reviewer had an agenda which he had declared by the end of the opening paragraph. 'Popular interest in the problem is unabated,' he explained, 'and a new generation of zoologists is shaming the un-enterprise of the last ... the pundits who did not stir from their armchairs.'

Gould, a man with 'an incorrigible interest in strange phenomena', had made a strong case for investigating the Loch and its alleged inhabitant. His book had been ignored by academics, 'apart from a few briefly derogatory reviews'. Constance Whyte had picked up where Gould had left off, making her own 'impassioned appeal for an adequate professional investigation' to confirm or refute the Monster's existence. The reviewer welcomed the third, revised edition of *More Than a Legend*, which he felt was the best 'general introduction'.

And so to the new arrivals. Dinsdale's book was 'very enthusiastic and readable' but his view that Loch Ness harboured 'a relict population of long-necked Plesiosaurs' owed more to 'courage than zoological knowledge'. Dinsdale's understanding of reptiles was 'wildly inaccurate' and he was also guilty of spelling mistakes, 'inordinate padding' and an uncritical acceptance of witnesses' testimony. The book could not be called authoritative.

This left Dr Maurice Burton, 'the first professional zoologist to write a full-length book on the Loch Ness Monster'. The book was the 'best constructed' of the four but was dangerously partisan. Burton now apparently believed that the 'Monster' was an artefact created by vegetable matter and marsh gas but still vacillated, unable to dismiss 'a hard core of evidence which may point to an unknown and zoological explanation'. The reviewer nailed up a 'disturbing' example of Burton's fickleness. Just six months earlier, Burton had been 'convinced beyond all doubt' that the Surgeon's Photograph was genuine and that there was 'no argument about the reality of the Loch Ness Monster, nor any doubt of its being a large animate body'. Now, it was only a large tree root.

The reviewer was identified in *Oryx* only by his initials, D.W.T., but anyone with more than a passing interest in the Monster, or who had skimmed the newspapers during the previous few months, would have known instantly that this was Dr Denys W. Tucker, lately of the Natural History Museum.

Tucker had already crossed swords with Burton and Whyte in the Letters column of *New Scientist* a few months earlier.[34] He had laid into Burton for his malleability and, latterly, his intense negativity about the Monster. Burton had no idea what he was talking about, because he had never seen the animal himself. Tucker had:

I, a professional marine biologist of respectable experience, did see a large hump travelling across flat calm water between Inchnacardoch and Glendoe on 22 March 1959, and do quite unashamedly assert that it belonged to an unknown animal . . . And at present, for the life of me, I cannot imagine any other animal than an Elasmosaur which would fit the bill.

Burton fought back. He had indeed 'toyed' with the plesiosaur notion, even before Whyte's book resuscitated it in 1957, but accumulating evidence now compelled him to reject it. Vegetable mats, or gas-inflated sheets of algae, sometimes extraordinarily lifelike and 'capable of shooting across the surface at surprising speed', were the real explanation.

This provoked Constance Whyte to wade in. Vegetable mats were just the latest of Dr Burton's ideas and, like their predecessors, would soon be 'more or less old hat'.[39]

Peter Baker agreed: Burton's vegetable creations could not explain the observed behaviour of the Monster, and there was only one place where the mystery could be solved – Loch Ness itself. The last word in this dog-eat-dog exchange went to the vertebrate palaeontologist Dr L. B. Tarlo, who wrote as though he had a score to settle with his former colleague at the Natural History Museum, Dr Denys Tucker. Tarlo was glad that Burton had at last abandoned the notion of a plesiosaur, but this made it 'all the more regrettable that Dr Tucker should continue to champion this lost cause'. Tucker claimed to have made 'an intensive study of the literature'; why, then, had he consistently failed to acknowledge all the work which Tarlo and others had done to reconstruct the habits of the long-necked plesiosaurs from their skeletal structure? These animals 'were poorly equipped for diving after their food and consequently spent most of their life at the surface' – where they would be obvious to all. 'If [Tucker] persists in his adherence to the plesiosaur theory,' Tarlo complained, 'surely it is up to him to dispute these conclusions and not simply ignore them?'[40]

Last exits

Tucker had the last word in *Oryx*, but by then – in terms of both career and credibility – he was writing from beyond the grave. In the summer of 1960, Tucker's battle with the Natural History Museum had spilled over into the newspapers. *The Times* and others had portrayed him as a wronged visionary, respected by his international peers if not his employers, who had stood his ground for his scientifically grounded beliefs in the Monster's existence. The Museum had behaved outrageously throughout, trying to gag him and then threatening to dismiss him if he persisted in pursuing his Monster, even off the Museum's premises and during his holidays. Tucker had refused to recant, and had now paid a terrible price for his principles. The Museum's Trustees had sacked him and banned him from the premises.[41]

Tucker had threatened an outcry in the press if the Museum's Trustees tried to fire him, but by now the editors had moved on to hotter topics. His trade union also failed to snatch him back from the jaws of unemployment. This left Sir Cyril Black, MP for Wimbledon, who took his constituent's case to the House of Commons on 3 February 1961. According to Hansard, Tucker was in the limelight from 4.01 p.m. to 4.30 p.m., but his 29 minutes of fame did not rescue either his reputation or his job prospects.[42]

Cyril Black had done his homework on his constituent. This international expert, the youngest DSc in the Museum and a 'brilliant colleague', had been dismissed after 11 years' exemplary service by the Museum's incoming Director, Dr Morrison-Scott. Significantly, Morrison-Scott had already left a trail of 'precipitate and unpleasant summary sackings' during his previous job at the Science Museum. Tucker's appeal against wrongful dismissal was entirely justified.

The counter-argument was put by Sir Edward Boyle, Financial Secretary to the Treasury, who had also done his homework. Having read all the documents, Sir Edward hesitated to point out that his Honourable Friend, the Member for Wimbledon, had not stated the whole truth when he described Dr Tucker as 'a man of enthusiasm and zeal'. Dr Tucker also had 'certain less admirable qualities' and had provoked a string of warnings from his bosses at the Museum. He had sealed his own fate by writing 'a thoroughly offensive memorandum' in April 1960. The Tucker dossier was waiting on Morrison-Scott's desk when he took up post as Director. Sir Edward concluded: 'I can understand the unwillingness of the Director to grant Dr Tucker the privilege of admission to the library and the study collection.'

Having heard both sides of the story, the special pleading from Wimbledon collapsed and the House adjourned on time for tea. This left Dr Denys Tucker in limbo, surrounded by the ruins of his career and all but unemployable. Opinion remained divided about the identity of the true villain. Some continued to argue that Tucker had been martyred for his belief in the Monster. Others felt just as strongly that the Monster was just a convenient scapegoat.

*

The spring of 1961 saw another dedicated Monster hunter bow out of the chase: Dinsdale's 'anonymous gentleman' observer, and the naturalist-sleuth who had broken cover to write to Peter Scott about the creatures that he had watched below the Horseshoe Scree and near Lochend, tailing the *Finola*. Torquil MacLeod had made little fuss when he broke away from his sick bed and chemotherapy and struggled 500 miles to London so that Scott's élite group could hear at first-hand what he had witnessed.

Scott's intuition to settle MacLeod's travel expenses immediately turned out to be well founded. Just a month later, on 14 May, Scott received a handwritten letter from Elisabeth MacLeod which informed him that Torquil had lost his 'long and courageous battle against his illness'.[43] He had died peacefully at home at Altourie, high above the northern end of the Loch.

Torquil was not the only stoical MacLeod. His widow finished her letter to Scott:

I hope that the work he was able to do on the mystery of the Loch Ness Monster will be helpful in the final classification of the animal.

Expeditionary Force

Many will remember the opening half of 1962 for the first footage of The Beatles in their definitive line-up (John, Paul, George and Ringo) performing live at the Cavern Club in Liverpool. There were other distractions, notably a chill in American–Soviet relations which would soon plummet into a new Ice Age, when Russian nuclear-capable missiles were photographed at San Cristóbal in western Cuba.

None of this interfered with David James's preparations for his brainchild, Operation Loch Ness, to be launched on 12 October. He wrote to Scott in mid-July, suggesting that they set up the 'Bureau for the Investigation of Loch Ness Phenomena'.[44] The Executive Committee would consist of James, another Scottish MP (William Anstruther-Gray, of Berwick and East Lothian), Richard Fitter, Constance Whyte and Scott. James was also working on Norman Collins, who had created *Dick Barton: Special Agent* and *Woman's Hour* while Controller of the Light Programme at the BBC and had then defected to co-found Associated Television (ATV).

In late July, James organised an 'Afternoon Study Conference'. Like Scott's informal advisory group a year earlier, this was held in a Committee

Room at the House of Commons, but any resemblance ended there. This was now James's show. In addition to the Bureau's Directors, James invited someone with no knowledge of Loch Ness or zoology, but who commanded near-universal admiration in Britain. Lieutenant Colonel H. G. 'Blondie' Hasler had a navy record that made Scott's and James's look pedestrian. Hasler had led the 'Cockleshell Heroes', a dozen Marine Commandos who paddled their explosives-laden canoes into Bordeaux Harbour and wrecked several Axis ships in December 1942. Hasler, an old friend of James, was delighted to while away a couple of weeks cruising up and down Loch Ness in his yacht *Jester*; his chances of nailing the Monster might be remote, but he would bring publicity and popular credibility to the mission.[45]

Again, Scott's diary got in the way. His scribbled exchange with Mike Garside, written across the top of James's letter notifying him of the conference, went as follows:

Scott: Can I go? *Garside*: No. BBC. *Scott*: Oh well. Send apology.

Plans for Operation Loch Ness had now crystallised with military precision.[46] At 08.39 Eastern Siberia Time on 13 October, James and his Land Rover, loaded with the photographic and radio equipment, would be waiting at Inverness station to meet the 18-strong team off the sleeper from London. Rotas had been drawn up to cover the four observation stations. James and his friend Clem Skelton (a lapsed Catholic monk and now a wizard with cameras and all things mechanical) would be supernumerary, visiting the sites in rotation while constantly on call to be summoned by walkie-talkie to sightings or emergencies.

The days were short – sunset on the 27th was at 16.21 – but the moon was full during the first week, and the Monster's nocturnal stirrings would be revealed by ex-RAF searchlights stationed north of Urquhart Bay and across the water on the Loch's eastern shore at Whitefield.

That was not all. Radar – 'highly desirable' at the searchlight bases – was still under negotiation, as was Admiralty clearance for James to use 'small' explosive charges to bring the animals to the surface. In the bag, however, were Admiralty-issue hydrophones (underwater microphones), ex-artillery range-finders, and mechanical sound generators usually used to set off acoustically triggered mines from a safe distance. An additional 'stimulant to appearance' adapted a recipe from Maurice Burton's expedition to the Loch in 1960: the 'gunny-bag', full of mashed fish and aniseed, to spread an 'attractive slick' around the river mouths that opened into Urquhart Bay.

From Saturday 13 to Saturday 27 October inclusive, Loch Ness would ac-
quire the trappings of a battle zone. Peter Baker, preparing to take another
group of Cambridge students on a combined photographic/sonar survey of
the Loch, was impressed. 'Please let me know when the blasting will take
place,' he wrote to James, 'as I should very much like to be there.'[47]

Results

The 1962 assault on the mystery of Loch Ness began in early July, with the
band of Cambridge students led by Peter Baker and Mark Eastwood.[48] The
shore watch with binoculars and cameras was supplemented by an ambitious underwater search strategy. The Loch was swept from end to end with
a 'sonar curtain' which spanned its width and hung right down to its floor.
The 'curtain' was produced by echo sounders carried by three boats which
moved in formation down the Loch and, in theory, would leave no hiding
place for any large aquatic animals (Figure 4).

The Cambridge Expedition clocked up nearly 500 hours of watching and
six full-length sonar sweeps of the Loch. The rewards were three sightings
(one of a distant pole-like object) and three strong sonar contacts that
seemed too big to be fish. Baker and Eastwood also concluded that the
classical multi-humped sightings were caused by the wakes of boats. They
wrote an upbeat piece for the *Observer* but had to concede that their findings again fell short of hard evidence for the Monster.[49]

Baker's Expedition and Colonel Hasler's mission in *Jester* were just
warm-up acts for Operation Loch Ness. On schedule, Loch Ness became a
vast *son et lumière* show for two weeks in October. Its dark and enigmatic
waters 'rang like a bell' with the mine-detonating noise machine; shivered
with the shock of gelignite sticks detonated on the shoreline; and stank of
rotting fish and aniseed.

The strategy paid off. Three different types of unidentified object were
observed.[50] A vertical 'finger shape', rising six to eight feet out of the water,
was caught in the beam of the searchlight wielded by Mr M. Spear (judged
'a truthful witness') about half a mile out in Urquhart Bay, before disappearing. A dark 'dome' had been spotted a couple of times, also at long
range.

Most convincing was a 'long, low, indeterminate shape' about ten feet in
length, seen barely 200 yards off Temple Pier and in good light on the afternoon of 19 October. Alerted by 'a vast concourse of salmon' thrashing and
leaping on the surface, several observers enjoyed 'an excellent view of the

Monster "feathering" through the water', apparently in pursuit of the fish. The spectacle lasted some minutes and 'a brief few moments' were captured on film by John Luff, an ex-navy man. The sight left a deep impression on Luff, who in all his years at sea had never met anything like it. 'If I had seen such a thing during active service in a destroyer', he said, 'I would have altered course.'[51]

Figure 4 'Sonar curtain', combining the signals of echo-sounders on boats moving in formation up the Loch. This arrangement was used to sweep Loch Ness from end to end during the Cambridge Expedition of July 1962, led by Peter Baker and Mark Eastwood. Reproduced by kind permission of *The Observer* (3 June 1962).

All the written evidence was reviewed by experts, together with a frame-by-frame analysis of Luff's film conducted by the RAF's Central Reconnaissance Establishment (whose cooperation was obtained through more string-pulling by James). During the war, the CRE had honed its skills on photographs taken from Spitfires at anything between 60,000 feet and roof level; their notable successes included working out that nondescript shapes and the shadows they cast were in fact top-secret radar emplacements or launch sites for V1 and V2 rockets. The CRE's verdict on Luff's film was that it showed about eight feet of something dark and glistening, which was 'not a wave effect and had some solidity'.[51]

The experts who examined the findings of Operation Loch Ness provided further evidence that James was determined to carve his own path through the mysteries of Loch Ness. None of them had been members of Scott's hand-picked panel, with their Regius Professorships, FRSs and fondness for clinging to old dogma. Those eminent people were quietly sidelined by James. Instead, he assembled his own 'Independent Panel'.[52]

The 'impartial chairman' was Adrian Head, a London barrister. The world of science was represented, respectably enough, by Dr N.H. Marshall, a marine biologist at the Natural History Museum. The Panel was completed by two experienced naturalists. John Robson, a member of the Otter Committee, was James's deer-stalking companion who later became his biographer.[2] John Buxton, a noted angler and wildlife photographer, was a keen 'observer of natural phenomena'.

The Panel met in late November 1962 to plough through 140 pages of material gathered by Operation Loch Ness and the report from the CRE. They concluded:

We find that there is some unidentified animate object in Loch Ness which, if it be an animal, reptile or mollusc of any known order, is of such size to be worthy of careful scientific investigation. If it is not of a known order, it represents a challenge which is only capable of being resolved by controlled investigation using careful scientific principles.[52]

This was essentially the verdict that Scott had tried, unsuccessfully, to persuade his own expert group to endorse. Unfortunately, the scientific establishment would regard James's panel as essentially a bunch of amateurs. Scott could say in all honesty that they had reached their conclusion uninfluenced by him. Indeed, it was only as a special favour that David James let Scott see the experts' draft report. His covering letter read, 'Technically speaking, I shouldn't send this to you, but I do'.[53]

In his reply, Scott pointed out that the Panel's draft statement needed to be rewritten because it would undoubtedly be widely quoted and was not 'zoologically expert' as it stood. The Panel's statement about 'an animal, reptile or mollusc' implied that reptiles and molluscs are not animals, and omitted a couple of other candidates. Scott suggested changing the text to '. . . a mammal, reptile, batracian [amphibian], fish or mollusc . . .'[54]

James's experts accepted Scott's suggestion except for 'batracian', which they dropped. Colonel Lane, progenitor of the giant salamander theory (1933), would have been dismayed. So would an American professor who was destined to visit Loch Ness in 1965 and would resuscitate the notion that the Monster was a massive amphibian.

Nation shall speak unto Nation

The findings of James's committee were kept strictly under wraps until the New Year. Meanwhile, James had brought in Norman Collins to be Chairman of the Loch Ness Phenomena Investigation Bureau, as it was now known. Collins had pulled his own strings at ATV and had found money for a 20-minute documentary about Operation Loch Ness, to be broadcast to coincide with the public announcement of the committee's verdict.

If anyone at the BBC was worried that ATV were trying to poach one of their prime assets, there was no cause for concern. Peter Scott, so used to the limelight on television and to having the nation in the palm of his hand, now found himself playing a supporting role to David James. He was summoned to Elstree Studios, Hertfordshire, on 18 February 1963 for a rehearsal in the morning and filming in the afternoon, and was presented with a script written by James.[55] Later that week, Scott received a letter from a director at ATV: 'Mr David James has asked me to write to you to ask you not to speak to the press until after the transmission date of Sunday 24th, when a statement regarding the findings will be made.'[56]

The programme, *Report on the Loch Ness Monster*, went out on the appointed day. Its impact was limited, because the broadcast was restricted to the Northern English and Grampian Regions; also, the Cuban Missile Crisis had reached its climax during the second week of Operation Loch Ness.

Compared with *Faraway Look*, the *Report* was drab and unpolished. It began and ended with James, with Scott and others chipping in on his cue. The opening looked contrived –

JAMES: Good evening. My name is David James and I believe in the Loch Ness Monster. My reasons for this are the same as for those who believe in the Rings of Saturn. I have never seen them but I accept the credentials of those who have . . . Probably it would be best if I asked my old wartime friend, Peter Scott, who is a naturalist, to give you the background . . .

SCOTT: David, what prompted you, a busy politician, to lead an expedition to Loch Ness?

– and so did the ending:

JAMES: . . . Well, ladies and gentlemen, that is our story, supported by nearly 1,400 years of legend . . . Are we right in believing that there is a large unidentified species present in Loch Ness? . . . And if our experiments prove that this is likely, isn't it high time that this fascinating zoological whodunnit were solved for all time?[55]

The day after the broadcast, the experts' verdict that Loch Ness harboured 'a large unidentified animate object' that could be 'a mammal, reptile, fish or mollusc' was unveiled to the press. The findings of Operation Loch Ness were condensed into a ten-page report published by the *Inverness Courier*. A cut-down version appeared in the *Courier* under the triumphant headline, 'There is a Monster in Loch Ness – Expedition's findings accepted by experts', with an I-told-you-so subtitle, 'Highlanders vindicated'.[50]

The Loch Ness Phenomena Investigation Bureau had made its mark under the inspired leadership of that dynamic adventurer David James MP. And this was just the start. Plans were already taking shape for an even larger-scale assault on the Loch in the summer, when the astonishing creature behind all those tantalising sightings, photographs and sonar tracings would finally be revealed.

Two and a half years earlier, Peter Scott had sought out James because he needed someone with influence and energy to help him fight for a proper scientific investigation of the Loch. James had charged ahead with great energy, but in a direction that was now diverging from Scott's plan. And Scott, increasingly pulled into his worldwide conservation work, was falling behind.

In spring 1963, Peter Scott had become one of those celebrities who could never be upstaged. But now it seemed that David James had done exactly that.

8

So near, yet so far

The phylum *Mollusca* is diverse and encompasses octopuses and squid, cockles and mussels, slugs and snails. As none of these looks obviously like the Monster, the interested reader might be left wondering how molluscs insinuated themselves into the list of candidate animals drafted by David James's Independent Panel in early 1963.

The idea originally came from F. W. Holiday, generally known as 'Ted' (Plate 23). Before he met the Monster, Holiday was a journalist who wrote articles and books about fishing and lived with his mother in Pembrokeshire. As a boy, he had been infected by Monster fever but did not make the pilgrimage to the Loch until late August 1962, when he was 41. His expedition was frugal compared with David James's extravaganza a couple of months later. After a two-day drive from West Wales in his dilapidated Ford Anglia van, he pitched up on the eastern shore of Loch Ness opposite Urquhart Castle. Inside the van were books about the Monster, 10x50 binoculars, a Rolleiflex camera, fishing rods slung under the roof, and an army surplus mattress with two tartan blankets and a nylon sleeping bag.[1]

Holiday had his first, massive stroke of luck just two days later. At 6 a.m. on 24 August, he was looking down on the Loch from several hundred feet up the mountainside above Foyers. In the mouth of the River Foyers he noticed a dark rounded shape at the centre of a huge circular wave and his binoculars showed a glistening black hump which he estimated to be over 40 feet long. To his eye, it was like the creature which 'gives female gardeners moments of shuddering horror by its very beastliness', only much bigger. The Monster was a gigantic aquatic slug. He rechristened it 'The Great Orm', from the Swedish *Sjö-orm* ('sea serpent') and set out to tell the world.[2]

His first target was Peter Scott. In a persuasive letter, he argued that 95 per cent of sightings, including the photographs taken by Hugh Gray and R. K. Wilson, were consistent with a giant slug.[3] Being a journalist, Holiday was good with words. This might explain why, in a scribbled note to David James, Scott described Holiday as 'a pretty sensible sort of chap' and his giant slug hypothesis as 'an ingenious argument'.[4] As a result, molluscs made it on to the LNPIB's shortlist of candidate Monsters.

The editor of *The Field* was also convinced, and on 1 November 1962 published Holiday's article on the Great Orm, the monstrous slug that lay behind the legend of the creature that inhabited Loch Ness.[5]

1963: Gelignite and gliders

Ted Holiday returned to chase the Orm the following August. The trip was a washout, with just one potential sighting which turned out to be a dead stag lying at the water's edge. This was not his only disappointment, as all the letters about his Orm/slug hypothesis that he had fired off to various universities had been met with cursory dismissal or silence.[6]

By contrast, business had boomed for David James's Second Expedition during the first two weeks of June 1963.[7] The Expedition Headquarters was a group of caravans in Fraser's Field at Achnahannet, two miles south of Urquhart Castle. Fifty-four volunteers – twice as many as in the previous year – arrived as per instructions with cutlery, a bottle opener, waterproofs and gumboots. Directed by their Group Commanders, they watched the Loch between dawn and dusk from ten observation sites dotted along the 70 miles of shoreline. The sites were served by mobile observation units which had 'Loch Ness Investigation' painted on the side and a camera plat-form on the roof; otherwise, they looked just like black London cabs, which is what they were.

Improved technologies had been brought in. The loudest was 'plaster-blasting', using six-ounce charges of gelignite taped to boulders along stretches of shore that belonged to understanding landowners. David James later maintained that the explosions were responsible for the 40 sightings recorded that fortnight, the biggest peak since the original frenzy of 1933–34.[7]

Much quieter were the three gliders which flew up from Nympsfield in Gloucestershire and circled above the Loch throughout the fortnight. The squadron leader was none other than Peter Scott, who had taken up the sport in 1956 and pursued it with his usual flair (British champion, 1963). Scott reckoned that on a clear day they could see down to 15 feet beneath the surface, even in Loch Ness. Unfortunately, they spotted nothing.[8]

Down on the ground, 40 eyewitnesses (only four of whom were with the LNPIB) filled in the official sighting report form for 'David James's Expe-dition'. Just one useful photograph was taken, showing a black object at a range of two miles. The RAF Central Reconnaissance Establishment kindly analysed the image and estimated the object to be 17 feet long.[9] Maurice Burton begged to differ ('a sheet of hessian wallowing in the water').[10]

Afterwards, James beefed up the trip's achievements in an article for the *Observer*, which had provided some funding. The new sightings, together with the 'very consistent descriptions' from the past, confirmed that a plesiosaur-like animal inhabited the Loch.[3] Ted Holiday was one of those impressed and went to visit the Bureau's office near Victoria, London SW1. There he found James surrounded by maps, charts and drawings of custom-designed camera rigs being built for 'the forthcoming hunt'. Excited, Holiday signed up to join the 1964 expedition.[11]

Others were less enchanted. In March, Constance Whyte had written another agitated letter to Scott, complaining that James had no scientific training and that his flights of fancy needed to be reined in. Scott agreed, but regretfully explained that his WWF duties (not to mention the 50 paintings he had promised for a forthcoming exhibition) left him no chance to intervene.[12]

Meanwhile, trouble was also brewing 500 miles south of the Great Glen. In Brighton Kemptown, increasing numbers of constituents felt that David James MP was spending too much time on the wrong side of the border.

1964: Long lenses and long knives

The Third David James-LNPIB Expedition of 1964 was the first to run through the summer, from Whitsun to October. When Ted Holiday arrived on 15 May, he found Fraser's Field filled with caravans, Elsan chemical toilets, a throng of volunteers, a television film crew and a buzz of excitement. He was promptly pressed into service, pulling trout out of the Loch as backing footage for the film crew. Then he was led to the front line by Clem Skelton, ex-Harrow, ex-RAF and ex-monk, and now the technical guru who trained the volunteers to use the camera rigs. Following the cable of a field telephone, they climbed up to the battlements of Urquhart Castle, where the technical drawings in David James's office had been transformed into the most formidable optic weaponry ever to target British wildlife.[13]

The camera rig was built around a 35mm cine-camera carrying a massive 36-inch (900mm) telephoto lens. With its 28x magnification, the lens could resolve a football a mile away on the opposite shore. The camera was fed by a magazine that could hold 1,000 feet of film, enough for 11 minutes of continuous filming. On either side, at the end of a yard-long steel arm, was a still camera with a 20-inch (15x) lens. Like Clem Skelton, the cameras were ex-RAF. Their kind had seen active service in the belly of the photo-reconnaissance Spitfires that had stared down from 40,000 feet on the V2 rockets at Peenemünde.

The camera rig was anchored to a 'pan-and-tilt' mounting on a heavy metal tripod; by turning hand wheels, the rig could be swung like an artil-lery piece to bear on a target anywhere in a wide arc across the Loch. All three cameras were operated by a single switch. While the cine-camera was running, both still cameras fired automatically every two seconds, taking stereo pairs of photographs from which the size and range of the object being filmed could be calculated (Plate 24).

Across the Loch at Sandy Beach, an identical rig was set up on a scaf-folding tower 12 feet above the ground. The mobile observation units were no longer London cabs but Bedford vans, painted Rural Green from Wool-worths. On the roof was a platform large enough for a folding beach chair and a tripod carrying a cine-camera with a 17-inch lens.

On arrival, Holiday had smelled optimism in the air, and the summer's first sighting followed that evening: a dark, 15-foot hump off Achnahannet, which instantly converted a sceptical schoolmaster into a fervent believer. The next morning, however, sea mist crept in, cutting the visibility to a hundred yards. This set the scene for the rest of the season; the entire six months saw just 15 days of clear and calm 'Monster conditions'.[14]

The gruelling monotony of watching from 5.30 a.m. to 10.30 p.m. was made bearable by the excitement of the mission and the colourful charac-ters swept up by it. Clem Skelton and his dog Horrible were complemented by Ivor Newby, naturalist, diver and folk singer, who arrived by canoe with his dog Laddie, having paddled two miles across the Loch. Treated with reverence by everyone was a man from the very start of the story: Alex Campbell, 'the redoubtable water bailiff who has seen the Orm many times'.[14]

That summer, 18 sightings were logged by the Bureau, one by Alex Campbell and only three others by LNPIB personnel. There were no pho-tographs. Back home in Pembrokeshire, Ted Holiday followed the 'mostly routine' reports from the LNPIB's London office with deepening pessi-mism. 'By mid-October,' he wrote, 'we all knew that the 1964 attempt had been unsuccessful and that all that remained was to conduct the autopsy'.[15]

David James tried to turn the autopsy into a Lazarus-style resuscitation. In another article for the *Observer* entitled 'Fair-weather Monster', he pre-dicted better weather and better luck next year.[16] He had another reason to put a brave face on things, because suddenly he had plenty of time for the Monster. In the General Election of 15 October 1964, James lost the previously safe Conservative seat of Brighton Kemptown. The margin was agonisingly narrow (just seven votes) but even seven recounts could not rescue James from a humiliating defeat. He took it hard – 'it felt as bad as being recaptured after escaping' – and retreated to Torosay Castle to regroup.[17]

James and his supporters blamed cock-up and conspiracy in Kemptown, rather than any concerns that he had neglected his constituents. However, 1964 had not been a good year for Brighton. At Whitsun weekend, the town had been turned into a battleground by warring factions of Mods and Rockers. Where was their MP? Up at Loch Ness with his Expedition.

As 1964 came to an end, David James ex-MP might have remembered his gut feeling when Peter Scott invited him to join the search for the Monster – that he had entered the House to make a reputation and not to destroy one.

1965: A near-miss and a big hit

Another year, another Expedition. 'Bigger and better' was Ted Holiday's verdict, largely because David James, freed of the need to spend time down south, had thrown all his energy into it.

The HQ in Fraser's Field now looked almost permanent, with its cluster of caravans. Following local grumbles about the desecration of Urquhart Castle, the camera rig was now installed on a wooden platform on the edge of the field, protected from the uninitiated by a sign that read, 'PRIVATE – NO VISITORS'. A Bedford observation van was usually parked six miles to the north beside the Clansman Hotel. Together, these two posts covered almost half the Loch; the other mobile units took care of the rest.[18]

A week after he arrived, Holiday had his second sighting, in the 'wonderful Scottish gloaming' of 15 June. For almost an hour, he watched a ridged, mustard-coloured hump moving down the Loch a mile out from the Clansman. It was tracked simultaneously from the opposite shore by two fishermen, one of them a policeman. Sadly, none of them had a camera.[19]

This was particularly bad luck, because the summer's weather turned out to be the worst on record and the mismatch between effort and reward was especially cruel: 150 days of watching with the camera rigs manned for 2,200 hours, for just nine sightings and no photographs. Even two businessmen, improbably standing on the steps of the YMCA in the centre of Inverness, had better luck. They were amazed to see a three-humped creature with a thick neck, 'ridged like a tractor tyre', proceeding down the rain-swollen River Ness towards the sea.[20]

A thousand visitors called in that year, but at 2/6d a head did little to keep the operation afloat. The *News of the World* chipped in, living up to its reputation for sensational headlines and sloppy reporting. James wrote to Peter Scott, inquiring if the World Wildlife Fund, now saluted across the world for its campaign to save endangered species, could help out with £500. Scott replied with a regretful 'No' that shut off any further approaches.[21]

However, against all that gloom, there was one significant stroke of good fortune, and one that would transform the LNIPB's prospects.

London had lost its charm for the well-travelled and 'surfeited' American tourist who arrived in September 1965, searching for 'something new, refreshing, different'.[22] His quest was unrewarded until he spotted a tourist poster on Oxford Street and an impulse sent him to see if the Scottish Highlands really were 'remote and breathtakingly beautiful'.

Two days later, sitting on a mountainside above Urquhart Bay, he could only agree. On the way down, he spotted something that changed his life: not the Monster, but a green Bedford van with a hefty tripod-mounted camera on its roof. At HQ, he and Bureau staff exchanged credentials. He saw dedicated adventurers searching for something incredible. They saw an answer to some of their prayers.

Roy Mackal was tall, blond, confident and forty years old (Plate 25). He was also Associate Professor in the Department of Biochemistry at the world-renowned University of Chicago. He spent several days at Loch Ness, learning about the Monster and convincing himself that the LNPIB was not 'an agglomeration of crazies', then rounded off his vacation on the Isle of Mull. There, he was captivated by David James, 'a medium, sandy man, given to honest woolens'. Mackal returned to Chicago, having found the 'new, refreshing, different' thing he had been looking for. He left behind the promise to help the LNPIB by finding American money and – crucially – by bringing top-class science to the hunt for the Monster.[22]

He was as good as his word. The Chicago Zoo and Aquarium declined to throw money at something that might not exist, but the Catholic Girls Junior College of Springfield, Illinois, and others dug into their purses. Mackal wangled an invitation for James to address the Adventurers' Club in Chicago. Founded in 1911 'to provide a home and hearth for those who have left the beaten path and made for adventure', the Club's members included conquering heroes such as Roald Amundsen, Sir Edmund Hillary and Thor Heyerdahl. James's newly flexible diary allowed him to address the Club in their traditional haunt (St Hubert's Old English Grill) on 15 January 1966. He was immediately admitted as a Member, to 'sit at the Long Table and receive the blessings of the Great God Wahoo', and $5,000 swiftly followed.[23]

Meanwhile, Mackal drew up Operation Bootstrap to give the LNPIB the scientific backbone it so desperately needed. His search strategy would include aerial surveillance, midget submarines, mobile sonar searches and underwater microphones. As well as photographs and film, tissue samples were essential, to identify whether the Monster was reptile, mammal, fish or invertebrate. To this end, Mackal devised biopsy harpoons, to be fired underwater from a submarine or above the surface from a crossbow or a fearsome .303 Greener harpoon rifle (Plate 25). Specimens could be captured by baiting, seine netting or 'compression shocking' with explosive charges as used to implode submarines. Although apparently unaware of the Fisheries (Dynamite) Act, Mackal reluctantly abandoned compression shocking, because an imploded Monster and a lot of equally dead fish might not best serve the cause.[24]

As Operation Bootstrap took shape and excitement grew, one fact was sidelined. Professor Mackal's expertise lay in the curious syringe-like 'bacteriophage' viruses that prey on bacteria, killing their victims with a lethal injection of DNA. As far as Monster-sized animals – perhaps 100 million times bigger – were concerned, Mackal was just a well-read amateur.

1966: You win some, you lose some

February 1966 started enigmatically, with David James scribbling 'Keep it quiet!' across the copy of a letter which he sent to Peter Scott.[25] The letter was from Lord Shackleton, Minister of Defence for the RAF, and 'it' was a 2,000-word Photographic Interpretation Report by the Joint Air Reconnaissance Intelligence Centre (JARIC), whose wartime predecessors at the Central Reconnaissance Establishment had prised secrets out of photographs from high-flying Spitfires. JARIC had now risen to a new challenge: a dark object just one-third of a millimetre across, on several hundred frames of 16mm cine-film captured in 1960 by Tim Dinsdale's 'magic lens'.

James had pulled his ex-service strings to persuade JARIC's experts to take a break from Cold War surveillance and help out with the Monster. The verdict on Dinsdale's film was essentially what James wanted to hear. The dark hump was 'probably animate', looked quite different from the boat which Dinsdale had filmed afterwards, and was moving too fast (10–11 mph) to be a boat. The object was about six feet across and probably 12–16 feet long.

Scott kept it quiet, but James did not. A six-page booklet, 'Report on a film taken by Tim Dinsdale, with an introduction by David James MBE DSC', price 6d, was published triumphantly by the LNPIB shortly after-wards.[26] Here was vindication of Dinsdale's film and proof that the LNPIB was chasing the real thing. Predictably, Dinsdale described the JARIC report as 'a brilliant piece of detective work of real significance'. Equally predictably, Maurice Burton saw nothing in its 'tenuous conclusion' to change his opinion about 'Dinsdale's sighting of a red-brown motor-boat'.[27]

It was a good year for Dinsdale. His annual pilgrimages to the Loch, firstly in a 'lonely hide' on the eastern shore and then on an oystercatcher-crowded gravel island in the mouth of the River Foyers, had borne fruit. His second book, *The Leviathans*, had just been published and, with brand-new cine equipment donated by Kodak and his 16-foot cabin cruiser *Water Horse*, he was the undisputed Laird of the Loch.[28]

Meanwhile, Ted Holiday's ideas about the nature of the beast were ma-turing in an interesting direction, away from a gigantic slug and towards an odd creature called *Tullimonstrum* (Figure 5). This fat, worm-like species with fins and an elongated proboscis was known only as fossils and had been extinct for even longer (305 million years) than the plesiosaurs (a mere 65 million years). Nevertheless, Holiday convinced himself that *Tul-limonstrum* fitted the Monster's description like a glove, and began writing a book about it.[29]

Roy Mackal was there in spirit throughout the summer, having accepted the post of Scientific Director of the LNPIB which James had offered him. Holiday was ambivalent about Mackal's role, partly for patriotic reasons: 'This was "our" monster and it was "our" privilege to put it on the map.' But he conceded the usefulness of American science and especially American money, which 'saved the entire study from dropping in its tracks from want of cash'.[30]

Money was David James's main preoccupation. The Fifth 'David James Expedition' did pretty well: more volunteers and visitors than ever before, 4,000 hours of watching and 29 sightings 'of tolerable authenticity'. However, that big-bang photograph still eluded them, while the threat of bankruptcy was their constant companion. James's response was to pitch even harder to potential donors, with ever-glossier Expedition Guides packed with the juiciest sightings, the JARIC report and other nuggets.[31] Science might be important, but without a good story there would be no money and that would be that.

Figure 5 Reconstruction of *Tullimonstrum gregarium*, an enigmatic
species known only from fossils in the Mazon County coal-measures near
Chicago. It was named after Francis J. Tully, an amateur geologist from
Lockport, Illinois, who discovered the type specimen in 1958. The scale-bar
is 6 inches (15 cm) long. Illustration by Ray Loadman.

Mackal breezed into London in September, and James introduced him
to his fellow Directors, including Peter Scott ('Director of the World Wild
Life Fund [sic] and a noted naturalist', as Mackal observed) and Richard
Fitter ('an ornithologist at the London Zoo'). Dinsdale was also there, as
'honoured guest'.[32] Then Mackal was whisked up to the Loch for a press
conference at HQ. The resulting newspaper headlines ('The monster is
a giant sea-slug or squid' . . . 'Scientists say Loch Ness Monster could be
Caught with a Giant Hot-Plate') suggested that this was not public engagement's finest hour. Mackal completed his tour of inspection by 'trying to
participate in everything', which included watching a convincing Monster
turn into a cormorant.[33]

On returning to Chicago, Mackal found his university chiefs 'surprisingly
sophisticated'. In other words, they would let him indulge his zoological
fantasy as long as it did not eat into university time or bring the institution
into disrepute. By now, Mackal was convinced the Monster existed. As to
its identity, he had 'no favourite view . . . but a very slight bias toward one
kind of animal – molluscs'.[34]

The year 1966 also saw a rift opening between the LNPIB and people
whom it could ill afford to lose. Peter Scott visited the Loch for two days
in mid-July to film an episode of *Look* entitled 'On the track of unknown

animals'. Those watching on 9 December 1966 saw Scott kitted out in a scuba diving suit, emerging from dark waters beside the Belgian crypto-zoologist Bernard Heuvelmans, similarly clad; not transmitted were the sound of Scott's teeth chattering and his plea to 'get me out of here PDQ'.[35]

Scott's concerns went beyond the coldness of Loch Ness. In November, he wrote to Richard Fitter, 'I am a bit worried about the unscientificness of the Bureau.' James's reports were extravagant, far outstripped the evidence and drew unwarranted conclusions. Scott had done his best to edit James's most recent offering into shape, but confided to Fitter that 'with my mod-ifications, it may be just acceptable, but only just'. He added ominously, 'David's uncritical approach is running all of us very near to being regarded as the lunatic fringe, and I may find it necessary to pull out.'[36]

Even more dismayed was Constance Whyte, who had already written to tell Scott that she was resigning as a Director of the Bureau. James had taken the enterprise down a path that she no longer wished to follow.[37] She dumped all her research notes and correspondence into a battered suitcase, put it in the attic and turned her back on her Monster.

1967: Testament of Youth

The Beatles released two albums during the Summer of Love in 1967, and both had a message for the Monster hunters. *Magical Mystery Tour* summed up the whole operation perfectly, while the cover of *Sgt. Pepper's Lonely Hearts Club Band* carried the picture of an old friend. Not the Mon-ster, which was still unknown outside the Great Glen in 1927 (when the eponymous Sergeant taught the Band to play), but a man who epitomised the Fab Four's new-found mysticism. Appropriately sandwiched between Hindu guru Sri Yukteswar and Mae West, Aleister Crowley (alias the Beast of Boleskine) glowers out from the back row of onlookers.

Fraser's Field 'wore a new look', according to Ted Holiday.[38] The visitors' centre, mobile workshop and five caravans (ranging from 'The Black Hole' to 'The Ritz') had all been painted a soothing green. The place was heaving; 150 volunteers and 30,000 visitors passed through that summer.

Roy Mackal's handiwork was evident in the new boat and equipment bought with the $20,000 that he had charmed out of Field Enterprises Inc., a big Chicago-based publishing house. Mackal himself did not arrive until September. Instead of launching his Operation Bootstrap, he spent two weeks interviewing eyewitnesses old and new. He was still firmly wedded to the notion of a giant mollusc, even though he had been told some months earlier that anything without a backbone was a non-starter. That intelligence came from Ralph Buchsbaum, who might have known a thing or two as he was Professor of Invertebrate Biology at the University of Pittsburgh.[39]

In the midst of all that activity, only one thing was missing. Like a failing heart, the Bureau was going flat out but, even with a hefty injection of dollars, its output was not sufficient to sustain life for much longer. Recycling years-old evidence such as Dinsdale's film was not going to keep them going; they needed the Monster.

It fell to one of the youngest volunteers to save the day. Dick Raynor had left school at 15 and was two years into an engineering traineeship in Scunthorpe.[40] 'Closing in on the Loch Ness Monster' in *Reader's Digest* of March 1967 made him want to join the hunt and, in early June, he drove up to the Loch in his clapped-out Ford Popular. En route, he collected two wealthy American lads of draftable age who may or may not have been of an age to be called away from Monster-hunting to fulfil their patriotic duties in Vietnam. The first of the Ford's cylinders that was doomed to fail did so at Scotch Corner; mercifully, the second hung on until they were within sight of Urquhart Castle.

On arrival, Raynor dumped his belongings in the Black Hole and was pushed straight into Clem Skelton's crash course in operating the various cameras – including the coded message that everyone longed to scribble on a piece of paper and give to a passing tourist to take to HQ. 'Camera jam at X' told those in the know that the Monster had been sighted at X, while hopefully protecting X from being overrun by excited visitors.

Then it was down to business: up before dawn to start the watch, fortified by coffee and porridge. It was tough, but huge fun. One evening, Raynor was mesmerised to hear about the 30-foot creature that surfaced in Borlum Bay, from the lips of the man who had seen it with his own eyes 30 years earlier. Alex Campbell held his audience entranced for a full half-hour, imitating with his hand the nervous head movements of the Monster as it looked towards the two drifters emerging from the Caledonian Canal.

Raynor had his (and the Bureau's) lucky break just before midday on 13 June. From the top of a Bedford observation van parked near Abriachan, he spotted a long wake crossing glassy-calm water in Dores Day, about 2,000 yards away. Through binoculars, he made out a dark object at the apex of the wake. He filmed it for two minutes before it disappeared; the sequence helpfully included the steamer *Scot II* crossing the field, giving both authenticity and scale (Plate 26).[41]

Back at HQ, he set the hares running. Clem Skelton swept up in his Jaguar Mark VII and whisked the camera away, and Raynor found himself an instant celebrity. The film was developed and sent to JARIC, which concluded that the object causing the wake was travelling at 4–5 mph and that the part visible above the surface was probably about seven feet long.

Raynor did not see his film until 4 January 1968, at the Bureau's delayed 1967 Christmas party. Held in the Chatham Rooms near Victoria, this was billed as 'Film show, buffet, ceilidh – bring guitars etc – piano provided', and members only were invited because 'classified material' was going to be revealed.[42] The 'classified material' was Raynor's film. When it appeared on the screen, he felt disappointed; it had looked much grander on the day through 12x binoculars. But David James rated it alongside Dinsdale's film and the best thing to come out of the LNPIB's efforts to date.[40]

1968: Echoes and reflections

During his fortnight of Monster-hunting in July 1968, Dick Raynor (now a trainee ship's engineer) was promoted to Group Commander and deployed on night drifting. This took a three-man crew out on the Loch after dark to pursue David James's hunch that the Monster was nocturnal. One man was ready with a camera and a massive Pentax flash that could turn night into day, while the others carried a powerful crossbow and the Greener harpoon rifle, both armed with the biopsy head designed by Roy Mackal. All these outings were bitterly cold and completely unrewarded.

That year, Raynor got to know the key personalities of the Monster community. Up on the hill at Strone, in a house with a grandstand view across Urquhart Bay, lived Wing Commander Basil Cary and his wife Winifred, known as Freddie. Both were staunch supporters of the Monster hunt, and both were formidable. The wing commander had been involved in the air strike against the *Bismarck*, while Freddie had played a secret role with the Special Operations Executive, laying traps for newly trained agents to filter out those likely to blow their cover behind enemy lines. Freddie had grown up around the Loch, and had seen the Monster several times. She was renowned locally as a dowser, able to locate missing objects by swinging a non-metallic pendulum over a map. According to legend, she could pinpoint with 100 per cent accuracy which pub Basil was in, and did not need to give her name when she rang the landlord with the command, 'Send him home!'[40]

On the front line at Achnahannet, Raynor met Roy Mackal – charming, highly intelligent and with hints of an intriguing past. Mackal let slip that he had rubbed shoulders with hoodlums at Frank Sinatra's wedding and that his basement was littered with functional rocket parts from the pre-NASA era. Also from Chicago was Holly Arnold, a 21-year-old who had done a high school project on the Monster and decided to trek up while visiting London. On arrival, a bizarre thing happened. Looking down the Loch for the first time, she felt that she had come home. She pestered David James for a job; he somehow made her temporary visa permanent and found some money to create the post of Bureau Secretary for her.[43]

Arnold's impression was of 'a wonderful collection of one-offs'. Quentin Riley had lived with the Inuit and found that the Eskimo Sun Dance was ideal for getting volunteers out of bed before dawn. Sir Peter Ogilvie-Wedderburn ('Og' to the troops) was famous for his splendid but frayed kilt and flask of ready-mixed gin and tonic. Lionel Leslie, an Irish cousin of David James, had explored four continents, done time in a real Parisian artist's garret, and now led a Highland bohemian life as a sculptor near Torosay Castle on Mull.

Just as colourful were some of the volunteers, who included housewives, hippies, professors, priests (lapsed and otherwise), lawyers, students and dropouts. Arnold did her best to protect the Bureau from the attentions of some of the odder members of the public. She kept their correspondence in an unlabelled file that she called 'The Crazies'.[43]

The year 1968 was a disappointing one, with just 14 sightings and only two from LNPIB personnel. However, you would not have believed it from the upbeat piece in *The Times* ('Questing the Beast'); the full-page picture spread in the *Scotsman* ('The Great Monster Hunt'), showing James behind a camera with 'an extra powerful telephoto lens'; or James's hype-rich Annual Report.[44]

Embedded in the Report was a three-page non sequitur about an LNPIB expedition to chase Irish lake monsters in the improbably small (half a mile long, 100 yards wide) Loch Fadda in Connemara, Co. Galway. The expeditionary force included Roy Mackal, Lionel Leslie, Ivor Newby and two businessmen from Field Enterprises Inc. who wanted to see how their money was spent. Newby brought his legendary amphibious car, which looked like a pale blue Triumph Herald until it drove into the water and began surging across the surface at several knots. That was when Newby remembered to insert an all-important bung; when he forgot, things went less smoothly.[45]

Loch Fadda yielded no secrets, other than a hint of 'something' breaking the surface when Leslie detonated seven pounds of gelignite on the shore. Neither did they find anything in Lough Nahooin, a few miles away. It would have been a miracle if they had. Lough Nahooin was barely twice the size of a football pitch and only a few feet deep.

Nonetheless, the trip provided the chance to try out the harpoon gun (Plate 25) and some plastic Monster netting, and to feature in an Irish television documentary. Ivor Newby celebrated the occasion by composing a lament, 'The Beast of Connemara', which he sang unaccompanied by the lakeside; this became the opening theme for the documentary, which took the same name in Gaelic (*Ollpheist chonamara*).

Luckily, the two businessmen from Field Enterprises had a great time and were most understanding.[45]

There was one ray of hope in 1968: a sonar survey led by Professor Gordon Tucker and Dr Hugh Braithwaite of the Electrical and Electronic Engineering Department at Birmingham University. Tucker, well respected in the field, had developed a new digital apparatus and took up James's suggestion (prompted by Peter Baker) to try it out in Loch Ness.[46]

The unit was installed below Temple Pier and directed its beam out across Urquhart Bay. On 28 August, it picked up several large objects that appeared to rise and fall in deep water at vertical speeds (over 400 feet per minute) that would rupture the swim bladders of fish. Some of the objects were estimated to be 20 feet or more in length. Tucker and Braithwaite quickly submitted their findings to *Nature*, which initially made encouraging noises but finally rejected their offering.

Braithwaite rewrote the paper and sent it to *New Scientist*, widely read but not peer reviewed and a downmarket home for original research. When *New Scientist* published the piece in December 1968,[47] *Nature* added insult to injury with an anonymous editorial about it, entitled 'Monster by Sonar', which concluded that 'there is little reason to take seriously the claims of Dr Braithwaite and Professor Tucker to have found a monster'.[48] Mackal was incensed at the 'inscrutable' behaviour and Peter Baker wrote to the editor in vigorous defence of the findings, but the damage was done. David James was unusually balanced in his Annual Report:

It is a temptation to suppose that these [the sonar tracings] must be the fabulous Loch Ness Monsters, now observed for the first time in their underwater activities . . . A great deal of investigation . . . is needed before definite conclusions can be drawn.[49]

And 1968 was also a year of ambitions realised. Ted Holiday published his book, *The Great Orm of Loch Ness*, which aimed 'to challenge some of the most generally accepted zoological concepts'.[50] Basically, the Great Orm was the descendant of 'certain unique fossils in America'. The Loch Ness Monster was one, and so were dragons in Scotland, Babylon and elsewhere, and even the 'dreadful and terrible' Fourth Beast described in Revelation.

Holiday had chanced upon the 'unique fossils' in the bulletin of the Field Museum of Natural History in Chicago. The centre spread showed a picture of 'a unique and wormlike creature . . . resembling a ribbon worm . . . with a tiny head, a long swanlike neck and a long, torpedo-like body ending in a powerful tail' (Figure 5). This was an artist's reconstruction of *Tullimonstrum*, named after amateur geologist Francis Tully. In 1958, while exploring a coal mine near Chicago, Tully cracked open a shale nodule which he had pulled out of a tip heap and became the first person to set eyes on the bizarre fossil. Experts at the Field Museum dated it to the tropical Late Carboniferous Period, 300 million years ago, and described its structure in detail – but had no idea what kind of animal it had been.[51]

Tullimonstrum plugged the gap in Holiday's hypothesis, but left him to explain how this tropical animal had survived 300 million years with no fossil records anywhere else, and transported itself into a small, extremely cold Scottish loch. Moreover, *Tullimonstrum* was not as much of a monster as its name implied. The biggest ever found is only 14 inches long, roughly one-fiftieth the size of the Great Orm which Holiday had seen at Foyers. Ted Holiday was undeterred. He had seen the Monster and knew what it was.

David James was also realising a big ambition. In April, the Special Branch had concluded that a Communist-inspired dirty tricks campaign had been mounted against him in Brighton. In May, the exonerated James was being groomed to replace the Conservative MP for North Dorset, who was standing down before the General Election in 1970. In June, James was selected as the Conservative candidate in waiting.[52]

Throughout his charm offensive in Dorset, James presented himself as a war hero, publisher, adventurer and former MP. There was no mention of the Monster that had once devoured so much of his time and kept him away from his constituents.

1969: Audio, no visuals

Summer 1969 was memorable for Neil Armstrong's giant step for mankind, and three days of Peace and Music (and much else) at Woodstock. It was also a good year for the Monster, which joined the Moon in the top ten subjects for which the office of *Encyclopaedia Britannica* was consulted. Interest ran so high that Rupert Gould's book was dragged out of its 35-year hibernation and reprinted alongside the bestsellers by Tim Dinsdale and Ted Holiday.

There were now eight caravans at Achnahannet, grouped around a wooden visitors' centre with a long-necked, two-humped plesiosaur-like creature painted on its front. The 52,000 visitors suggested excellent health, but other vital signs were less encouraging. As usual, the money was about to run out and, paradoxically, the harder the LNPIB looked for the Monster, the less they seemed to find.

Dick Raynor spent some days scouring Inverness for defunct television tubes, to be used to force Monsters to the surface.[40] Explosions had been banned in and on the Loch, but not *im*plosions. A high-vacuum television tube, weakened by cutting into the glass with a file and dropped into the water in a weighted bag, acted like a depth charge and imploded at a depth of 150 feet with a bang that could be heard at the surface. Unfortunately, the Monster was as unmoved as it had been by patch-blasting.

Luckily, a new weapon was about to be unleashed. Dan Taylor was an earnest and stubborn 24-year-old from Atlanta, Georgia, who had long been fascinated by the Monster.[53] He was also a do-it-yourself fanatic and a US Navy submariner. On 1 June 1969, he arrived at HQ with a truck bearing *Viperfish*, a 20-foot, two-ton submarine which he had built for £8,000 in his parents' garden (Plate 27). By happy coincidence, *Viperfish* was the same colour as the submarine in which the Beatles had all lived in 1966.

Taylor and his vessel were welcomed with 'unprecedented publicity' and predictable headlines. The *Daily Mail* ran 'They're all hunting Nessie in a yellow submarine', above a photograph of Taylor peering out of his conning tower, with David James, Clem Skelton, the kilted Sir Peter Ogilvie-Wedderburn and Holly Arnold lined up in front. Concerns were expressed about *Viperfish*'s tiny plastic propellers, its thrusters held on with jubilee clips, the hull which felt like papier mâché, and the absence of lights to see outside. Taylor explained that his vessel was made of 600 layers of fibreglass, but let slip that his life insurance company had cancelled his policy.[54]

Viperfish arrived unfinished and did not take to the water until August. By then, she was fully equipped for the hunt, with twin biopsy harpoons fired by compressed air.[53] Those worried about the ethical problems posed by sampling a Monster might have been comforted to learn that the whole issue had been thoroughly debated in the House of Lords; assured that 'the submarine operations have no aggressive intent', their Lordships had granted permission to proceed.[55] Mackal hoped that the Monster would be curious and come in close – ideally very close, because the range of the harpoon was only a few feet beyond the submarine's nose.

Finally, the world's press lined up to watch *Viperfish* being towed out into Urquhart Bay. Initially, she refused to submerge, then sank quickly, and several minutes later shot out of the water 'like a flying fish'. The submarine had nose-dived into the mud 60 feet down, and Taylor had to blow the ballast to get free. *Viperfish* went on to complete 45 dives and Taylor had some exciting moments – such as when an invisible force spun the submarine through 180 degrees while it was resting on the bottom. Sadly, though, there was nothing to see or biopsy.[53]

Later, another submarine joined the chase. The three-man *Pisces* was bigger and better, and proof that technology can gallop ahead too fast for its own good. Its original mission was to tow a five-ton model Monster around various scenic locations for a film entitled *The Secret Life of* (brace yourself) *Sherlock Holmes*. All went well until the director decided that the beast looked better without its humps; unfortunately, these contained its buoyancy tanks and the model Monster went the way of all aquatic things that are heavier than water.

Happily for the LNPIB, *Pisces* was liberated to join the hunt for the real thing. During its 40-plus dives totalling 200 hours, *Pisces* explored vast trenches over 900 feet deep (150 feet below the official bottom of the Loch); encountered a powerful current which swirled it right round; and logged a large sonar contact, 50 feet above the bottom, which moved away as the submarine closed in. Positive results: nil.[56]

That summer, hopes were again pinned on sonar. Tucker and Braithwaite returned from Birmingham, recorded more suggestive but non-diagnostic contacts, and decided against trying to publish their findings. Their task was made harder by a rival expedition, mounted by Independent Television News (ITN) and the *Daily Mail*. This involved a minesweeper, a tethered observation balloon and a high-powered sonar unit that used an audible frequency. As well as interfering with Tucker's recordings, this produced a loud 'ping' that could be heard across the Loch. Positive results: nil.[57] A further disincentive was a savaging in the *Daily Mail* by veteran reporter Vincent Mulchrone, who poured vitriol on Tucker, Braithwaite and the whole Monstrous affair: 'Nessie is a myth, a delusion, a tourist bait, a fraud.'[58] Mulchrone was not Monster-naive; his recent exclusive, reporting the discovery of the first genuine skeletal remains of a Monster, had caused great excitement. So had the revelation that the bone originally belonged to a whale and had been nicked from someone's rockery by three lads.[59]

The third prong of the sonar attack was more promising. It was wielded by Bob Love, a brilliant American electrical engineer and underwater expert. After some weeks, Love picked up a large contact off Foyers on 10 October. He tracked it for three minutes at a range of several hundred yards as it traced out a broad looping course between 200 and 500 feet below the surface. Sadly, that was his only interesting result.[60]

As usual, Tim Dinsdale and David James were much in evidence. Dinsdale spent 82 days in his cabin cruiser, *Water Horse*, camouflage-draped and bristling with cameras, radios, fishing rods and a parabolic reflector-cum-microphone. Positive results: nil. David James took charge of a drifter trailing 1,800-foot 'long lines', each carrying 100 baited hooks to tempt passing Monsters. Results: several trout caught and three hooks lost, cause unknown.

In July, *TV Times* published a long interview with David James, featuring drawings of the 'adapted plesiosaur' which still survived in Loch Ness 'over 70 million years after its official extinction'.[61] The absence of hard evidence of the 'adapted plesiosaur' did not deter James from writing an Annual Report that was even richer in hyperbole than its predecessors.[62] Once more, sonar came to the rescue; Bob Love's contact might have been nothing better than suggestive, but the three-minute tracing clinched it for Field Enterprises. This was lucky, as the year which had seen the Monster right up there with the Moon landing could claim only 14 sightings, no photographs and not even a shred of tissue.

1970: All change

The new decade ushered in new people, new ideas, new technology and a change welcomed by signwriters and typesetters: the LNPIB dropped its 'P' for 'Phenomena' and became simply the Loch Ness Investigation Bureau. By contrast, the operation itself mushroomed. To keep everything in order, Tim Dinsdale was appointed Director of Surface Photography, and Bob Love his Underwater counterpart.[63]

Dinsdale took his responsibilities seriously. He recruited Wing Commander Ken Wallis, distinguished bomber pilot and designer-builder of the tiny, open-cockpit one-man helicopters known as 'autogyros'. Minus his handlebar moustache so that he could double for Sean Connery, Wallis had flown *Little Nellie* in his most outrageous mission, outmanoeuvring ordinary helicopters and an erupting volcano in the James Bond film *You Only Live Twice* (1967). Now, he swapped rockets for cameras and spent two happy weeks flying over the Loch (Plate 28). He saw nothing, but the time was well spent. As well as excellent publicity for the LNIB, the sortie was useful practice for Wallis' later participation in the hunt for another elusive life form, the fugitive Lord Lucan.[64]

Meanwhile, Roy Mackal was mounting his 'Big Expedition', an all-American assault on the Monster, inspired by an invitation early that year to join 'The First Monster-Hunting Seminar', organised by Harold E. Edgerton, Professor of Electronic Engineering at the Massachusetts Institute of Technology.[65] 'Papa Flash' Edgerton had invented the 'strobe' flash which made light using electronics rather than by burning out a magnesium filament in a flash bulb. This device revolutionised underwater photography and won him the friendship and admiration of Jacques Cousteau. Edgerton's clever cameras were the best in the world for penetrating the peat-brown waters of Loch Ness, but he had already promised them to Cousteau that summer. Mackal was therefore left to improvise.

The seminar at MIT opened up another avenue, suggested by zoologists from the Marine Biology Laboratory on Rhode Island. Hydrophones (underwater microphones) could be used to record 'bioacoustic calls' of aquatic animals and potentially match them against the library of sounds produced by fish and mammals. Mackal was impressed by field trials of a hydrophone in Chicago. Dunked into his kitchen sink, it allowed him to eavesdrop on neighbours' conversations, transmitted by the plumbing. Then, in the city's aquarium, he tapped into the songs of dolphins, and, out on Lake Michigan, the mating calls of drumfish.[66]

Mackal remained determined to obtain a tissue sample, if not a whole Monster. Abandoning the crossbow and the harpoon gun, he now envisaged 'saturation bombing' the area where a Monster was spotted. Dozens of biopsy darts released from the air would slice through the water, take a sample if they hit something and float back to the surface for recovery. The idea never got off the ground, because of cost and Wing Commander Wallis' refusal to contravene air law by dropping things out of an aircraft. Meanwhile, Bob Love had designed his own apparatus to catch a Monster. Baited with 'highly aromatic' kippered herring (which eels were known to die for), the 18 x 6 x 6-foot cage made of high-impact plastic tubing carried a Citizens' Band radio to alert the hunters that the trap had shut.[67]

Field Enterprises Inc. funded several cameras and hydrophones, together with sound-activated tape recorders and waterproof 55-gallon drums in which they would float. Other equipment was built in Mackal's apartment, to the consternation of his neighbours. The whole lot – a ton in weight and filling 50 crates – was shipped out to the Loch at Field's expense in early July. Mackal followed later: 'I arrived in September with my wife to instigate the specimen collection program.'[68]

The Big Expedition required large numbers of marker floats, which cleaned the Inverness branch of Woolworths out of orange and white footballs. Many compromises were necessary. Edgerton's state-of-the-art camera was replaced by an Instamatic in a waterproof casing, triggered by movement of an arm carrying bait. The baited arm had to be wedged in the neutral position while the assembly was lowered through several hundred feet of water, to stop it firing the camera prematurely. The best wedge was found to be a cylinder of Polo Mints glued together, which took about 15 minutes to dissolve underwater. Unfortunately, the Instamatics confirmed that they were not good enough. All the exposures showed an impenetrable blackness, which would swallow up all but the brightest Monster.[69]

The 'specimen collection program' also had to be rationalised. Bob Love's Monster trap could withstand a 1,400-pound swipe from a 15-foot, 1,000-pound animal but it remained on the drawing board. Mackal's version was conical, just five feet across and six feet high, and made of chicken wire. He did not catch anything.[70]

Meanwhile, Dinsdale was out in *Water Horse*, and Bob Love was chasing sonar echoes beneath the waves. So was Hugh Braithwaite, but without Gordon Tucker, who had drifted away after Vincent Mulchrone's diatribe in the *Daily Mail*. Dinsdale recorded one sighting, of a strange shape 'like a thick telegraph pole' rising out of the water, which sank as he approached.[71] On shore, the usual party atmosphere prevailed, culminating in a celebratory meal at the Drumnadrochit Hotel which featured a locally sourced menu: Consommé Clem – Grilled Mollusc with Asparagus Achnahannet – Sighting Soufflé – Ness-Café.[72]

It was business as usual, and nobody was prepared for a new wave of invading Americans who swept in later that autumn.

The application of science

Another participant at Harold Edgerton's Monster-hunting seminar at MIT was a 48-year-old patent lawyer called Robert Rines, representing the 'Academy of Applied Science' in Boston (Plate 29). Rines now turned up in late September with Martin Klein, president and chief inventor of Klein Associates, specialist sonar manufacturers. They brought new tools to hunt the Monster – a battery of 'sensory attractants' and Klein's new sonar scanner.

Rines explained that the attractants – scents, flavours and chemical compounds that signalled food, company and sex – had been developed in collaboration with university and industrial laboratories. Many ingredients were secret; those revealed included salmon oil, essence of kipper and sex hormones from fish. Also undisclosed was how these compounds were chosen, given that the nature of the Monster and its primal urges were entirely unknown. The attractants' efficacy remained uncertain. Investigators who accidentally anointed themselves with the substances were not eaten or molested, and there was no evidence of Monsters in any state of excitement. Further enticements were provided by playing underwater 'recorded sounds of captive eels, salmon and dolphins, engaged in various acts ranging from feeding to love play'. These too failed to do anything detectable for the Monsters, although 'pulse-like emissions' were later picked up by hydrophones in the areas where the suggestive sounds had been broadcast.[73]

Rines's team were luckier with sonar. Klein had invented a 'side-scanning' emitter/detector that looked like a narrow sledge and radiated a fan of signals in a plane perpendicular to its long axis. Mounted vertically under water, the device could detect objects in the horizontal plane up to a mile away. Towed behind Tim Dinsdale's *Water Horse*, it looked straight down and out to either side. The sonar mapping revealed an underwater picture of Urquhart Bay which was much more exciting than previously believed, with ridges and canyons up to 400 feet deep. There were also shoals of fish, more than enough to feed a population of large animals.[74]

Within a month of arriving, Rines called a press conference to announce their provisional findings. In contrast to the hostility and cynicism that Gordon Tucker had endured from the *Daily Mail* the previous year, Rines was besieged by journalists. Locals, including the postman, were offered bribes to steal photographs of the sonar traces, and the photographer resorted to sleeping with the prints under his pillow. When finally revealed, the images seemed no more exciting than the ones Braithwaite and Tucker had published in *New Scientist*, but the reception was ecstatic. In sharp contrast to the Brits, the Americans had come, seen and were poised to conquer.[75]

Not to be outdone, Roy Mackal had one last fling with his hydrophones in the final stormy days of October. For weeks, they had plumbed the depths and found them to be acoustically dull. Now, listening in the tamer surroundings of Urquhart Bay, they heard something new: an irregular clicking that melted away whenever a boat approached and picked up again when the engine noise subsided. They detected the same sound at other sites, together with other new noises: a rhythmic knocking, and occasional heavy swishing. Mackal was thrilled:

Suddenly we realised that these were not mechanical sounds but calls – produced by living creatures in the water below. Alone in a small boat with darkness falling it was an awesome feeling . . . Minute after minute calls continued to be recorded. At last we had it – an extended recording of calls from unknown animals in Loch Ness.[76]

But with that tantalising cliff-hanger, it was time for Professor Mackal to go home. After all, the new academic year was already well under way back at the University of Chicago.

Handover

The grand finale of the 1970 Monster season was a violent night with 100–mph winds that tore across Fraser's Field and wrote off two caravans. This was also the end of an era.

Ted Holiday was not alone in believing that David James had 'turned fortuity into an organised operation'.[77] Unfortunately, although well-oiled and blessed with equipment, personnel and the determination to succeed, the operation had failed to deliver. It had also inflicted collateral damage. Constance Whyte, doyenne of serious Monster hunters, had turned her back on the mission that she had inspired because, in her view, David James had been obsessed by publicity and money rather than gathering solid evidence. Peter Scott had also drifted away, worried that he would be consigned to the 'lunatic fringe' if he were too closely associated with James and his hyperbole. Since filming Look with Bernard Heuvelmans in July 1966, Scott had not been back to the Loch.

Now, James was about to move on, to pastures old. In the General Election of 18 June 1970, he was elected Conservative MP for North Dorset. His campaign had focused on his parliamentary experience and the achievements that brought him to This Is Your Life. The Monstrous entanglement that had once dragged him away from his duties was expunged from his record, and from now on he had to be conspicuously active down south in Dorset and Westminster. His return to the House after six years in the wilderness was not going to be easy: there were still sniggers in the corridors of Westminster about him and his Monster.[78]

*

The LNIB faced an uncertain future, too. The good news: the manufacturers of Polo Mints, delighted with the publicity about their crucial role in underwater Monster photography, would provide an unlimited free supply for the coming season. The bad news: Field Enterprises Inc., who had bankrolled the LN(P)IB to the tune of $100,000 over four years, decided that a promissory Monster was no longer good enough. With regret, they withdrew their support.[79]

All this left a potentially life-threatening vacuum: no leader, no money, no Monster. It was a golden opportunity for Robert Rines of the Academy of Applied Science to ride in from Boston and save the show. And steal it.

9

Serious science

During 1971, the LNIB slipped off Peter Scott's list of things that mattered. In late August, he received David James's report to the Directors, which was a mixed bag: a grant of £5,000 from the Highlands & Islands Development Board and over 50,000 visitors expected, but a bitter tussle with Inverness-shire County Council over planning permission for a permanent HQ-cum-visitor attraction. Scott scribbled 'Quite interesting' above the letterhead, and it was filed away.[1]

He was left cold by Monstrous things that once would have stirred him. A worryingly close encounter with an animal at least six feet across had been reported by experienced diver Robert 'Brock' Badger, just 20 yards offshore, and a photograph taken by a Frank Searle showed a two-humped object at close range. Searle, whom James described as 'a naturally suspicious hermit', had lived in a tent by the shore for three years and clocked up over 10,000 hours of Monster-watching. James, excited by the photograph, had written 'Strictly Private – not for reproduction' across the back of the print he sent Scott. That was also filed away without comment.[2]

Both James and Scott were frantically busy with their respective careers, which might partly explain why Scott's office in Slimbridge logged no correspondence about the Monster during 1972. After an 18-month void, the Monster surfaced again in an irate letter written to Scott in April 1973. A ten-year-old boy had saved up his pocket money for his £1 junior membership and sent it to David James at the LNIB office in Victoria. Nobody had bothered to reply, even after polite reminders. The lad's mother now demanded an explanation for the LNIB's 'disgusting display of inefficiency and bad manners'.[3] Scott wrote a placatory response, but was too busy with his Presidency of the World Wildlife Fund and jetting around the world for the International Union for the Conservation of Nature to dig any deeper.

And he was going up in the world as well as around it. *Sir* Peter Scott had recently been ennobled for 'services to the conservation of wild animals'.

RIP LNIB

If Scott had inquired, he would have found that the LNIB's 'shocking display' was a symptom of a much deeper malaise. The organisation had collapsed into terminal decline and was heading for its last rites. Money was running out fast and the Council had finally rejected their planning application.

Dick Raynor's two-line report on 20 October 1972 was the last entry in the LNIB's log book. Shortly afterwards, he and Holly Arnold faced the sad task of laying the LNIB to rest and disposing of its assets. The photographs, documents and maps that had gripped the attention of over a quarter of a million people were taken down and given to the Great Glen Exhibition in Fort Augustus. The caravans were sold by sealed auction, mostly to travelling people; some of them did not travel very far before the rust underneath proved to be structural rather than cosmetic. The green Bedford vans, still with 'Loch Ness Investigation' painted on the side, were also sold off. Raynor spotted one a few years later in Dingwall, its insignia intact.

Finally, Raynor dismantled the camera rig that had stood proud at Achnahannet. The photographic equipment was crated up and stored temporarily at Boleskine House, now owned by Jimmy Page, lead guitarist with Led Zeppelin and wild boy (but an amateur compared with his idol, Aleister Crowley). Later, David James drove over from Torosay in his Volvo and took the lot down to London to get what he could for it. Being 30-year-old RAF-surplus stock which had seen heavy use, this was not much.[4]

Surprise, surprise

Scott would probably not have had time to do anything useful, even if he had known. On 18 November 1973, he received Norman Collins' last letter as Chairman of the LNIB. They were broke; there would be no summer expedition in 1974 and they could not even afford to print a farewell newsletter. The LNIB would continue to exist in name, in case money miraculously appeared from somewhere. The shopfront still stood, but the premises were now empty. A one-line item might have puzzled Scott, had he noticed it. Dr Robert Rines of the Academy of Applied Science, Boston, had been invited to join the Board.[5] Scott made no attempt to follow up the lead.

For Scott, 1974 was another entirely Monster-free year, and 1975 began as it would continue – hellishly busy. The Severn Wildfowl Trust that he had set up 30 years earlier at Slimbridge had matured into the Wildfowl and

Wetlands Trust and started a family, opening three new Slimbridge-style centres in England and Scotland. Meanwhile, Scott had been appointed Chancellor of Birmingham University. Wearing both his Chancellor's robes and his IUCN hat, he organised a big international symposium on campus, entitled 'Man and his Environment'. Apparently, he had left Loch Ness behind and moved on to bigger and better things. Then, at the end of March, the Monster lumbered out of oblivion and back into Scott's life. This time, it would push him off his fence and in among the believers.

The moment of conversion came when David James sent him a copy of a book entitled *The Loch Ness Story*, published a few months earlier.[6] Its cover carried a bright art naif view of a hump cruising past Urquhart Castle. The author was Nicholas Witchell, identified in the blurb as a law student and editor of a student newspaper. Inside were faces that Scott did not recognise. A fresh-faced lad in an anorak turned out to be the author, while a middle-aged man wearing heavily framed glasses was named as Dr Robert H. Rines, President of the Academy of Applied Science.

Witchell began his book dramatically with three sentences that must have made Scott sit up straight:

At about twenty to two in the early hours of the morning of the 8th August 1972, a large animal glided silently through the black waters of a Scottish loch. Its journey was possibly its normal nightly excursion except that on this occasion its path took it to within a few feet of the camera sitting on the loch floor. Pictures of the animal were taken and history made –

One of the underwater photographs was reproduced, and it was stunning. 'The flipper-like appendage of a large, unknown animal in Loch Ness' could have belonged to the plesiosaur drawn for Scott's article in the *Observer* in 1961.

This bolt from the blue left Scott grappling with some big questions. Who was Rines, and what was the Academy of Applied Science? What did the other photographs show? And how had Scott managed to miss all this?

Whizz kid

When he first saw Loch Ness in September 1970, Robert H. Rines was 48 years old, divorced and a partner in a successful law firm in Boston.[7] The business had been founded by his father, a patent lawyer and inventor who had passed on his fascination for gadgets. Robert's first patent application, for a pocket-sized folding cutlery set, failed only because someone else had already invented it. In his defence, he was just eight years old and had a greater distraction in the violin. Three years later, music took him to a summer school in Maine, where a visiting dignitary heard him play and asked if they could perform a violin duet. The man who played second fiddle to Rines on that occasion was a certain Albert Einstein.

Having notched up his first proper patent at the age of 19, Rines went on to study physics and engineering at the Massachusetts Institute of Technology (MIT). He graduated high in his class in 1943 and was siphoned off into MIT's radiation research laboratory and then the US Army Signal Corps. While working on a top-secret microwave system to detect enemy aircraft up to 200 miles away, Rines filed patents for electronic 'lenses' to focus microwave signals.

After the war, he followed his father into the legal profession and before long was an up and coming patent lawyer at Rines & Rines in Boston. In 1964, he wrote *Create or Perish*, a textbook on patent law which was based on lectures that he gave at MIT. He also founded the Academy of Applied Science, a 'non-profit, tax-exempt, scientific and educational corporation' based in Belmont, near Boston. Funded by Rines and his brother-in-law (another partner in the family firm), the AAS's mission was to 'stimulate invention, innovation and other creative endeavours [and] the interest of youth in the applied sciences'. However, its most conspicuous spend supported the 'creative endeavours' of older people. As well as Rines's own trips to Loch Ness, the AAS paid for expeditions to find Bigfoot and the legendary fleet of King Jehoshaphat which, according to an authoritative source (1 Kings 22:48), came to grief somewhere in the Red Sea.[8]

In October 1970, Rines flew home from a productive summer at Loch Ness, excited by the state-of-the-art sonar surveys which had lit up the dark waters of Urquhart Bay. As well as the unsuspected ravines cut deep into the floor of the Bay and the shoals of fish to sustain piscivorous Monsters, Martin Klein's side-scanning probe had detected transient large contacts that might be the Monsters themselves.

Back in Boston, Rines contacted Harold Edgerton at MIT, who was intrigued by the sonar results and bemused by the technical difficulties of working in an environment that could have been designed to prevent photography. Edgerton was busy but promised to rise to the challenge. And, as the inventor of the electronic flash unit that photographed the D-Day beaches from 3,000 feet on the pitch-black night before the invasion,[9] he did.

Rines returned to Urquhart Bay the following summer with $100,000-worth of high-intensity Edgerton strobe flashes coupled to underwater cameras that could be triggered automatically at set intervals. The apparatus made Roy Mackal's Polo Mint-Instamatic setup look like the work of an amateur, but photographed nothing of interest. However, the trip was memorable for other reasons. Rines and his secretary, Carol Williamson (who later became his wife), were treated to a spectacle that the laws of chance had withheld from millions of other hopefuls.

On the afternoon of 23 June 1971, they visited Wing Commander Basil Cary and his wife Freddie in their house overlooking the south side of Urquhart Bay. Over tea, they spotted a disturbance about half a mile offshore. Through Basil Cary's elderly brass telescope, Rines saw 'a large, darkish hump like the back of an elephant', which he estimated to be 20 feet long.[10]

That was the moment that changed Rines's life. It also led nicely into 1972, the year in which the Monster finally made serious scientists sit up and take notice.

Pay dirt

In April 1972, Frank Searle was not the only person scanning the Loch. Across the water, 19-year-old Nicholas Witchell (Plate 30) was living in a wooden hut that he had built on the hillside below the Carys' house.

Witchell had finished school the previous summer and saved up enough from office work and paper rounds to spend six months looking for the Monster before starting university in October. This was his third visit to Loch Ness. The first time, aged 17 and inspired by a school science project, he and two friends had camped on the shore south of Dores. After a few days, the friends got bored with not spotting the Monster and drifted away; thereafter, Witchell came alone. Now, he spent most of each day gazing out over the Loch, armed with binoculars and a long-lensed Bolex cine-camera lent to him by the LNIB.[11]

The Monster never revealed itself to Witchell but it still knocked his plans for life off course. He was due to read Law at Leeds, but soon found that he preferred writing. So when Constance Whyte refused to do another edition of *More Than a Legend*, Witchell jumped at the chance to bring the story up to date. As it turned out, that summer would provide the most exciting material in the 1,400-year history of the Monster – and the best opening sentence of any book about it.

*

Rines was in buoyant spirits as he settled into Monster-hunting that June. Harold Edgerton, about to fly out of Boston, received a telegram which began 'HITTING PAY DIRT' and ended 'CAN YOU POSSIBLY PASS THROUGH DRUMNADROCHIT EN ROUTE TO GREECE TO HELP?'[12]

Unfortunately, Edgerton had a rendezvous in the Gulf of Patras to explore wrecks from the Battle of Lepanto (1571), and had to leave Rines to his own devices. These were an underwater camera which took photographs automatically at intervals and a sonar unit that recorded the size, range and speed of anything caught on film. The sonar unit was a commercial depth-sounder, mounted sideways on a metal frame to fire its beam horizontally. The rig was lowered from a boat on to the lake bottom, and the echo signals were fed up to the boat and recorded on a paper chart. The camera and flash were set on the lake bed further out, pointing towards the area targeted by the sonar beam. The camera was basic: a 16mm movie camera set to expose a frame every 15 seconds, loaded with 50 feet (just over 2,000 frames) of Kodachrome 25 colour transparency film. The ten-millimetre wide-angle lens brought anything over a yard away into focus. Mounted beside the camera was a strobe powerful enough to cut several yards into the gloom, synchronised with the shutter.[13]

Urquhart Bay, rich in both salmon and sightings, seemed a promising place to snap a Monster. The best spot was identified by an 'optimum search pattern model', a complex mathematical method which had helped to find an American hydrogen bomb that fell into the Mediterranean in 1966. The model pinpointed 'grid element 2903', 40 feet down and 60 yards off Temple Pier on the northern side of the Bay. Here, the AAA's new sonar map showed a ridge that would provide a grandstand view of anything big coming up from the depths in search of fish.[14] This was where Rines and his team laid their trap and settled down each night to watch the paper crawling out of the chart recorder.

Their moment came early in the flat-calm morning of 8 August. Rines later described the underwater scene in detail.[15] The sonar rig lay on the Loch's bed 30 feet under a small fishing boat named Narwhal, anchored 30 yards offshore. The camera/flash unit was on the bottom 50 feet beneath the Nan, a motor launch moored about 40 yards further out (Figure 6). At 1.40 a.m., the surface suddenly came alive with salmon jumping. On Narwhal, the sonar chart showed the pinpoint echoes of the fish turning into streaks as they fled. At 1.45 a.m., a 'big, black trace', much heavier than the fishes' echoes, began to take shape about 40 yards away.

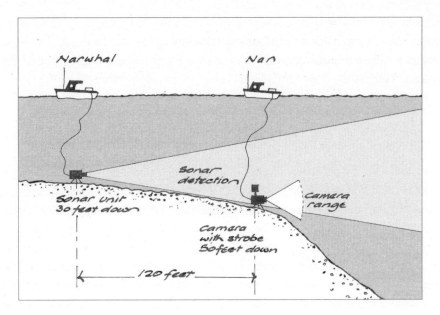

Figure 6 Photographic and sonar equipment deployed on 7 August 1972 at
'Grid Element 2903', near Temple Pier in Urquhart Bay. The layout is
as described by Robert Rines. Illustration by Ray Loadman.

Peter Davies, skipper of *Narwhal*, was shaken by the 'sheer size of the echo trace'. Frightened by the realisation that a 'very large animal' was just 30 feet below him, he climbed into the tender and paddled over to *Nan* to tell Rines what was happening. He rowed Rines back to *Narwhal*, where the chart still displayed the heavy black smudge. As they watched, a breeze sprang up, *Narwhal* swung on her moorings and the contact melted away.[16]

The contact had appeared close to where the camera was pointed, but had anything been photographed? At first light, the camera was pulled up and the film was shipped directly to Eastman Kodak's headquarters in New York. Only four of the 2,000 frames on the film showed anything other than blackness. Taken between 1.45 and 1.55 a.m., while the 'large object' was being drawn on *Narwhal*'s chart recorder, these revealed 'the hazy outlines of something large and solid'. To see what lay behind the haze, the frames were sent to the Jet Propulsion Laboratory (JPL) in Pasadena, California, to be run through the computerised enhancement process used to sharpen up images beamed back from NASA space missions.[17]

Meanwhile, half a dozen sonar experts had pored over the paper strip torn from the chart recorder on *Narwhal*. They concluded that the heavy black trace that appeared just after the fish scattered must be a 'large marine

animal' (or possibly two), 20–30 feet in size and bearing 'echo protuberances' consistent with humps or appendages up to ten feet long. At least
one large animal had swum up to the underwater ridge, within range of the
waiting camera.[18]

Gilding the lily

Nowadays, manipulating digital images is child's play. In 1972, when a
mainframe computer filled a room and a hard disc was too heavy for a
man to lift, image processing was genuinely space-age technology. Rines
entrusted the image-sharpening to Alan Gillespie, a PhD student in Geology at the California Institute of Technology. Gillespie had developed his
own technique and was allowed to use the computers at the nearby JPL
out of hours. In Gillespie's process, a tiny beam of light was swept across
the original colour transparency and its intensity measured on the other
side of the film using a highly sensitive photocell. Each frame was broken
down into two million spots of light, analogous to the pixels that make up
modern digital images. Filters were used to remove colour bias from each
spot and enhance contrast, stripping away the 'fog' of peat particles to yield
a high-definition black and white image.[19]

The basic technique had worked wonders with the fuzzy images transmitted back from Mars by the Mariner missions, unmasking features that
had been invisible in the original. Nonetheless, sceptics detected a whiff
of witchcraft whenever images were manipulated, even by computer, and
suspicions were bound to be raised by anything emanating from the Loch
that had spawned so many hoaxes. Nicholas Witchell took pains to reassure
his readers that 'there can be no suspicion that the computer process fraud-
ulently "touches up" a picture'. JPL's experts put it more eloquently: 'it can
never make Bach sound like Beethoven.'[20]

After processing, three frames remained so cryptic that they were kept
under wraps for further analysis. The enhanced fourth frame showed an
extraordinary object: clean-cut, diamond-shaped and with a central rib,
joined through a sturdy root to a mottled surface that filled the frame (Plate
31). The structure was massive, estimated by Rines to be six feet long and
at least two feet across. It looked like the rear right flipper of an aquatic
animal – but as Harold Lyman of the New England Aquarium in Boston
put it: 'General shape and form of flipper does not fit anything known
today.'[21]

This sharpened image also proved that computers could turn a sow's ear into a silk purse. Steeped in fog, the original frame showed a vague diagonal line corresponding to the central rib, but only an eye of immense faith could have made out anything like a flipper.

Telling the world

Rines's triumph after just three years must have been galling for the LNIB, limping towards the end of its twelfth barren season and facing the likelihood that it would not survive the winter. Even worse, Rines's plans did not include the LNIB. He would not announce his discovery at Loch Ness or even London, but in Boston. The Academy of Applied Science had fixed a press conference for 2 November, jointly with the Boston Scientific Society and the American Institute of Electrical Engineering. The LNIB was listed as a partner but would play no active role. In the meantime, Rines told them, they must say nothing.[22]

Four days before the press conference, Holly Arnold sent out a brief cyclostyled report, confidential and restricted to LNIB members; it turned out that this was the last report that they ever received.[23] The LNIB's work over the summer would be presented at the annual Christmas Party in the Chatham Rooms, Victoria, on 13 January 1973. The high point would be the astonishing underwater photographs which Dr Rines was about to unveil in Boston. Nobody from the LNIB had yet seen them, but Rines said they showed a massive flipper and a 'tail-like structure' several feet long. Various other images being analysed at JPL would be disclosed in due course.

For someone who had captured the closest pictures ever of the Loch Ness Monster, Rines's publication strategy was odd. Serious zoological journals would have leaped at the chance to publish the images, but Rines targeted *Time*, the American news magazine, and MIT's in-house journal, *Technology Review*. On the Monster's side of the Atlantic, he chose only the *Photographic Journal*.[24] This was sent only to members of the Royal Photographic Society and its existence was unknown to 99.9 per cent of the British population. These were not places to publish quality science. None of them used peer review and scientists would see them only by accident. This meant that no card-carrying zoologist could inspect Rines's photographs before they were published.

Rines's astounding flipper photograph made a big splash in Boston and, to a lesser extent, across the USA. Surprisingly few ripples reached Britain. The many who failed to notice included Peter Scott, who had somehow dropped off the LNIB's mailing list. Over two years passed before he finally saw the picture that, if someone else had taken it, might have saved the LNIB from extinction.

A tough act to follow

Fired by the success of the previous summer, Rines returned to Loch Ness in July 1973. He brought another $100,000-worth of improved underwater cameras, loaded with ultra-fast black and white film and triggered whenever an object over five feet across entered the beam of a coupled sonar detector. He chartered Basil Cary's yacht *Smuggler* for three months and experimented with the underwater rig in Urquhart Bay. The apparatus was seen by millions on BBC TV's *Blue Peter* and exposed thousands of feet of film, but failed to photograph anything of note. However, twelve sonar incursions of 'large objects' were logged, similar to the contacts recorded simultaneously with the 'flipper' photographs the previous year.[25]

The summer was busy for others. Nicholas Witchell was frantically writing his book, armed with the contents of Constance Whyte's suitcase of Monster archives. Meanwhile, an 'enthusiastic Oriental team' of divers arrived from Japan, under the intermittent leadership of Yoshio Ko. In real life, Mr Ko was an impresario who was at home staging Tom Jones concerts and Muhammad Ali prize fights in Tokyo. In Loch Ness, he was out of his depth. The Japanese expedition found only a lake which was bigger, deeper and colder than they had been led to believe.[26]

The season was more productive for Frank Searle, the naturally secretive hermit (Plate 32). His tent, glorying in the title 'Visitor Centre', was now firmly on the trail of must-see attractions recommended by the Tourist Information Office in Inverness. 'My simple tent had become a kind of Mecca,' he wrote. Up to a thousand visitors per week called in. The vast majority went away again, although a string of attractive young women (he called each one his 'Girl Friday') were sufficiently taken by Searle's ready smile, weather-beaten features and Clark Gable moustache to stay on and share his vigil, coffee and bed.[27]

Rines had ambitious plans for the next summer, 1974. He revealed these during one of his evening lectures on patent law at MIT, held from 7 to 9 p.m. in the dining hall of MacGregor House. Four massive concrete blocks were being cast, which would sink into the silt 90 feet down and provide a rock-solid foundation for an improved, sonar-triggered camera. Rines confided that an unpublished picture showing 'the whole body of a second animal' was currently being processed at 'an unnamed lab in California [which] does Venus', and was presumably JPL.[28]

Unfortunately, the concrete piles and the new photo failed to materialise and Rines's summer residency at Loch Ness drew another blank. Meanwhile, Witchell had finished his book, due to be published in the autumn. Down on the shore, Frank Searle notched up two more spectacular Monster snapshots (Plate 33), 26,000 visitors and another brace of Girls Friday.[29]

Searle partly filled the void left by the demise of the LNIB. So did Rip Hepple, one of the LNIB Group Commanders who had helped to build up the visitors' centre at Achnahannet. Hepple issued 'Nessletter No. 1', with news of sightings and happenings on the Loch and expressed the hope that this would be 'the first of many'. That was in January 1974. Nessletter No. 120 was sent out in January 1995.[30]

In from the cold

Leafing through Witchell's book in late March 1975, Peter Scott found more questions than answers. On further scrutiny, the 'amazing' flipper photograph looked too good to be true. Could it be a fake? Why were the other underwater photographs not shown? Robert Rines was evidently no stranger on the shore of Loch Ness, but was he trustworthy? Scott put his concerns in a letter to David James, adding that what Witchell had written about the so-called Academy of Applied Science was 'not entirely reassuring'.[31]

James was busy with the Common Market Referendum and took nearly a month to reply, rather brusquely. Rines was 'a well-known and respected patent lawyer . . . who has indulged in research projects of this nature for many years'. The photos were unquestionably genuine. Scott should try to meet Rines during his forthcoming annual pilgrimage to Loch Ness.[32]

Scott duly wrote to Dr Robert H. Rines at the Academy of Applied Science in Belmont, near Boston:[33]

As you will know, I have long had an interest in the possibility of large animals in Loch Ness . . . In some regrettable way, I had not seen your fantastically interesting photographs of 7 August 1973 [sic] until the publication of Nicholas Witchell's book. I am rather astonished at the lack of serious attention given to your photographs and feel that this situation should be remedied at the highest level.

Rines's reply began, 'Dear Sir Scott' and explained that the flipper photograph came from a series which was undergoing computer enhancement at JPL. 'Serious attention' had in fact come from JPL and 'several other American scientists'. Rines had 'tried to avoid publicity', which might explain why Scott had missed all this. Rines would be in London with his wife and new baby son in early June, and would be delighted to meet Scott and David Jones [sic]. The letter was sent from the Rines family law firm, not the Academy of Applied Science. It ended, 'Cordially, RINES AND RINES' and was pp'd by a secretary.[34]

Scott and Rines met for the first time in the second week of June, at Slimbridge. They got on well. The Rines family's thank-you note, from Bob, Carol and baby Justice, was on a jaunty Nessie postcard sent from Tychat, the house which they had recently bought on the hillside overlooking Temple Pier and Urquhart Bay. Rines promised to tell JPL to send Scott all the photos – and of course would let him know if they found anything new at the Loch during the summer.[35]

Scott's suspicions had not been allayed by meeting Rines face to face. A month later, he wrote to his wise counsel, Sir Solly Zuckerman, who was about to visit MIT. Scott explained the background and went straight to the point. Any evidence suggesting that the photographs and sonar charts could be 'a deliberate, elaborate and skilful forgery' would be 'helpful'. Specifically, Scott asked 'to what extent can we depend on the integrity of Dr Rines and his associates?'.[36]

Zuckerman replied promptly. He was spending a whole day with the President of MIT and would 'make the necessary enquiries'. On returning to England, he reported back to Scott in a letter marked 'Confidential'. The President of MIT had never heard of the Academy of Applied Science but knew about Rines. 'I showed him your letter', wrote Zuckerman, 'and the gist of his reply was, "Tread warily".'[37]

But by then it was too late, because Rines had found something new.

Full frontal

The Monster was now firmly back in Scott's life. In August, he joined Richard Fitter and Tim Dinsdale to interview Alan Wilkins, who had seen the Monster a month earlier; Constance Whyte was also persuaded to attend. Wilkins and his son had watched a dark object cruising up the Loch a mile off Invermoriston. Through 10x50 binoculars, up to three humps were clearly visible.[38] Wilkins was Head of Classics at Annan Academy near Dumfries, and a convincing and matter-of-fact witness. His account inspired Scott to sketch a series of bulky, long-necked plesiosaur-like animals with one, two or three humps (Figure 7).[39]

Later that month, Scott went to stay with Bob and Carol Rines at Tychat. He inspected the underwater operation at first hand, reliving the moment in 1966 when *On the Track of Unknown Animals*[40] had left him shivering in the Loch in his wet suit (Plate 34).

Then, on 1 October, he found himself on a plane to Boston, because Rines had told him about two even more astonishing underwater photographs of the Monster. These were not the unseen frames that had been incubating for three years at JPL, but new images taken just a few weeks earlier.

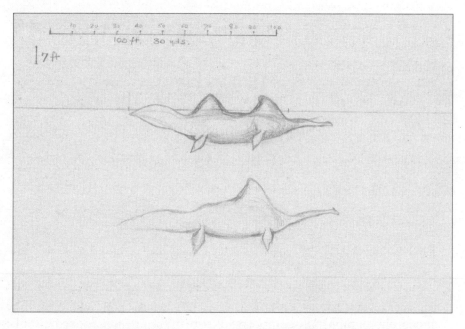

Figure 7 Sketches by Peter Scott of hypothetical humped Monsters,
drawn in August 1975. They were inspired by his meeting with Alan Wilkins, who
had reported watching a creature that showed one, two or three humps
as it swam across Loch Ness. Wilkins was the classicist who approved the
derivation of Scott's scientific name for the Monster, *Nessiteras rhombopteryx*.
Image reproduced by kind permission of Dafila Scott and the Syndics
of the University of Cambridge.

＊

All summer, Rines had stuck to his favourite hunting ground off Temple Pier, just a few hundred yards below his house. A new sonar-triggered camera rig (another $100,000) was placed on the loch bed 80 feet below the control boat, *Hunter*. Rines also dug out 'Old Faithful', the camera/strobe unit which had taken the flipper photograph in 1972. Suspended in mid-water about 40 feet under *Hunter*, 'Old Faithful' pointed towards the target area of the new sonar unit and fired automatically every 12 seconds.[41]

The old technology proved its worth on the morning of 20 June. The clever new camera saw nothing because freak turbulence covered it with silt from the bottom. Luckily, the camera/strobe dangling 40 feet above it was not obscured, and three frames exposed during that session captured something other than black water. One was just the underside of *Hunter*, but the other two were much more extraordinary.

The first, taken at 4.32 a.m., showed a bizarre object hanging in the water, with a long, slender structure curving up from a bulbous shape that carried two angular protuberances (Plate 35). The second photograph, from 11.45 the next morning, was even more cryptic: a grotesque shape with two stalk-like projections, harshly lit and pockmarked with deep shadows (Plate 36).

Rines was convinced that both frames had captured parts of the Monster. The first had caught its front end, with the upper torso, parts of two front flippers and the graceful curve of its neck. The total visible length of the creature was calculated at 17 feet. Tantalisingly, the head had missed the flash and could not be made out. Within the grotesque shape of the second photograph, Rines could see the left-right symmetry of the creature's face. It had been looking straight into the camera, and the deepest pool of shadow was its open mouth. The two stout projections were on the top of its head, just like the 'antennae' which various witnesses had described.[42]

These new additions to Rines's Monster picture gallery soon acquired nicknames – the 'whole-body' and the 'gargoyle head'. Combined with the 'flipper' photographs, the composite picture of the creature looked just like a long-necked plesiosaur, such as the *Elasmosaurus* which Denys Tucker had suggested in 1960. The bizarre structure of the head was a surprise but nobody knew what a plesiosaur's face looked like because the soft tissues which gave it shape did not survive fossilisation. However, plesiosaurs' nostrils were set high on the skull, raising the possibility that the 'antennae' were in fact breathing tubes.[43]

Rines was ecstatic about the new photographs. Scott was intrigued, but later admitted that he found them much less convincing than the 'flipper' images, which he now accepted as showing parts of a living, plesiosaur-like animal.[44] Nonetheless, he was carried along by Rines's enthusiasm. They began hatching plans to confront the scientific establishment with Rines's complete package of evidence, and to push for a concerted investigation of the Monster.

This time, the new evidence would be presented in Britain and published in a quality scientific journal. Scott suggested a scientific symposium in Edinburgh, followed by a high-profile meeting at the House of Commons to be co-ordinated by David James, backed up by an article in *Nature*. Rines agreed and Scott flew home to start making arrangements.

Rines had sent copies of the new images to zoologists at the Smithsonian Institution in Washington, DC, the Royal Ontario Museum, and the Natural History Museum in London. Otherwise, he insisted on absolute secrecy until the results were unveiled at the scientific symposium. He omitted to tell Scott that someone else outside the Academy of Applied Science had already seen the secret new photographs. The other party was not a high-powered zoologist but a third-year student at Leeds University. Nicholas Witchell's *The Loch Ness Story* had sold well in hardback, and Penguin had signed him up for an updated paperback edition. After Rines told Witchell about the new photographs, Penguin saw an opportunity and flew Witchell out to Boston to negotiate their publication in his book – the first opportunity for the public to see these extraordinary pictures.[45]

Rines gave his consent as long as he could set the date of publication. This was a magnificent coup for the 22-year-old Witchell. It was also a disaster waiting to happen.

Meetings of minds

Through the autumn of 1975 Scott's plans for his three-pronged attack gelled nicely: the scientific symposium, then David James's meeting in the House of Commons, and a final flourish that would be seen by every serious scientist on the planet – an article in *Nature*.

To get the symposium moving, Scott had dangled some strange but enticing bait in front of the Royal Society of Edinburgh, the most prestigious scientific body in Scotland. World-class zoologists would be invited to review the evidence and plan further research into the Loch and its Monster. Scott would be in the chair, bringing his own agenda of protecting endangered animals. It was an attractive package, and the Society agreed to support the symposium and to host it in their handsome building near Princes Street.

Edinburgh University signed up, too, followed by Heriot-Watt, the city's second university. The date was set for Monday, 8 December.

Scott worked hard on the programme, squeezing in time between meetings in London, Slimbridge and Lausanne. Rines and Edgerton, who styled themselves as 'instrumentation people', would present their search strategy and what they had found. The photographs and sonar traces would be interpreted and debated by respected zoologists. George Zug, Director of Zoology at the Smithsonian Institute, and Christopher McGowan, Associate Curator of Vertebrate Palaeontology at the Royal Ontario Museum, had been sent the new photographs and declared themselves excited. Zug wrote to Scott: 'I believe three frames clearly depict parts of the animal fondly known as Nessie.'[46] In the opposite corner would be five senior scientists from the Natural History Museum, including the Keepers of Zoology, Fish, Palaeontology and Fossil Reptiles. They could not see anything in the new images to suggest a large animal. Instead, they suggested that the 'gargoyle head' might belong to a dead horse, while the 'body-neck' could be a branch or perhaps a massive swarm of phantom midge larvae. Without wishing to question the integrity of the investigators, a hoax perpetrated by a third party could not be excluded.[47]

Weighing up the evidence and arguments would be an audience of nearly two hundred: Professors, Fellows of the Royal Societies of Edinburgh and London; Directors, Chairs and Presidents of august scientific and conservation organisations; and others with a long-standing interest, including Constance Whyte, Peter Baker, Dick Raynor and Holly Arnold.[48]

Tim Dinsdale watched the plans taking shape and wrote approvingly to Scott: 'The great Drama is drawing to its conclusion, as the once empty stage is now alive with characters awaiting the curtain's rise for the final Act.'[49] Unfortunately, the 'final Act' did not follow the script.

Conservation piece

To Peter Scott, President of the World Wildlife Fund, the Monster was more than a new species. As a prime example of a precariously poised species that could easily be snuffed out, it could be turned into an icon for endangered animals – even more powerful than the giant panda which he had drawn for the WWF's logo. The Monster needed protection and the timing was perfect for it to be a test case for the Conservation of Wild Creatures and Wild Plants Act, recently passed by the UK Parliament. More speculatively, Scott drafted some eye-catching artwork to include the Monster in the WWF's 'Save One Species' appeal (Figure 8).[50] His stance won the grateful support of the Zurich-based World Federation for the Protection of Animals:

We are touched by your concern for the well-being of the animals located in Loch Ness and would like you to know that WFPA is prepared to commit itself to total protection of this apparently shy and sentient organism.[51]

However, Scott was left with a practical problem. This new species, of which not even the tiniest fragment had ever been studied in the laboratory, had to have an identity. As a minimum, it needed a proper scientific name and everyone had to know what it looked like.

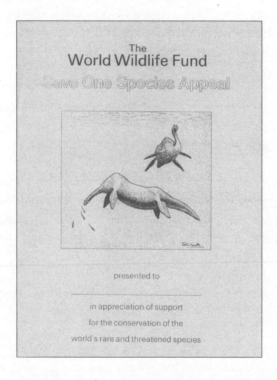

Figure 8 Artwork drafted by Peter Scott to include the Loch Ness Monster in the series of animals highlighted by the World Wildlife Fund for their 1975 'Save One Species Appeal'. Reproduced from the Peter Scott Archives by kind permission of Dafila Scott and the Syndics of the University of Cambridge.

Scott already had the ideal contact to help concoct a Greco-Latin name in the style of Linnaeus: Alan Wilkins, eyewitness, believer and Head of Classics. Assisted by a Greek dictionary, Scott bounced ideas off him and eventually came up with a name that Wilkins applauded as 'excellent'.[52]

Nessiteras rhombopteryx was a neo-classical Greek translation of the Monster's diagnostic features, which made no assumptions about what sort of animal it might be. *Nessiteras* comes from *Nessi-* ('of Ness') and *teras*, meaning a 'wonder' or 'marvel', and *rhombopteryx* from *rhombo-* ('diamond-shaped') and *pteryx* ('wing' or 'fin', as in the 'finger-winged' pterodactyl). The name had linguistic credibility and sounded good. Scott decided to make the naming of the creature the main thrust of the paper he was drafting for *Nature*, arguing that a species could not be granted statutory protection without a formal scientific name.

To some, this looked contrived. Dr R. G. Melville, Secretary of the International Commission on Zoological Nomenclature, wrote to warn Scott that many zoologists might not be convinced that he had 'a tangible species worth naming'.[53] Melville went on:

Egg on face is not an agreeable sensation . . . I think you might well ask whether it is the right thing to do. Your reputation among other zoologists might not be enhanced. If naming the Loch Ness Monster is going to bring ridicule on the head of the namer, why be the one to attract it?

Sir Arthur Landsborough Thomson, past Chairman of the Natural History Museum's Trustees, was even blunter.[54] In a letter headed 'Not for quotation' he demanded to know 'Who says?' that the Monster had to have a scientific name to be legally protected. What Scott needed above all was the flesh and bone of a type specimen. Without that, the Monster 'not merely beggars description, but buggers it'.

But by then, Scott's article was with *Nature*, and it was too late.

A fair likeness

Scott had said publicly that the Monster 'looks rather like a plesiosaur', and the sketches and doodles of the last few months now coalesced into his definitive impression: hunchbacked and stouter than reconstructions of *Elasmosaurus*, with a long slender neck, a fat tail and the hallmark diamond-shaped flippers. His famous oil painting, later known as *Courtship in Loch Ness*, shows Scott's talents as a wildlife artist in a surreal setting.

Pictured as though cruising through honey, two massive but graceful Monsters hang just below the surface. Beneath them, the water darkens steadily, while the only hints of the world above are the dappling of sunlight on the creatures' upper parts and a streak of brightness where the hump of one breaks through (Plate 37).

Scott painted the picture for David James, but unveiled it in front of another old friend.[55] In late November, David Attenborough was summoned to the Natural History Museum, where Scott produced his painting and asked: 'What do you think?' Attenborough put on his glasses for a better look and both lenses promptly fell out on to the canvas. This was not because Attenborough's eyes were out on stalks; the spectacles were new and the frames were loose. His vision restored, Attenborough said that he found the picture 'rather exciting' and 'courageous'.

Until then, Attenborough had not been on Scott's guest list for the Edinburgh symposium. An invitation quickly followed on 28 November. This left just ten days for Attenborough to rearrange his diary – and for the whole thing to career off the rails.

Like a sieve

Rines had wanted publicity, and now he had it. From the safe haven of Boston, it was just a spectator sport. Across the Atlantic, Scott was in the thick of a Monstrous media feeding frenzy and was losing control.

The first sign of trouble had arrived a week earlier, in an apologetic letter from Nicholas Witchell.[56] He explained about his paperback and Rines's new photographs, and let slip that Rines had instructed Penguin to publish it on Sunday, 7 December, the day before the Edinburgh symposium. Worse still, Witchell had allowed a trusted friend, now a reporter for the *Yorkshire Post*, to flick through an advance copy. Unfortunately, Witchell's trust was misplaced, and details of the secret new photographs would hit the headlines before they were formally unveiled at the symposium. Witchell's letter ended, 'I am extremely sorry if today's events have taken the lid off.'

Scott was furious, mainly because Rines had not bothered to tell him any of this. He could see his carefully laid plans beginning to unravel, because other forces were lifting the same lid. Meanwhile, various journalists had tried to prise the confidential report out of the five experts at the Natural History Museum. The experts revealed nothing, so the pressmen followed their instincts and invented the story that the Museum thought the photographs were dubious. This provoked Rines to fire off an incensed letter to

Dr Gordon Sheals, Keeper of Zoology, expressing his 'dismay at your obvious handicap' in interpreting the images. Sheals replied that they had kept everything confidential, adding that the press onslaught over the Monster had brought the Museum to a 'virtual standstill'.[57]

On 30 November, a scathing article appeared in the *Sunday Times* entitled, 'Nessie, this is your best show yet'. The newspaper's Insight team had dug up some dirt on 'the monster's latest revival on the Fleet Street stage . . . ironically led by an American patent lawyer'. Rines was the founder, bankroller and beneficiary of the Academy of Applied Science ('less impressive . . . than its name suggests'), and was hawking around a set of underwater Monster pictures to *Time* and *National Geographic* with an asking price of $100,000. There were 'considerable doubts' about the mysterious computer enhancement process, especially as Alan Gillespie had worked his magic on enlarged, high-contrast copies of the original images – which he had never been shown. And why were the Monster photographs so fuzzy that they had to be sharpened up by computer, when the same camera had apparently taken clear pictures of ordinary Loch fish such as salmon and eels?

Finally, Insight noted, the whole business could turn into a wealth-creation scheme for British bookmakers as well as Rines. There had been an outbreak of suspiciously large bets that the Monster's existence would be proved within a year. The first bets, at odds of 100–1, were placed in Reading (where Witchell's paperback was being printed in 'complete secrecy'). In no time, more than £20,000 had been bet, forcing Ladbroke's to shorten the odds to 6–1. The Insight team sought the wisdom of Jimmy 'The Greek' Snyder in Las Vegas, America's most celebrated tipster. Snyder retorted, 'I know that Goddam Monster doesn't exist', and refused to lower his odds below 100–1.

By now, the 'confidential' scientific case to be revealed at the Edinburgh symposium was as watertight as Ivor Newby's legendary amphibious car when he forgot to put the bung in. The Insight article about 'that Goddam Monster' was probably the last straw for the Royal Society of Edinburgh. They should have done the gentlemanly thing and told Scott to his face. Instead, he only found out on Monday, 1 December, when a reporter rang him for his reaction to the news, on the front page of several newspapers, that all three Scottish sponsors had pulled out of his symposium.[58]

Scott later wrote to a friend that life was too short to worry about such things.[59] When his initial fury and frustration subsided, he settled down to the task of cramming the symposium's day and a half of concentrated science into the two-hour evening session which David James had organised in the House of Commons for 10 December.

Strangely, Rines was not dismayed at the change in plans. Indeed, he was 'delighted' that the symposium was being switched to London, where his team would stay at the President Hotel 'in great secrecy'.[60]

Nicholas Witchell had written a breathless Preface for his new book:

This paperback edition . . . is being rushed out in the autumn of 1975 . . . when the world is about to witness one of the greatest and most significant discoveries of the 20th century. Detailed colour photographs of the head and body of the 'Monster' have been taken by a highly respected American scientific team. They have set the zoological world, and will shortly set the whole scientific and lay worlds, ablaze with excitement.[61]

He had also snapped off one of Scott's three prongs of attack. Luckily, there was still hope for the House of Commons meeting and for the most powerful prong, the paper by Scott and Rines which was about to appear in *Nature*.

Nurture at *Nature*

Take a room full of scientists anywhere in the last 50 years and ask them to name the world's top scientific journal. The answer: *Nature*, and with good reason. Leafing through its pages, you are so close to the cutting edge that you have to watch your fingers. Here were reported some of the greatest discoveries of the twentieth century, including the neutron, the coelacanth, the double helix of DNA, lasers and pulsars. The journal was notoriously tough and mostly fair; some rebuffed authors took rejection badly, threatening to kill the editor or set fire to themselves outside his office. With over 95 per cent of submissions rejected, getting a paper into *Nature* was a glittering prize that could make a scientist's career.

Nature had never taken the Monster seriously, because there was no hard evidence that it existed. In late 1933, a correspondent spent a morning at the Loch and concluded that the story had no scientific basis. He blamed 'fertile (if unconscious) imagination' and the gullibility of people like Rupert Gould. He also hinted that Hugh Gray's recent photograph of a 'twenty-foot water-serpent' had been touched up.[62] Denial had remained *Nature*'s policy ever since, until the issue of 11 December 1975. Incredibly, this devoted three of its hallowed pages to the Monster and splashed Rines's flipper photograph across the cover.[63]

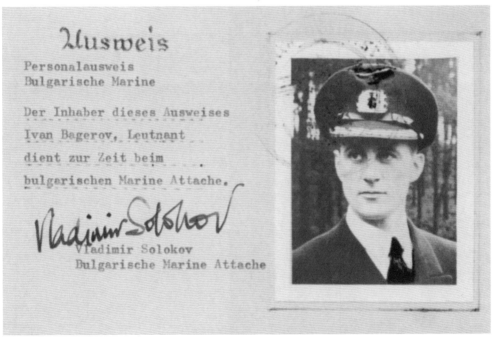

21. David James (1919–86), Conservative Member of Parliament for Brighton Kemptown (1959–64) and Director of the Loch Ness Phenomena Investigation Bureau. BOTTOM: reconstruction of the forged papers in the name of Lieutenant Ivan Bagerov, Bulgarian Navy, with which James escaped from the Marlag prisoner-of-war camp near Bremen in December 1943.

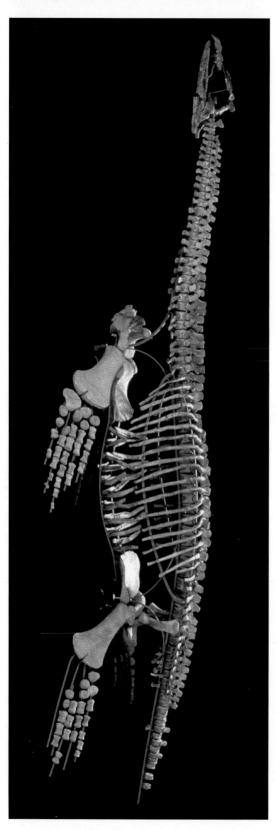

22. Skeleton of the elasmosaur *Cryptoclidus eurymerus*, found in the Oxford Clay beds near Peterborough, Northamptonshire. This marine reptile lived in the subtropical Middle Jurassic period, 160–5 million years ago. (The Trustees of the Natural History Museum, London)

23. F.W. 'Ted' Holiday (1920–79), journalist, angler and author of *The Great Orm of Loch Ness* (1968). Holiday suggested that the Monster was a giant slug-like invertebrate related to the extinct *Tullimonstrum*, an unclassifiable creature known only from fossils found in Carboniferous coal-measures south-west of Chicago. (Dick Rayner)

24. The camera rig at Achnahannet, with a cine-camera carrying a 36-inch telephoto lens mounted between two still cameras with 20-inch lenses. David James is third from the left. (Chris James)

25. Roy Mackal (1925–2013), molecular virologist and Associate Professor in the Department of Biochemistry at the University of Chicago. His analysis of data about the Monster led him to conclude that it was either a giant mollusc, related to sea-slugs, or a huge amphibian. Mackal (on the right) is pictured beside Loch Fadda, Connemara, in July 1968, with a biopsy harpoon which he designed; the blunt stops on either side of the biopsy head are to prevent excessive penetration of the Monster. (Ivor Newby)

26. Frame from the cine-film taken on 13 June 1967 by Dick Raynor, showing a wake crossing the Loch near Dores. A dark object appeared intermittently at the head of the wake. The passenger steamer *Scot II* is shown in the foreground. (Dick Raynor)

27. US Navy submariner Dan Taylor (behind) and his bright yellow, home-built craft *Viperfish*. In front, from l. to r.: Peter Davies, David James, Sir Peter Ogilvie-Wedderburn, Holly Arnold and Clem Skelton. (Ronald Spencer, Daily Mail and Solo Syndication)

28. Wing-Commander Ken Wallis, veteran of Bomber Command and *You Only Live Twice*, at the controls of his home-built *Wallis 117* autogyro, on the shore of Loch Ness in July 1970. Wallis was unsuccessful in finding both the Loch Ness Monster and Lord Lucan. (Ivor Newby)

29. Dr Robert H. Rines (1922–2009), American patent lawyer, inventor, composer and President of the Academy of Applied Science in Boston. Between 1970 and 2008, Rines made thirty trips to Loch Ness. The AAS expeditions which he led in 1972 and 1975 took the underwater photographs of the Monster that were published in December 1975 in *Nature*, the world's top scientific journal. (Nicholas Witchell)

30. Nicholas Witchell, who wrote *The Loch Ness Story* (1975) while still an undergraduate student, pictured on the shore of Loch Ness in summer 1976. (Nicholas Witchell)

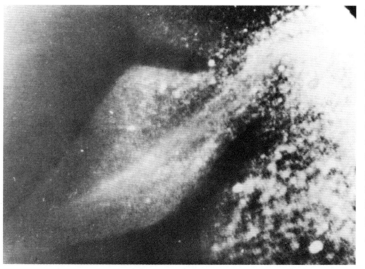

31. The 'flipper' photograph, taken by the underwater camera later known as 'Old Faithful' on the morning of 8 August 1972, near Temple Pier in Urquhart Bay. The original image (see Plate 46) was enhanced by Alan Gillespie using a computer programme developed at the Jet Propulsion Laboratory in Pasadena, California. (Academy of Applied Science, Belmont)

32. Frank Searle (1921–2005). The 'man who took his camera to bed', Searle spent 11 years living beside Loch Ness and produced a series of photographs purporting to show the Loch Ness Monster. (Highland and Islands Archives, Inverness)

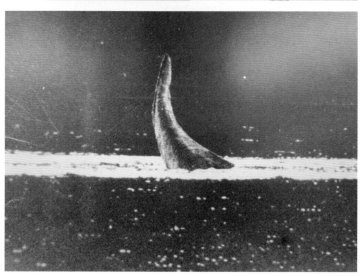

33. One of the many photographs produced by Frank Searle, supposedly showing the Loch Ness Monster. This one, taken in November 1975, was claimed to show the animal's tail (or flipper). (Highland and Island Archives, Inverness)

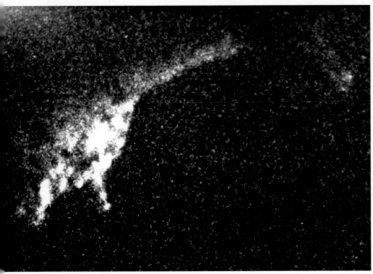

34. Robert Rines, Peter Scott (in diving gear) and Scott's personal assistant Michael Garside, photographed by Dick Raynor at Temple Pier, Urquhart Bay, in early summer 1975. (Dick Raynor)

35. The 'whole body' photograph taken by the automatic camera known as 'Old Faithful', near Temple Pier on the morning of 20 June 1975. (Academy of Applied Science, Belmont)

*

Nature should never have accepted the paper by Scott and Rines. Scott was supposedly the senior author, but he had not collected or analysed any data. The flipper photograph had already been published – normally a death sentence for a submission to *Nature* – while the 'gargoyle head' picture was withheld. Everything hinged on the image enhancement, but the process (which 'cannot alter shapes or otherwise falsify the record') was not explained and the original images were not shown. To many in the *Nature* office, this seemed more like a pre-Christmas joke than the hard-core science for which the journal was renowned and respected. Manuscript editor Fiona Selkirk watched the article causing bemused incredulity wherever it went, from the features editor who 'fell about laughing' to the typesetter who smirked when he saw the title.[64]

Yet the manuscript sped like a charmed arrow through *Nature*'s famously ruthless selection process. Its safe passage was assured by David 'Dai' Davies, the Editor-in-Chief. Davies, formerly a seismologist in Cambridge, had recently taken the helm at *Nature* and seemed determined to challenge scientists and their prejudices. Something about the Monster article must have moved the earth for him. He surprised everyone by locking it away in his filing cabinet and telling the business manager, Adrian Soar, to negotiate an exclusive tie-in deal with a newspaper. This also upset the old order. *The Times*, which ran a regular column 'From *Nature*', was outbid by the *Evening Standard*.[65]

Nature published 'Naming the Loch Ness Monster' as an anonymous item in the News section. This was not a research paper but a plea for the name *Nessiteras rhombopteryx* to be accepted, so that the Monster could be recognised as a true species and therefore protected legally. Scott and Rines were mentioned at the end, as 'a naturalist' and 'a prominent Boston lawyer [who] has looked for Nessie every summer since 1970', but were not listed as authors (Scott had written 'Where are our names?' in red ink across the proofs).[66] The flipper and whole-body photographs were displayed alongside Scott's drawings, which look plausible but are arguably works of art, not of science. Scott and Rines admitted that naming a new species without solid proof is 'unsatisfactory', but argued that the creatures needed urgent protection: 'Better safe than sorry.'

They speculated that 'Nessie' (the term used throughout the article) was a reptile, related to 'fossil marine forms' and 'living sea-serpents' such as *Megophias monstrosus*, named after a sighting in Massachusetts Bay in 1817 and also in the absence of physical remains. Scott and Rines further suggested that Nessies may have been landlocked in the Loch

12,000 years ago and were first noted during St Columba's visit in
AD 565.

To believers, seeing the Scott–Rines article in the holiest place in the scientific firmament was a triumph over the narrow-minded establishment, and
a challenge to look at the evidence with fresh eyes. Sceptics saw a piece of
nonsense which the new editor, in a moment of weakness or idiocy, had allowed to sneak in without peer review. *Nature* had debased itself and sunk
as low as the *Evening Standard*, which ran the story on the same day.

The question on everyone's lips was put to Scott on the morning of
10 December, when he and Rines faced the world's media at a press
conference under *Nature*'s distinctive banner (Plate 38). 'Is this a hoax?'
asked a reporter. Scott deferred to Rines, who said without hesitation, 'No,
these photographs are absolutely genuine.' Satisfied, the questioner left it
at that.[67]

Show and tell

The Bostonian flying circus that descended on London was completed
by Harold Edgerton, who jetted in from a working holiday in Greece on
Jacques Cousteau's *Calypso*. Cousteau, whose opinion of the Monster
was unhelpful ('I don't believe in that bullshit')[68] stayed behind. Judging
by the number of journalists who pushed notes under Rines's door in the
President Hotel, their hideaway was no longer secret.[69] Perhaps the press
believed Charles Wyckoff, another celebrated optical expert at MIT and a
member of the AAS, who described the proof of the Monster's existence as
'the greatest thing that has happened this century'.[70]

The Grand Committee Room in the House of Commons was a stately setting for the two-hour session on the evening of Wednesday, 10 December
1975. It was packed with Lords, Ladies, MPs and most of those who had
been invited to Edinburgh. Nigel Sitwell, the editor of *Wildlife* magazine,
later wrote that 'for most of those present, it was one of the most exciting and dramatic evenings of their lives'. Nicholas Witchell agreed, struck
by the audible gasps of surprise as the evidence unfolded on the screen.
Witchell sat beside a well-oiled but luckily non-flammable MP who was so
moved that a cigarette end fell unnoticed from his nerveless fingers and was
left to smoulder on the carpet.[71]

After opening comments by Lord Craigton, Chairman of the All-Party

Committee on Wildlife Conservation, David James welcomed 'our American guests who have done some very extensive and imaginative work over the last five years'. Peter Scott introduced himself as 'a painter with some claim to biology . . . but never a professional scientist' and explained that 'we have today published in *Nature* a description of the Monster' and had named the creature *Nessiteras rhombopteryx* in order to grant it legal protection.[72]

Rines, a powerful speaker 'like a Baptist minister', then set the scene.[73] Urquhart Bay, its floor riven with canyons and caves that nobody had suspected until the AAS mapped it properly with their clever sonar. 'Grid element 2903', the Monster-sighting hot spot off Temple Pier, pinpointed by computer analyses too complex to explain. A zoologically rich habitat: Roy Mackal had estimated that the Loch contained up to 13 million salmon, easily enough to feed 10–15 one-ton Monsters. Rines introduced Harold Edgerton who demonstrated how he had won his nickname 'Papa Flash', with strobe images of a dynamite cap beginning to explode, fish photographed from Cousteau's *Trieste* 36,000 feet under the sea, and a flash that blinded the audience as the camera set up on the stage took their photograph 'for posterity'.

Then Rines got down to business. The original flipper photograph might be familiar as it had been published in the *Photographic Journal* 'in 1972 or 1973', but here was another, taken 45 seconds earlier and never shown until now. The same massive diamond-shaped structure filled the frame, but was in a different position, proving that this belonged to a moving, living animal. Rines skated over the computer enhancement – 'JPL have a way of taking the fog out' – but dwelt on the sonar contacts which confirmed that 'big multiple trace objects' had approached the camera at the crucial moment. Six world-renowned sonar experts agreed that the record showed two animals, each with appendages 10–18 feet long.

Next came the whole-body and gargoyle photographs from June 1975. Around the time that the whole-body picture was taken, the control boat began rocking, evidently because 'something big had come in and had its picture taken'. At that point, Rines sat down to 'let the zoologists speculate on what we're seeing here'.

With minor variations, they sang off the same sheet. Zug of the Smithsonian believed that the photographs and sonar records showed 'an animate object or parts thereof', namely the recently described *N. rhombopteryx*, previously known as the Loch Ness Monster. The species of animal could not, however, be identified. McGowan of the Royal Ontario Museum and A. W. Crompton, Professor of Biology at Harvard, both agreed that the findings suggested 'the presence of large aquatic animals in Loch Ness', which must now be studied.[74]

Further supporting evidence was then presented. The usual gallery of photographs and Tim Dinsdale's film were shown, followed by eyewitness accounts from Mr Lowrie of the *Finola*, Scott's classics expert Alan Wilkins, and (on film) 'perhaps the most famous eyewitness of all'. Alex Campbell, now 'a spare old Scotsman' but still charismatic, again described his best ever sighting, early on a bright, flat-calm morning in 1934. The 30-foot humped creature with its six-foot serpentine neck appeared in Borlum Bay, looking agitatedly around, and then sank when two trawlers appeared from the Caledonian Canal. Campbell recounted how he ran home to consult his book on prehistoric animals. 'The nearest approach,' he said, 'was the picture of the plesiosaur.'[75]

Scott then summed up and gave his personal view. After fifteen years of sitting on the fence, he had made his mind up. 'The underwater pictures leave no further doubt in my mind,' he said, 'that large underwater animals exist in Loch Ness.' [76]

Question time followed.[77] A zoologist from London wondered what Dr McGowan thought of Scott's drawing of *N. rhombopteryx*. McGowan admitted that palaeontologists were 'usually a bit fussy about wanting a lump of bone in their hand', without which it was 'extremely foolish' to speculate on the animal's identity. Humphrey Greenwood of the Natural History Museum wanted to know why other underwater photographs, allegedly showing the creature's belly (and even freshwater parasites attached to it), had not been shown. Answer: they had not been fully processed. Greenwood responded politely that the event had been 'beautifully presented but – if I may be so bold – unscientific'.

A Scottish Professor of Biology challenged Mackal's estimate of 13 million salmon, which would make the Ness 'the best salmon river in the world'. Maybe 50,000? David Attenborough weighed in, concerned that the central rib of the flipper made it look more like a fish's fin than a plesiosaur's paddle. He was promptly put right by an expert in plesiosaur locomotion from Tübingen University, who insisted that it looked exactly like a plesiosaur flipper 'with meat on'. A freelance journalist wondered what they were supposed to see in the gargoyle photo, and seemed unconvinced by the claim that it was the Monster's face with nostrils. The Professor of Physiology at King's College London felt that the evening's evidence had added 'only a little', and asked why no calibration shots of ordinary fish had been shown. The Professor had an interest to declare. He was Peter Baker, who 'in a former existence', had led the Oxford–Cambridge expeditions in 1960–62.

The 'Kensington Five' from the Natural History Museum held their most damaging fire until the end.[78] Gordon Sheals, Keeper of Zoology and the ringleader, suggested that the 'flipper' was just the caudal fin of a fish and that the estimated length of six feet was 'wildly out'. The enhanced 1972 photographs might look like a flipper, but there was no evidence whatsoever that the 1975 images showed the same object, or indeed anything living. Sheals aimed his last poisoned arrow straight at Scott:

I have to say that the publication of an article describing and formally naming an animal thought to be represented in these photographs is unfortunate and regrettable ... Scientifically, it has served no useful purpose and merely added to the clutter of dubious names already in the literature.

Rines held his tongue, leaving Scott to explain that the naming was intended to protect the species. Just as time ran out, Scott added that he found Dr Sheals' comment 'equally regrettable'.

Closing the conference, Lord Craigton remarked that the 'Palace of Westminster has been put to good use this evening'. Not everyone agreed.

Season's greetings

Scott was ready for repercussions from the House of Commons meeting but was quite unprepared for what happened two days later. Friday, 12 December was a day of high ceremony with Sir Peter Scott, Chancellor of Birmingham University, presiding over the Graduation Ceremony. It ended with a brief news item on ITV that left Scott 'horrified'. Nicholas Fairbairn, the flamboyant MP for Perthshire and Kinross, had discovered that Scott's meticulously crafted *Nessiteras rhombopteryx* was an anagram of 'Monster hoax by Sir Peter S'.[79]

Scott immediately protested his innocence, but the press picked up the story with glee. The publication in the *Observer* of a hostile article, 'Nessiteras absurdum',[80] did not help. The next day, Dai Davies, clearly worried about *Nature*'s reputation, wrote to Scott about the 'most extraordinary and delightful coincidence' of the anagram. His tone was jocular but anxiety clearly showed through:

Many people have latched on and assumed it is indeed a hoax. Maybe a tiny letter from you with something like 'what an amazing coincidence but I swear it is no more' might be good. Assuming, that is, that you are not having us all on . . .'[81]

Scott's insistence that it was just a fluke was supported by Bob Rines, who rang from Boston a couple of days later to say that he too had been playing around with the letters and had come up with 'Yes, both pix are monsters. R'.[82] But this particular genie was out of its bottle and making the most of its new-found freedom.

In the run-up to Christmas, Scott's mailbag was swollen by letters expressing doubts, anger, regrets and commiseration.

An angry one came from Peter O'Connor, purveyor of the dubious Monster-in-the-shallows photograph and now styling himself 'The Taxidermist of Europe' on the High Street, Luton. As O'Connor had already named the Monster *Nessiesaurus o'connori*, he demanded that corrections be printed if Scott attempted to use his new nomenclature. He added ominously that 'my future actions depend on your co-operation'. A lesser man might have told the Taxidermist of Europe to get stuffed. Scott replied politely: 'Do not worry. If your name satisfies the International Code of Zoological Nomenclature, ours will become a synonym.'[83]

Placatory letters arrived from some of the Kensington Five at the Natural History Museum. Humphrey Greenwood apologised about tempers becoming 'a little frayed', explaining that taxonomists enjoyed 'spirited debate'. Alan Charig also regretted Gordon Sheals' 'oration', saying that he was usually a likeable chap. Charig still felt that naming the Monster was undesirable but admitted: 'Emotionally, I hope that Nessie does exist; rationally, I am far from convinced.'[84]

Charig also thanked Scott for his 'Nessie'. This was a Monster drawn by Scott under the legend, 'A Nessie for Alan from Peter, at Christmas 1975.' Scott had done four different designs of a Monster in various poses (Figure 9), photocopied and hand-coloured them and sent them out to three dozen people from the LNIB, AAS, *Nature* and – as a peace offering – the Natural History Museum. Above his Nessie to Gordon Sheals, ringleader of the Kensington Five, Scott wrote 'I harbour no ill-will'.[85] Sheals did not reply.

On Christmas Eve, Scott wrote to a friend that Christmas was looking pretty miserable, except for the 'magical sight' of 300 Bewick's swans feeding under the floodlights outside his window at Slimbridge. He ended by remarking that Rines's recent photographs were unconvincing, but that the 1972 flipper picture 'in my view can only be an animal'.[86]

That was also the view of seven-year-old Susan Young in Stroud, a few miles away across the new M5 motorway. 'Dear Sir Peter Scott,' she wrote in pencil, 'I thought it looked like an Elasmosaurus instead of a Plesiosaurus. I know this because we were doing about Dinosaurs in school. With lots of love, Susan.' Over the page was a PS in exuberant crayon: 'I hope you have a HAPPY CHRISTMAS.'

Scott wrote back to thank her for her kind letter and express solidarity. 'You are quite right. I too believe that it is an Elasmosaurus.'[87] And he did his best to have that HAPPY CHRISTMAS.

Figure 9 'A Nessie ' drawn by Peter Scott for Dick Raynor, Christmas 1975. One of a series of 'Christmas Nessies' which Scott sent to various friends and foes after the House of Commons conference on the Monster. Reproduced by kind permission of Dick Raynor and Dafila Scott.

10

No hiding place?

Even if Scott's Christmas was happy, its immediate aftermath was not. On 27 December, a harbinger of the hostility to come appeared in *The Times* under the title, 'Why *Nessiteras rhombopteryx* is such an unlikely monster'. The author was Adrian Desmond, a noted dinosaur expert who was on a fellowship at Harvard, just a bone's throw from MIT. He attacked all the arguments that the Monster could be a plesiosaur and was particularly concerned about the gargoyle head, which 'bears little resemblance to the rest of creation'. *Nature*, which was 'daily forced to turn down excellent hard-core scientific papers', should never have published such stuff. An animal could not be proved to exist without a type specimen. Carl Linnaeus, the father of taxonomy, had made the same mistake when he invented *Homo caudatus*, the cat-tailed man which existed only in tall stories from exotic lands. 'As a palaeontologist,' Desmond explained, 'I need bones.'[1]

There were more omens in a letter from Denys Tucker, written on New Year's Eve:[2]

Please pay attention, because I'm serious and I know what I'm talking about – I've been kept out of any sort of a job for 15 years. The nasty little Establishment jackal pack has started snuffling around in the undergrowth and this time it's after you. The anagram has had some of them rolling in the aisles.

Scott needed no reminding about that. The morning after the House of Commons conference, the *Daily Express* had announced, 'Naturalist Sir Peter Scott laid his reputation on the line yesterday by declaring that the Loch Ness Monster should be awarded a zoological name'. The headline – 'Sir Peter says it can't be a hoax' – was too close for comfort to 'Monster hoax by Sir Peter S', which broke the next day.

Tucker made another pointed comment:

Rines, presumably, is so stupid that he doesn't know when his camera is operating in mid-water and when it is sitting on the bottom.

He did not explain the remark, but its significance became clear when separate letters from Dick Raynor and Holly Arnold arrived a few days later. Both wanted to correct an error regarding the 1972 photographs on which Scott had nailed his faith. At the House, Rines had claimed that the sonar unit and camera were sitting on the floor of Urquhart Bay. In fact, both were dangling in mid-water and – according to Raynor, who had been on *Narwhal* until two hours before the flipper photograph was taken – swinging 'quite considerably'. The implications were obvious. The sonar unit and camera might not have been pointing at the target area of 'grid element 2903' when the large sonar contact apparently closed in and the photograph was taken. Indeed, there was no way of knowing what either had captured.[3]

More worrying tidings arrived in a confidential letter from across the Atlantic. Alan Gillespie, who had computer-enhanced the 1972 images, was concerned that something odd had happened to the flipper photographs after they left his hands at JPL. 'I was never convinced,' he wrote, 'that the picture published by Witchell as the Rines/Edgerton photograph was the one I worked on.' The processed image that he had sent Rines was 'less convincing' than the versions published in Witchell's book and in *Nature*. Gillespie enclosed copies of the images which had rolled off JPL's computer, but did not suggest what he or Scott should do with the information.[4]

To cap it all, David Attenborough wrote to say that he had followed up his question at the House and, according to other experts, plesiosaur paddles had no central rib. The 'flipper' in the photograph might belong to a fish, but did not seem to match the creature that Scott believed it to be.[5]

At this point, the advice of the President of MIT to 'tread warily' when dealing with Rines might have returned to haunt Scott, especially as another act of faith was taking shape on his desk. This was a big article for *Wildlife* magazine entitled, 'Why I believe in Nessie'. In it, he described his moment of conversion: 'And then I saw Bob Rines' new photographic evidence.'[6]

Bad press

As expected, the correspondence column of *Nature* bristled with indignation, dismay and astonishment at the decision to publish the Scott–Rines article.

Gordon Corbett, one of the Kensington Five at the Natural History Museum, was coldly furious. The paper could 'mislead laymen' but did not withstand proper scientific scrutiny. Until the enhancement process was validated, the photographs were uninterpretable. *Nature* had debased itself and was dragging zoological taxonomy and the reputation of scientific publishing in Britain back to the bad old days of 'uncritical mythology'.[7]

Hostile fire also came from Professor L. P. Halstead, a plesiosaur expert at Reading University. Halstead was renowned for the length of his publication list and his fearlessness in standing up to everyone from the Nigerian secret police to the Vice-Chancellor of Reading. He challenged every plesiosaur-related statement in the *Nature* paper, and proposed an ingenious explanation. Norse longboats were known to have had diamond-shaped paddles and wooden posts carved to resemble the heads of fantastic animals. Rines had probably photographed a Viking shipwreck.[8]

Scott defended himself against these and other attacks as best he could, pouring particular scorn on Halstead's Viking boat. Presumably because this was zoology rather than 'instrumentation', Rines played no part in 'Sir Peter Scott replies', printed in *Nature* a couple of weeks later.[9]

The mailbag brought some comfort. Constance Whyte offered sympathy for the 'harassing' times he had endured, especially as 'the denial by the zoological establishment of facts that are beyond argument is a disservice to the cause of TRUTH'. Scott agreed, adding that 'the anagram was a coincidence we could have done without.'[10] There was light relief from another attempt to name the Monster. The artist's impression of *Plesiophonus harmonicus* looked rather like *N. rhombopteryx* except for its tartan markings and inbuilt sonar system for avoiding scientific instruments. The article was published in the spoof *Journal of Irreproducible Results*.[11]

A conciliatory letter from Adrian Desmond, author of the 'unlikely monster' article in *The Times*, might also have helped to lift Scott's spirits. Desmond still did not think that a sustainable colony of Monsters could survive in the impoverished waters of the Loch, but he threw in a crumb of hope. It was possible that some plesiosaurs had generated enough heat by vigorous swimming to be warm-blooded and therefore might have survived in water cold enough to kill conventional reptiles. Desmond also admitted:

I should like nothing better than to switch camps. I earnestly hope to lose my case . . . I can only reiterate my sincere hope that I am wrong.[12]

Distillates of wisdom

Ted Holiday had also been busy. Pursuing his fixation on invertebrates, his article, 'The case for a spineless Monster', appeared in *The Field* in early February.[13] He had rejected *Tullimonstrum* in favour of the worm *Sabellaria*, which had a beautiful diamond-shaped flange. Admittedly, *Sabellaria* was only a few millimetres long, but an enlarged version would look the part. A role model might be *Megascolides*, a 12-foot worm that sprayed defensive slime so energetically that anyone handling it had to wear protective clothing.

There was more backbone in coverage of the Monster in the March issue of *Wildlife*. The cover boasted Scott's haunting portrait of two cruising Monsters and the strapline, 'For the first time – the full story behind the Nessie photos'. Nigel Sitwell, the magazine's editor, summarised the evidence presented at the 'highly charged' session in the House of Commons: the unsuspected underwater caverns 'capacious enough to provide refuge for large animals'; grid element 2903 in Urquhart Bay, where Monsters were photographed, exactly as the computer had predicted; and the assertion by Alan Gillespie that 'I detect no evidence of fraud'.[14]

'Why I believe in Nessie, by Sir Peter Scott' followed, complete with his drawings of *N. rhombopteryx*. Scott's personal journey to belief had begun with *More Than a Legend* but was only fulfilled 15 years later when 'I saw Bob Rines' new photographic evidence'. And now –

My personal belief is that the new evidence, taken with the old, suggests there are indeed large animals in Loch Ness. My own guess, from all reports, is that they might look rather like plesiosaurs.

Scott did not mention the 'delightful coincidence' of the anagram, but was ready for any cynics who suspected that this was 'a hoax perpetrated by the people who took the photographs':

Having got to know Dr. Rines and his colleagues, I am prepared to stake all my knowledge of human relationships in stating that they are not hoaxers.

Roy Mackal's long-awaited book, *The Monsters of Loch Ness*, was published that summer.[15] At 400 pages, this was the biggest book ever about the Monster, partly because it was padded out with appendices about eels and salmon, designs for Monster traps and the entire chapter on Loch Ness lifted from Sir John Murray's Bathymetrical Survey of 1910. For anyone wondering what a professor of molecular virology looked like, Mackal was pictured in front of an electron microscope in his lab in Chicago and sighting down a crossbow loaded with his Monster-biopsy device.

The book contained some surprises. The Loch was 150 feet deeper than the Bathymetrical Survey had found, and the famous landmark of the Horseshoe Crag, generally believed to lie between Foyers and Fort Augustus, had been transported to the western shore. The photographs taken by Hugh Gray, Lachlan Stuart and Robert Rines were all 'positive evidence', although Mackal believed that Stuart's picture of three humps showed three Monsters swimming in formation rather than the back of a single creature. Frank Searle's greatest coups were just floating logs, while the Surgeon's Photograph was an unidentified diving bird. The two humps photographed by bank manager Peter Macnab were extremely dubious. Mackal devoted four pages to pointing out oddities in the image which suggested that it had been doctored, but stopped short of accusing Macnab of fraud.[16]

Mackal had also been ruthless with eyewitness accounts. Without explaining how, he whittled down over 3,000 sightings to just 269 'valid' reports which filled a 40-page spreadsheet. The last, at 09.20 on 6 August 1969 (observers: Mr and Mrs Geoffrey Craven and two children), occurred some 1,400 years after the first (AD 565, precise time uncertain; observers: St Columba, Lugne Mocumin, others). Sighting no. 45 was of a 30-foot humped creature with a five-foot neck recorded by 'Mr A.C.' at 09.30 on 22 September 1933; no. 78 was a 30-foot humped animal with a six-foot neck that turned from side to side, seen by Alexander Campbell in May 1934.[17]

Anyone who thought the monster looked like one of Peter Scott's plesiosaurs was in for a shock. Mackal had dumped not only the Surgeon's Photograph but also the long neck seen by many witnesses, including Alex Campbell (alias A.C.). Possession of a long neck was just one of 32 reported features of the Monster which Mackal analysed. Some of the others might have been news to seasoned Monster watchers, such as a 'pulsing neck region', 'no head–neck differentiation' and making a barking or hissing sound. Mackal simply added up the total score out of 32 for each of his main candidates, namely seals, sea cows, plesiosaurs, amphibians, followed by giant eels. Top came amphibians (28/32, a commendable 88 per cent) and giant eels (25/32, or 78 per cent). He ruled out plesiosaurs and the rest because 'they do not have the correct configuration'.[18]

Mackal still felt drawn to the 'wondrous world of molluscs', having mentally scaled up a pair of sea slugs that he kept in a tank in his office. Unfortunately, there were no obvious solutions to the problems of inadequate size and stiffness. Sea slugs are even smaller than Holiday's 14-inch *Tullimonstrum* and, realistically, a 20-foot animal would need a backbone to be able to nip across the A82.

Mackal's favoured candidate was a giant amphibian. This was an obvious choice for an aquatic creature that ventured out on land – where, he

reckoned, it had been seen 178 times in 1,400 years. Even the giant salamander was not giant enough, but fossil amphibians could be. Particularly
seductive was *Eogyrinus*, a 15-foot brute from the Carboniferous period
that had looked more like a crocodile than a salamander. It had a humped
back, a good-looking tail and odd little feet that, in poor light, could pass
for flippers. Being an amphibian, it may well have had 'a pulsing throat
region', but could never claim to have a long neck. In fact, Mackal's reconstruction of *Eogyrinus* appears to have no cervical vertebrae, and therefore
no neck at all.[19]

Figure 10 'Hypothetical amphibian, based on Loch Ness data', which was
Roy Mackal's favoured candidate for the Loch Ness Monster. From
The Monsters of Loch Ness, p. 214–5. Illustration by Ray Loadman.

As Mackal remarked, 'nature encourages incredible adaptations'. So
did the artist who drew the 'hypothetical amphibian, based on Loch Ness
data', for Mackal's book. The creature looks like a corpulent newt. Male and
female are shown, the latter slightly smaller and facing the other way. On
the assumption that 300 million years is time enough to grow a neck, the
artist has given them one, together with diamond-shaped flippers à la Rines
(Figure 10).

Mackal and his publishers had high hopes for his book, with an initial
print run of 25,000 copies. However, he confided in a letter to Scott that
his pursuit of the Monster had come at 'tremendous personal cost'. His
colleagues had turned hostile and the University of Chicago had recently
taken away his academic tenure. To stay there, he would now have to raise
grants to cover his salary as well as the costs of his virus research.[20]

But as the letter from Denys Tucker had recently reminded Scott, Mackal was not the first scientist to have laid down his career for the Monster.

At the end of March, Bob Rines wrote an excited letter to Scott about the plans taking shape for the 1976 AAS expedition. This would be bigger and better than ever before, with a rig carrying sonar-triggered, wide-angle strobe cameras to take stereo pairs of images of the whole monster rather than a part of it. The new kit would be tested in the seal and shark tanks at the New England Aquarium in Boston. Also, Chris McGowan from Ontario had obtained funding to scour the Loch's floor for Monstrous bones, using Martin Klein's new high-resolution, three-channel sonar. Finally, the *New York Times* was sponsoring the whole operation. Hopefully Scott could call in over the summer?[21]

On the day Rines put pen to paper, Scott was in the Natural History Museum as a guest of the British Underwater Association. Their Spring Conference included papers on the interpretation of Rines's photographs and sonar traces.[22] A sonar expert had taken his equipment up to Urquhart Bay; depending on viewpoint, his conclusions were either interesting or worrying. Large sonar signals could be caused by sediment stirred up from the lake bed, rocky outcrops, boats' wakes and even the 'thermocline', the invisible boundary between the relatively warm surface layer and the permanently cold deep water. As a further complication, movement artefacts would be expected if the sonar unit was swinging in the water, not anchored on the bottom.

The expert was asked what he made of Rines's moving sonar target, picked up just before the flipper photograph was taken. He replied, 'I'm not entirely convinced that it was moving.'

Long hot summer

The warm-up for the Monster season 1976 was monitored with amusement by Frank Searle. Someone claimed to have recorded the Monster's mating call from the Loch; played back on the radio, it sounded like bleeps from a sputnik. Searle was suspicious and, sure enough, a toy manufacturer pitching for Christmas soon admitted the hoax. Realising that sex was a non-starter, a piano-maker wondered whether the Monster might be moved by Beethoven, performed on one of his own creations which he carted up from Birmingham on a trailer. There was no detectable response – or to Elvis, which someone else tried.[23]

The approach of spring also induced mating behaviour in Searle. He put up a sign in the Inverness Tourist Office, advertising for a Girl Friday to join him for a year or two. Job description: 'Public Relations Officer, typist, letter-writer, camera maintenance, cook, girlfriend – the lot.'[24] As usual, there were takers.

Things were going well for Searle. The County Council had just granted planning permission for the 'Frank Searle Loch Ness Investigation Centre', based in his blue caravan at Lower Foyers. This was a poke in the eye for the late and unlamented LNIB, which Searle loathed. Also, his book, *Nessie: Seven Years in Search of the Monster*, had just been published. Illustrated with his increasingly fanciful photographs, this presented a uniquely Searle-centric view of the Monster. To Searle's disgust, the publisher had cut the manuscript to avoid being sued by Dinsdale, Scott and Witchell but some 'vicious and derogatory comments' had been left in. Searle mistrusted 'Doctor' Rines, who had acquired that title after 'a brief visit to a college in Taiwan' to which his father had made a substantial donation in 1969. His so-called Academy also looked dodgy. Searle had sent a Christmas card to 'Mr R. Rines, c/o The Academy of Applied Science, Boston, USA', only to have it returned marked 'Insufficient Address'. By contrast, the letter sent by a girl in Philadelphia to 'Frank Searle, Shores of Loch Ness, Scotland' had reached its intended recipient without delay.[25]

Denounced as a hoaxer by all those whose guts he hated, Searle had an increasingly strong following at home and abroad. He was hunted down by the American magazine *Argosy*, which described him as 'far from crazy' and reported his exploits in breathless detail. These boiled down to 19 years as a soldier; seven years by the Loch; 22,000 hours of continuous Monster-watching; 26 sightings; and seven photographs.[26]

His greatest fans were naturally his Girls Friday, who were given every chance to see if he lived up to his reputation as 'the man who took his camera to bed'. Occasionally, his veneer of charm cracked, such as when visitors asked too many questions about exactly when, where and how he had taken his extraordinary photographs. Searle in a rage was not a pretty sight, as had been discovered by the young LNIB volunteer sent down to Foyers on a clandestine mission to photograph Searle's most recent pictures. 'For once in my life,' wrote Searle, 'I let my temper overrule my judgement . . . I grabbed him and punched him very hard several times.' Luckily for Searle, the young man did not bring charges and the LNIB folded shortly afterwards.[27]

Across the Atlantic, the AAS had brought out 'the first complete publication anywhere in the world of new and American evidence in one of the most fascinating scientific mysteries of the century'. It was a short piece in the MIT alumni magazine, *Technology Review*. The same issue carried a letter praising the editor for resisting peer review, which would reduce the magazine to a 'sterile technological journal' and was pointless, because educated readers knew full well that articles might 'reflect the bias of the author'. In that spirit, Rines laid out all his photographs for inspection.[28]

Plans for the summer were already advanced. The start-up meeting for the 'expedition to reveal the mystery – once and for all!' had taken place on 5 January, while Scott was battling with the fallout from *Nature* and the anagram. Rines, Harold Edgerton and Charles Wyckoff met for dinner in Edgerton's apartment near MIT, overlooking the Charles River. Shortly afterwards, the new 250-pound TV/camera rig was tested in the New England Aquarium and successfully took calibration shots of bemused fish. On a lake in New Hampshire, Martin Klein's high-resolution sonar proved capable of spotting mastodon bones dropped on to the bottom. Undeterred by the earlier failure of 'attractants' to tickle the Monster's senses back in 1970, Rines ordered extracts of the sex glands of eels and sea cows, together with tape-recordings of various sea creatures in agony or ecstasy.[29]

There were some wrinkles in the blanket. Firstly, the *New York Times* generously agreed to sponsor the expedition for $75,000, as long as they had world rights and a day's lead on any publications. The AAS–NYT expedition then discovered that it had picked up a parasite. *National Geographic,* which had turned down Rines's photographs because $100,000 seemed steep, were mounting their own expedition – not to chase the Monster, but a 'general survey'. They would be in Urquhart Bay at the same time, and, although they were not looking for the Monster, would set up their own underwater camera there too, just in case.[30]

The indigenous peoples of the Great Glen also needed consideration: 'The Beast was Scottish ... there could be British worries that we would usurp the Beast for Old Glory.' The AAS's solution? 'We had to ensure that there were British faces in the expedition crowd.'[31]

The Rines family took up residence in Tychat on 31 May 1976 and the AAS–NYT Expedition got under way. The camera rig was placed 200 feet off Temple Pier in 70 feet of water. Fish bait and hormones were trailed across the Loch from Urquhart Bay to Fort Augustus. McGowan and Klein began probing the Loch floor for the Monster's bones. As Dennis Meredith, editor of *Technological Review* and official chronicler of the Expedition wrote, 'Now we were ready for the Beast'.[32]

*

Nearby, *National Geographic*'s underwater camera was having no luck, but as they were not seeking Monsters this was to be expected. Instead, they photographed the Loch and interviewed eyewitnesses. Winifred Cary, wife of the wing commander, was a 'true believer' who had seen the Monster 18 times since the age of 12. However, the star witness was undoubtedly Alex Campbell, 'softly spoken and gentle of manner, not a man given to histrionics', who lived in the house in which he was born, 'a cottage throttled with flowers' by the River Oich. Campbell again described the extraordinary spectacle in Borlum Bay in mid-May 1934, with the 30-foot creature like a plesiosaur, nervously turning its head and submerging with a 'swoosh' when a trawler appeared.

National Geographic filled a 20-page full-colour spread with their story.[33] The AAS–NYT Expedition outdid them with a 170-page book, written by Dennis Meredith. Much space was devoted to the AAS's former glories (1972–75), described in prose that encompassed 50 shades of purple:

An animal loomed out of the blackness, sweeping past in a swirl of water so quickly that the camera caught only a tantalising, small slice of its surface . . . The camera was teased again an hour later when the visitor returned from the darkness . . .

The camera won the third round by capturing the visitor's image . . . a scientist's dream and a child's nightmare . . . a gnarled, gelatinous surface, suffocatingly close . . . an underwater close-up of the fabulous Loch Ness Monster.[34]

Over the summer, *National Geographic*'s photographic haul included a red high-heeled shoe and a salt cellar on the Loch floor, Alex Campbell and atmospheric landscapes. Meanwhile, the AAS/NYT expedition tried hard to repeat their triumph, taking over 108,000 underwater photographs. There were few fish, perhaps because of the record-breaking hot summer, and no Monster.

Luckily, as Meredith wrote, 'sonar proved to be the star of the 1976 expedition'. One contact was as big as the one that had caused all the excitement in 1972; it later turned out to be a 'delayed tracing' from a recently docked boat. More promising was a 30-foot target with strange 'parallel filaments' which Rines 'squeezed as far as he could for significance'. Squeezing even harder were Edgerton and Wyckoff, who were convinced enough to write '*N. rhombopteryx*' on the chart beside the enigmatic smear.[35]

Meredith's sonar 'star' might have been the discoveries made by Klein's high-resolution equipment at the north end of the Loch – not a Monster's bones but a series of stone circles and an aircraft, deep under water. The circles, quickly nicknamed 'Kleinhenge', looked like Pictish settlements from 3,000–4,500 years ago. Clearly, the standard wisdom that the Loch filled at the end of the Ice Age 12,000 years ago was all wrong. Were these baffling formations dwellings, or maybe some sort of giant calendar? Meredith hoped that, in time, they would 'speak of their origin to archaeologists'.[36]

Leaving new mysteries for others to solve, the AAS–NYT expedition members put on a brave face and drifted back to the USA at the end of July. The submerged aircraft turned out to be a Wellington bomber which had ditched on New Year's Eve 1940. Some time later, the stone circles did 'speak of their origin', but not to archaeologists. Dick Raynor remembered hearing that surplus rock from the building of the Caledonian Canal had been dumped near Lochend. Further inquiry confirmed the story and the diagnosis was later clinched by proving that one of the ancient stones from Kleinhenge matched the rock in the walls of the Canal.[37]

October 1976 saw the end of a significant era, heralded by the headline, 'North Dorset MP to quit'.[38] David James had bedded down convincingly in the rolling green hills of the West Country but now, with Margaret Thatcher poised to take over as leader of the Conservatives, it was time to answer the call of the ancestral home he had neglected for so long. A snapshot in the James family album shows his Volvo with a rolled-up rug on the roof, en route from Dorset to the Isle of Mull and Torosay Castle.

Clear water

A letter of introduction had arrived in Slimbridge at a bad time in early January 1976, when Scott was busy repairing Monster-inflicted damage. For Philippa's attention, he scribbled 'Interesting?' across the top. Her verdict was 'Interesting, but you should not undertake such a thing without meeting him and getting references.'

The letter was from an Adrian Shine, offering some lateral thinking about the Monster.[39] 'Lateral' meant about 50 miles to the south-west of Loch Ness, and a wilder and more beautiful lake. Lady Scott's caution might have stemmed from Shine's photograph on the front of a brochure entitled 'Loch Morar Expedition 75'. Bearded and in his mid-twenties, Shine seemed to be stuck in what looked like a camouflaged oil drum with windows. Beside him was a Union Jack and he was lifting a whisky bottle and a tumbler in a toast.

Shine's pitch was more serious. Morar was another Scottish Loch with a monster, known affectionately as 'Morag'. This was a humped creature with a long neck, putatively the same species as the one in Loch Ness. Loch Morar had 'a respectable pedigree of sightings', including an exciting encounter in August 1969 when two men shot at 'a huge aquatic animal' that had just rammed their boat.[40]

Shine argued that this was a more promising place to hunt Monsters than Loch Ness. Morar was smaller – just ten miles long – but even deeper, going down to 1,017 feet. It had once communicated with the open sea and drained into the Sound of Sleat south of Skye, celebrated for the sighting of a 50-foot sea serpent in 1872.[41] There was little human disturbance, with a single small road at its western end, serving the village of Morar. Crucially, the Loch's water was crystal-clear and ideal for underwater photography. On a bright day, a diver could look up through 50 feet of water and count the planks on the underside of a boat, while a camera 100 feet down could capture the outline of a swimmer on the surface.[42]

The first person to hunt Morag had been Elizabeth Montgomery Campbell, a volunteer with the LNPIB who had become frustrated with the obstructive waters of Loch Ness. The LNPIB's Christmas Party of 1969 inspired her to set up a splinter group to explore Loch Morar. Her decision was later applauded by Richard Fitter, who felt that this would be 'a much better bet for solving the mystery of the Loch Ness phenomenon'.[43]

Campbell and David Solomon, a fish ecologist from the University of London, set about raising funds and recruiting volunteers for the first expedition in summer 1970.[44] Conditions were more spartan than at Loch Ness, with camera stations in tents and the dormitory/HQ in the corrugated iron shed that passed for the village hall at Morar. However, the rewards soon became obvious. In a flight of fancy at the House of Commons meeting, Robert Rines had called Loch Ness 'a veritable Garden of Eden'; in reality, it was more like a brownfield site. By contrast, Morar was teeming with life, with a healthy food chain that could keep a colony of fish-eating Monsters in prime condition. And, in water as clear as any Highland spring, it was only a matter of time before they revealed themselves.

Word soon spread, attracting visitors such as Tim Dinsdale and Professor [sic] Robert Rines. Through 1971, the operation burgeoned and diversified into studying the Loch's ecology in the round, while waiting for serendipity to fill in the Monster-shaped void in the database. By the end of the summer of 1972, there remained just one missing ingredient: Morag. Campbell and Solomon decided to write a book and then called it a day.[44] This left the way open for Adrian Shine.

*

Shine grew up in Surrey and was, by his own appraisal, 'lazy and unpromis-ing' at school. A spell as unpaid bosun's mate on the tall ship *Sir Winston Churchill* in spring 1967 gave him the chance to appreciate ecology at first hand – looking out from the masthead towards Land's End and the burning wreck of the oil tanker *Torrey Canyon*, spilling its guts into the Atlantic. After school, various career options beckoned: India, chasing the Loch Ness monster, or rowing across the Atlantic. Someone had just done the Atlantic, so he went to India to retrace Kipling's footsteps.[45]

When Shine returned to England in spring 1970, Loch Ness appeared to be in the doldrums and he stayed south of the border. The publication in 1972 of *The Search for Morag* by Elizabeth Montgomery Campbell and David Solomon persuaded him to visit Scotland for the first time. He took a tent, a sea anchor, a powerful lamp and a camera up to Loch Morar, hired a boat and spent a week drifting across the water at night, watching and learning. Morag did not appear, but the place worked its magic on him. He said later, 'I came to conquer, but ended up captured.'[45]

Back in Surrey, he began building an underwater camera hide – *Machan*, the improbably small vessel in which (to celebrate its launch) he was later photographed with whisky and Union Jack. The name was not that of some splendid Pictish chief, but the Hindi word for the lair from which hunters watched for tigers. Its design was inspired by both Jacques Cousteau and Heath Robinson. Made of fibreglass, it was roughly spherical, 40 inches in diameter and just deep enough for a man to sit on the floor (sandwiched between ballast tanks) and stare up at the surface through plate-glass panels (Plate 39). Luckily, the curse of health and safety had not yet been visited upon the planet. *Machan*'s watertight seals worked best under at least 20 feet of water but not so well on the way down, and air was piped down a garden hose from a small electrical pump.

In summer 1974, Shine completed 30 dives in Machan 'without incident', going down 30 feet in areas likely to be favoured by Morag. Thanks to the clarity of the water, he could survey a wide area of surface by looking up from the bottom, and a large animal would immediately be obvious as a silhouette against the sky. The search strategy was valid and Machan was baited with irresistible fish oil; unfortunately, Morag refused to show.

The following summer, 1975, Shine assembled a small team of divers and the *Pequod*, a boat with a glass dome in the hull through which the lake bed could be scanned for bones or other Monstrous remains. Neither *Machan* nor *Pequod* delivered Morag, but, compared with the decades of effort at Loch Ness, these were still early days. Besides, another strand of research was turning out to be valuable and fascinating in its own right, whether or not Morag did the honourable thing and allowed itself to be photographed. The bigger picture was the Loch's ecosystem, which Montgomery and Solomon had begun to catalogue, from microscopic diatoms to otters and the newt that turned up in their drinking water. Shine now picked up from where they had left off.

His letter to Scott in January 1976 pointed out that, with Dinsdale the loner going his own way in *Water Horse*, this was the only organised British outfit actively seeking the Monster. Shine was planning to add underwater television cameras and hydrophones, courtesy of the Royal Navy, to his armamentarium. Would Scott lend his support?

Scott replied, expressing interest and encouragement but – preoccupied with clearing up after the other Monster – bided his time.

As time goes by

While the AAS–*New York Times* and *National Geographic* teams slugged it out in the opaque depths of Urquhart Bay, Shine continued to lay bare the secrets of the unspoiled and photogenic Loch Morar. His team of 20 divers now included Ivor Newby, who had defected from Loch Ness, and the man who had first photographed a living coelacanth. They explored the lake edge down to 100 feet or more and proved that low-light television cameras could clearly see a diver silhouetted against the sky. Seven sightings were recorded, including a brief glimpse of a hump reported by the dive leader, but no hard evidence – film footage or mortal remains – was forthcoming.[46]

At both Lochs, the Monster-free impasse continued through 1977, with some light relief at Loch Ness. This was thanks to Tony 'Doc' Shiels, a self-styled wizard and psychic from Cornwall, usually assisted by two lady witches who enlivened proceedings by swimming naked in the sea. Shiels had already photographed Morgawr, a Cornish sea monster from Falmouth who looked as though she was made of plasticine.[47]

Turning his attention to Loch Ness, Shiels summoned forth the Monster and took several photographs. In the most famous, the Monster apparently sports a grin, reminiscent of toothpaste commercials of the period. Uncharitable souls nicknamed it the 'Loch Ness Muppet' photograph and pointed out that surface ripples showed through the creature's neck, almost as if a picture had been cut out and pasted or projected on to a watery background to be rephotographed. Shiels explained that the Monster was a giant cephalopod (the 'elephant squid') and was see-through because it flits between our world and its own, parallel universe.[48]

Tim Dinsdale was more impressed. Writing to Peter Scott, he described Shiels's photographs as 'either the best surface stills ever obtained or just another fudged-up job'. Dinsdale had not met Shiels but had corresponded with him. Shiels's letters indicated 'a man of remarkable intellect, much character and a quite disarming sincerity'.[49]

The fourth edition of Tim Dinsdale's *Loch Ness Monster* (1982) features a facsimile of the sworn statement by Antony Nicol Shiels of Truro in the County of Cornwall that 'the aforementioned photographs are genuine'.[50] The book's cover shows the 'Muppet' Monster, complete with grin (Plate 40).

The year 1978 slipped by with no sensational breakthrough at either Loch, apart from finding molluscs and other invertebrates at the surprising depth of 1,000 feet in Morar. Possibly dragged down by Shiels's photographs, the Loch Ness Monster lost some ground. One expert lumped it in with other pseudoscientific obsessions including UFOs, ancient spacemen, Bigfoot, horoscopes and the Bermuda Triangle.[51] Depressingly, the expert was from the Smithsonian Institution in Washington, home of Rines's great supporter, George Zug.

Over at Loch Morar, this had been Shine's fifth unrewarded year for underwater photography. Realistically, only a tiny area could ever be covered visually, and Shine decided that the search must now shift to sonar. In recent years, the technology had moved on – and, used intelligently, sonar could look for large creatures in waters murkier than in Loch Morar.

Fusion and friction

David James made the first move to entice Shine into applying his Morar experience to crack the bigger problem at Loch Ness. In 1976, he drove down from Torosay to visit Shine at Loch Morar and made his pitch. 'It's like climbing Everest,' he explained. 'We need a younger man with bigger lungs.' As always, James was persuasive, and in no time Shine found himself in the smoke-filled Strangers' Bar at the House of Commons, and then in a committee room with a senior military man who was promising him divers, rations and equipment. Next, Peter Scott met Shine in London, and also turned on the charm. The search for the Loch Ness Monster desperately needed a solid scientific footing, like Shine's excellent operation at Loch

Morar. Scott had an additional motive. He confided to Shine, 'We need you to run a check on Dr. Rines.'[45]

In spring 1979, the remnants of the LNIB were absorbed by the Loch Morar Expedition to become the 'Loch Ness and Morar Project', directed by Shine and with James, Scott and Norman Collins (the former LNIB Chairman) as patrons. Operations continued at Loch Morar but this was now a testing ground for a new, British assault on that other, bigger Loch, where the Americans held sway.

In January 1979, Rines wrote confidentially to Scott with news from Loch Ness that was being sent only to 'trusted friends'.[52] Using Charles Wyckoff's super-sensitive film (ASA 30,000), a 'large object within relatively close range' had been photographed underwater. Also, Martin Klein's side-scanning sonar had spotted large targets that moved fast in mid-water and emitted ultrasonic noises – almost like a submarine, but none had been in the Loch at the time.

And the AAS had a new secret weapon – 'Project Bell' – being developed with MIT and the US Navy. Two dolphins were being trained in a lagoon in Isla Morada, Florida, to track turtles while wearing cameras and strobe flash units. It was assumed that their skills would be transferable to Scotland and that they would cope with the salt-free water of Loch Ness for 1–2 hours and return safely to the sea-water pen to be built for them there.

Rines's other 'trusted friends' might have received further details, or even sight, of the photographs of the 'large object'. Scott did not. Unfortunately, Project Bell had to be called off in late June; as explained in a mournful news item in *New Scientist* ('Ness hunter dies'), one of the dolphins did not complete its training.[53] This left the AAS in its default mode of business as usual and more of the same.

Plan B

Even with co-operative experts and high-end commercial equipment, Shine's sonar adventure took time to mature. It was 1981 before a floating sonar station ventured out on Loch Ness. This was a 40-foot catamaran named *John Murray*, after the ex-*Challenger* pioneer who had led the Bathymetrical Survey of 1897–1909. It was big and ugly, with an angular, futuristic superstructure and inflatable hulls which allowed it to drift along in silence. As the first long-term sonar survey of a large lake, this provided crucial information about real contacts and a baffling array of artefacts.[54]

John Murray's successor, *Monitor*, took over in 1984 and was moored in mid-Loch over the 600-foot basin off Horseshoe Crag. Urquhart Bay, the traditional hunting ground of the AAS, was avoided because it was a busy anchorage and full of sonar artefacts from signals bouncing off the steep sides of the Loch, the wakes of boats and even the mooring chains hanging in the water. *Monitor* was anchored in position using over a mile of rope and provided a stable platform from which the Loch's repertoire of sonar tricks could be unravelled.[55]

Over the next couple of years, Shine's team found that even a calm surface could hide unimaginable turmoil in the depths. Periodically, massive sonar signals appeared from nowhere and disappeared again. These turned out to be a huge underwater wave (a 'seiche'), described originally in Lake Geneva. The Bathymetrical Survey had found evidence of a seiche in Loch Ness in 1904,[56] which was confirmed in a *Nature* paper in 1972.[57] The wave could reach 130 feet in height but its crest was 200 feet beneath the surface, which rose by only a few inches. The false sonar contacts could appear to rise, dive and disappear – just like those which Hugh Braithwaite and Gordon Tucker of Birmingham University had reported in 1968, to the derision of *Nature*. On a smaller scale, as the British Underwater Association had told Scott in early 1976, dramatic sonar echoes could be produced by wobbles in the 'thermocline' boundary between the surface layer and deeper cold water.[58]

All this raised some interesting and potentially uncomfortable questions about previous sonar studies, notably those conducted in noisy, echo-infested Urquhart Bay. How much was fact, and how much artefact?

Shine always knew that he would not make friends by trying to demystify the Loch. One local non-friend was Frank Searle, who had generally turned nasty during the early 1980s and threatened to disembowel Shine with a knife. Searle was also thought to be responsible for daubing 'SHINE CON MAN' in large white letters across the noble stones of Urquhart Castle.[59]

A more subtle critic was Professor Henry Bauer of Virginia, crypto-zoologist, who believed that the Monster existed and that the human immunodeficiency virus (HIV) did not cause AIDS. Bauer felt that Shine's supporters had been foolish to claim, 'rather wishfully', that 'we are extending our influence and have become, in the eyes of more and more people, the authority on the controversial issue of the Loch Ness Monster'.[60] Bauer found this stance 'impertinently aggressive', especially as Shine and his people were amateurs. By contrast, Dr Rines's team provided a dazzling combination of the highest expertise in photography generally and underwater photography in particular, and in electronic gadgetry in general . . . a model of how one attacks a novel problem – a careful sifting and evaluation of data.

The opposing camps met in July 1987 at a symposium about the Monster, organised at the Royal Museum of Scotland in Edinburgh by the International Society of Cryptozoology and the Society for the History of Natural History.[61] Many familiar figures were there – Bob Rines, Roy Mackal, Henry Bauer, Tim Dinsdale and Adrian Shine – and all but one of them read scripts that had not changed significantly in a decade.

Shine alone presented new data: comprehensive ecological studies which suggested that Loch Ness was too impoverished to support large aquatic animals; careful dissection of sonar pitfalls in both Lochs; and cores drilled out from their beds which proved that Morar had once held marine life, whereas Loch Ness had never been open to the sea.[62]

Richard Fitter gave a paper on the history of the Monster, 'from St Columba to the LNIB', with a sad epitaph – 'no substantive advance' – to honour the LNIB's decade of endeavour.[63] (Dinsdale later amplified that statement, saying that the LNIB's strategy had been '99.9% futile').[64]

However, nobody there had reason to brag. Sightings continued to be reported by 'ordinary' people, but those with the technological firepower to prove the case – Rines, Shine and Dinsdale – had not come up with anything new about the Monster for 12 years.

Plumbing the depths

Shine had played a good hand at the Monster symposium, but his best trick was still up his sleeve. He was planning 'Operation Deepscan' to test, possibly to destruction, the hypothesis that Loch Ness harboured a Monster. A comb that no large animal could evade would be dragged through the water, end to end. The comb would be 900 yards across with broad, flat teeth that consisted of sonar beams directed at the lake floor from a line of boats advancing down the Loch.

Operation Deepscan was hugely ambitious but had precedents. A quarter of a century earlier, Peter Baker and the Oxford–Cambridge Expeditions of 1960 and 1962 had swept their sonar curtain several times down the length of the Loch. Their cue was followed in 1983, when $100,000-worth of clever sonar technology materialised in Urquhart Bay. This was the brainchild of two bright young things, Rikki Razdan from Boston and Alan Kielar from Buffalo, New York. They had set up an electronics company, and like Martin Klein, designed and built their equipment. Instead of trying to survey the Loch at large, they concentrated their resources – an array of 144 floating sonar units – on an 85 x 85-foot square of the Bay. Roy Mackal would have been interested in their automatic Monster-biopsy system, which fired darts at objects over five feet across that entered the sonar cage.[65]

Razdan and Keilar had come to the Great Glen hoping to be surprised but expecting to be disappointed, as they had spent weeks analysing sonar and photographic records from Loch Ness and were not impressed by any claim that a Monster had been detected. They waited patiently for three weeks but detected only a three-foot fish. After returning to the USA, they published their findings in a magazine with a giveaway title (*Skeptical Inquirer*) and argued that their own negative results, rounding off decades of inconclusive research, made the Monster's existence very unlikely. However, true sceptics could have responded by pointing out that their search, although commendably intensive, had covered only a millionth of the Loch's total volume and that cautious Monsters might well have avoided such an obviously abnormal stretch of water.

In their article, Razdan and Keilar also took the opportunity to set some records straight. Their main target was Dr Robert Rines and the Academy of Applied Science, and they pulled no punches. Their countrymen's claims were based on overinterpretation and possibly fabrication. The dramatic sonar 'contacts' that were supposed to be large animals swimming up from the depths were only artefacts, while the celebrated flipper photographs looked as though they had been touched up. To compound the felony, Razdan and Keilar quoted a rumour that the Monster-spotting site in Urquhart Bay ('grid element 2903') had not been identified by a brilliant computer analysis of statistical probabilities; instead, Rines had been told to look there by a local dowser, who waved a light on shore when the boat reached the right spot.

The piece by Razdan and Keilar was picked up by *Discover* magazine, which had a wider circulation. In advance of publication, the editor contacted Charles Wyckoff for his comments. Wyckoff was incensed and he, together with Rines, Edgerton and Alan Gillespie, responded with a furious defence of their methods and findings.[66] As on many previous occasions, the protests of those who had come to believe in the Monster fell on deaf or prejudiced ears. The editor of *Discover* refused to print their letter.

*

Operation Deepscan[67] had already tested the water when the Edinburgh Monster symposium took place in July 1987. The previous October, ten boats had lined up to sweep the northern basin of Loch Ness. All were fitted with identical echo-sounding units (Lowrance X-16), sending a 30-degree fan of signals vertically downwards at right angles to the direction of travel. The experiment was nearly wrecked by high winds, but was invaluable for fine-tuning the equipment and plans for the following year.

The real thing required a flotilla of boats, which were kindly provided by Caley Cruises in exchange for publicity; this was assured, as all the hotels around the Loch were full and the people from newspapers and television outnumbered those involved in the expedition. This was a far cry from trailing a sonar scanner behind a boat. On 9 October 1987, after five days of training, the 19 cruisers lined up at 50-yard intervals across the northern end of the Loch. Any unusual contacts on the Loch floor or in mid-water would be marked with a buoy, and a radio message sent to the response vessel *New Atlantis*, waiting behind the boats of the line with its high-resolution tracking sonar.

The flotilla cruised in formation down the middle of the Loch to Fort Augustus and returned the following day (Plate 41). Three unusual echoes were recorded on the first day; all had disappeared by the time *New Atlantis* reached the spot. Two were thought to be artefacts from sonar reflections, but the third remained a mystery. It was 'smaller than a whale but larger than a shark', over 500 feet down and apparently diving.[68]

Adrian Shine had thought that Operation Deepscan might be his swansong, but it was not to be. He had started chasing the Monster at the age of 23 and now, 14 years later, had spent over a third of his life in pursuit. During that time, he had evolved from an opportunist looking for the 'one in a million chance' into an investigator with research students, academic collaborators, a string of publications and a reputation as a serious scientist. He had also changed from a believer into a 'sympathetic sceptic', still hoping against the odds that his faith might be restored.[69]

Operation Deepscan should have been the ultimate acid test: if nothing was found, there *was* nothing to find. However, that large unidentified contact left everything in limbo. From its size, it could have been a submerged tree trunk, but that could not be confirmed. Also, comprehensive though it was, the sonar sweep left some wriggle room for Monsters to slip past. On the broadest part of the Loch, east of Urquhart Bay, the line of boats covered only one-third of its total width.

All along, Shine had taken pains to point out that Operation Deepscan was 'not a quest for the dragon of popular expectation'.[70] Predictably, most of the media missed the point. As the Monster had not been found, the whole thing had been a failure.

Marking time

In late 1987, Sir Peter Scott was an elder statesman and, in public perception, on a tier not far below royalty. He was also 79 years old, suffering from angina and under doctor's orders to live more healthily in order to avoid a recurrence of the mild heart attack which had spoiled Christmas 1984.

Scott had long since withdrawn from the front line but was still in touch with some of the Monster's devotees. A few months earlier, Tim Dinsdale had written gloomily, 'Nessie in the 80s has, if anything, been going backwards'.[71] By contrast, Bob Rines remained upbeat after his eighteenth visit (again fruitless) to the Loch. He was, however, irritated by Adrian Shine's claim that Operation Deepscan had found a rotting tree stump on the bottom of Urquhart Bay which was fished out by Dick Raynor and closely resembled the 'gargoyle' head photographed by the AAS in 1975.[72] Scott had only been copied into that letter, and did not bother to respond.

The following May, Peter Scott went on record about the Monster for the last time. In a letter to the editor of the *Western Morning News*, he wrote:

Although I feel that the chances of a large unknown animal living in Loch Ness are slim, I do not believe that possibility can yet be ruled out.

And by the way, he added, that anagram was pure coincidence.[73]

Last fling

Adrian Shine's Operation Deepscan was a difficult act to follow, but someone was bound to try. The person who took up the gauntlet was a man whose life had been knocked off course by the Monster, but who had done well from the collision.

Having forsaken law for the BBC, Nicholas Witchell had become even more familiar on the small screen than Peter Scott, initially as the Ireland Correspondent and then as the frontman on the *Six O'Clock* and *Breakfast News* programmes. The Monster had been good to him, and his book *The Loch Ness Story* had been a bestseller in all four editions. Witchell remained 'intrigued by the mystery', and although his book no longer began with a large animal gliding silently through black water, it was still clearly written by a believer.

In 1992, approaching the age of 40, Witchell decided that it was time to 'put something back'.[74] The result was a formidable, multidisciplinary collaboration, welded together by Witchell's powers of persuasion. The scientific establishment was represented by the Freshwater Biological Association, the Royal Geographical Society (of which Witchell had been elected a Fellow) and – coup of coups – the Natural History Museum. A key partner was the Norwegian technology company Simrad, famous for its civilian and military sonar equipment. Witchell came up with the name 'Project Urquhart', after the Castle which had greeted him each morning during the summer of 1972, on opening the shutters of his hut on the hillside at Strone.

The Project's mission statement had echoes of both Sir John Murray's Bathymetrical Survey and Shine's Operation Deepscan. Over two years, the Loch's floor and sides would be mapped in fine detail and a complete eco-logical survey conducted. Neil Chalmers, Director of the Natural History Museum, explained that science had neglected the Loch for too long and 'must now overcome its embarrassment'.[75]

This would have been music to the ears of Constance Whyte, who had pleaded for just such a survey 35 years earlier in *More Than a Legend*. However, some of the old tensions remained. John Lambshead, Head of the Museum's Department which specialised in worms, admitted that reputable scientists approached the Monster 'like you would approach a nuclear bomb. To be labelled as a Monster hunter can set your career back ten years'. Dr Lambshead had no interest in the Monster and would be busy enough looking for nematode worms – which occupied other fascinating ecological niches such as damp German beer mats and the penis of the blue whale.[76]

Newspapers also picked up a certain ambivalence about the real purpose of the Project. Headlines ranged from 'Telly Nick in Nessie proof bid' (*Sun*), to 'Witchell wary of Monsters as study begins' (*Sunday Times*) and 'Loch explorers claim Monster is not on agenda' (*Scotsman*).

Even more revealing was a confidential guide to help Project members sidestep tricky questions from the press; the document was 'Strictly Private and Not for Circulation' and was to be kept in a safe place.[77]

Q: What about Nessie? A: This is an ecological survey of the Loch.

Q: Isn't this just a glorified Monster hunt? A: No. See above. The Natural History Museum and Freshwater Biological Association do not take part in 'disguised' projects.

Phase 1 of Project Urquhart began in July 1992 by mapping the Loch from *Simrad*, a 150-ton research vessel. *Simrad*'s control room looked like 'a cross between a computer video arcade and the bridge of a do-it-yourself Starship Enterprise'.[78] Over three weeks, *Simrad* covered 500 miles and took 17 million readings with its 120 sonar beams, which radiated from beneath the hull like a spread fan.

The Bathymetrical Survey of 1903–4 had acquired data approximately 70,000 times more slowly, but had been remarkably accurate. There was no sign whatsoever of the canyons and undercut caves which the AAS's sophisticated sonar mapping had created in Urquhart Bay in 1971. *Simrad* also carried an ex-military sonar tracker, which locked on to moving targets and was normally used for hunting submarines. On the evening of 28 July, this device picked up a 'very strong' contact in mid-water between Invermoriston and Foyers and held it for two minutes. It was not identified, but an experienced sonar operator from Simrad thought this 'unlikely' to be one of the Loch's known fish species or a spurious echo.[79]

Phase 2 of the Project, a detailed examination of life in Loch Ness and 'potentially one of the most exciting such studies in Britain', was to follow in summer 1993. *Simrad* was replaced by the Scottish Marine Laboratory's research boat *Calanus* and a support vessel, both equipped with fish-finding sonar units. The actual fish population was mapped and compared with the numbers predicted from similar lakes, looking for a shortfall that might be explained by hungry Monsters. The preliminary results looked promising, but the project's funding ran dry before clear conclusions were reached; another four large sonar contacts were logged but remained unexplained. As hoped, a new animal species was discovered. This was a tiny nematode worm, presumably happier in the Loch than in some other peculiar places where nematodes live.[76]

Witchell said later that Project Urquhart 'got the Monster out of my system'. By the time the new millennium had begun, he had been persuaded that there was not enough food in the Loch to support a viable population of Monsters. After a third of a century as a believer, he lost faith.[74]

By contrast, Robert Rines still believed and returned to the Loch in 2000 for the twenty-first time. He was probably unaware that he was fulfilling a Cassandra-like prophecy made in 1957 by Constance Whyte:

There is a challenge to British scientists on their own doorstep and if it is not taken up, the initiative may pass to others – a team of wealthy enthusiasts may arrive from across the sea and solve the problem for us.[80]

Identity parade

By now, readers will have realised that the question 'What is the Loch Ness Monster?' has no simple answer, because the thousands of people who have seen, photographed or filmed their 'Monster' have not encountered the same phenomenon.

Before moving on to the big question of whether a large, unknown animal species inhabits Loch Ness, we need to consider the numerous pretenders – animal, vegetable, mineral and imaginary – which have managed to pass themselves off as the real thing. Looking at pseudo-Monsters in more detail will take us on a voyage of discovery to faraway places including a coal mine in Illinois, a Butlin's holiday camp and Paradise. Along the way, we shall meet the laws of physics, the circuitry of the human brain and distant echoes of the Great Glen Fault – and, of course, prehistoric animals whose shelf-life has outlasted their best-by date by 65 million years.

Tricks of the light

What you see at Loch Ness is not necessarily there. Mirages are not confined to the sands of the Sahara, but can pop up wherever layers of air at different temperatures are laid on top of each other. The resulting gradient in optical density can bend (refract) light like a lens. Refraction can make an object appear smaller or bigger, distorted or in the wrong place (Figure 11). A fine example was written up in the journal *Science* under the provocative title, 'Atmospheric refraction and lake monsters'. The test object was a two-foot wooden post on the edge of Lake Winnipeg, photographed from a few hundred yards away on a fine October afternoon. The image of the post performed optical gymnastics, first shrinking and then stretching to twice its true height – and all in the space of three minutes.[1]

Loch Ness is a breeding ground for optical illusions, especially on mirror-calm days when there is no wind to stir the layer of cold air lying on the water and ideally a light mist or heat haze. These conditions have long been recognised as the best for spotting the Monster. Mirages reported from the Loch have included snow-capped mountains, cliffs and even a steamer, plucked from somewhere and pasted in above the water.[2] On a smaller scale, mirages can turn everyday objects into shapes that defy identification, even by seasoned observers. On the calm, hazy morning of 3 September 1933, a 'strange object' shot out of the water in Borlum Bay a few hundred yards beyond the Abbey boathouse. It gave a fine show of its 30-foot body ('like two rowing boats end to end') and six-foot neck, ending in a small head ('like a cow's, but flattened') which turned alertly from side to side. The creature appeared nervous, and when two trawlers came in sight it dived with a great flurry of foam. The witness was no amateur, but a man who could boast two decades of first-hand experience of the Loch and its tricks: water bailiff Alex Campbell.

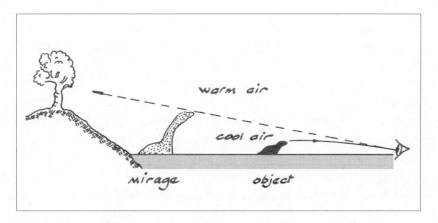

Figure 11 Mirages, optical illusions due to bending (refraction) of light across layers of air with differing optical density. An ordinary object such as a bird can be magnified and distorted into a grotesque shape. Illustration by Ray Loadman.

This dramatic spectacle, with an excellent view and a knowledgeable observer, had all the ingredients of a classic Monster sighting. However, a few weeks later, at the same time of day and under similar conditions, Campbell saw 'something very like' his Monster. This time, the mist melted away to reveal a group of cormorants with their leader standing up on the water and flapping its wings. In the uncertain light on the earlier occasion, refraction had magnified the birds 'out of all proportion to their proper size'. Campbell explained all this in a letter to his employer, the Ness Fishery Board, and added that mirages could also give floating gulls, barrels and bottles 'a very grotesque appearance'.[3]

Even without refraction errors, it can be difficult to judge distance and size on calm, featureless water. Ivor Newby, an experienced observer, remembers tracking a length of timber through binoculars as it floated down the Loch. He convinced himself that it was a railway sleeper several hundred yards away, and was startled when a duck flew into the field of view and landed beside the object. If it had been a railway sleeper, the duck's wing span would have been over 15 feet.[4]

Tricks of the mind

The witnesses who watched a single-engined plane crash into the sea 500 yards off Shoreham on the afternoon of 19 January 1934 agreed that it was white, with a distinctive red stripe. The coastguards were quickly on the scene but found no bodies or wreckage. The mystery deepened when it transpired that no aeroplane had been in the area that day. The eyewitnesses, whose integrity was never doubted, refused to change their statements. Verdict: nothing to concern air accident investigators, but a mass hallucination.[5]

Optical illusions are extraordinary renditions of real objects and can be photographed; hallucinations, being perceptions of things that are not there, are only in the mind and impossible to prove or disprove. Hallucinations that affect more than one person simultaneously were recognised and given diagnostic respectability in the late nineteenth century. Dr E. G. Boulenger of the London Zoo Aquarium was not necessarily being derogatory when he attributed the sudden flurry of Monster sightings in 1933–34 to mass hallucination.[6] This might or might not have explained why 50 passengers on two buses were convinced that they had spent 13 minutes watching a large two-humped animal playing in the Loch off Urquhart Castle on 28 October 1936.[7]

The ability of the human imagination to embroider banal objects was neatly demonstrated by Adrian Shine, who rigged up a contraption operated by strings to hoist a wooden post out of the water at Loch Ness. His inspiration was a sighting of a pillar-like object which had risen and sunk off the shore of the Loch, near where two innocent-looking lads were sitting on the beach beside a line that disappeared into the water. The witnesses of Shine's experiment knew that something was about to happen, but a third of them reported seeing objects that did not look like the post (Figure 12).[8]

Even more mischievous was Richard Frere, the naturalist and lover of the Loch who regarded the 'hunt for Nessie' as a lowbrow tourist stunt.

Frere demonstrated that the products of a vivid imagination can quickly infect others to produce a remarkably detailed and consistent multi-witness account of something that was not there. He and a friend set up a camera in a layby at Lochend, where the Loch was flat-calm except for the wakes of three trawlers which had just disappeared around a headland. They told the crowd which quickly gathered that they had seen a large humped animal. When Frere and his co-conspirator sneaked away 20 minutes later, they left an excited throng of eyewitnesses haggling over the details – number of humps and length of neck – of the 20-foot animal that they had watched.

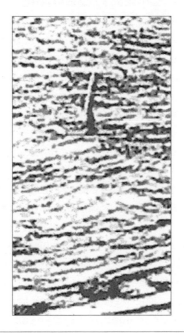

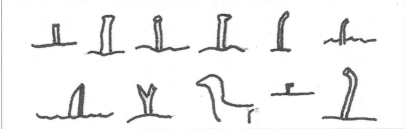

Figure 12 Eyewitnesses' sketches (bottom) of what they observed when a square-section 18 x 18-inch wooden post (top) was hoisted out of Loch Ness 450 feet away. One-third of them believed that they had seen something other than a post, including complex shapes. From Shine A. J. Postscript: surgeon or sturgeon? *Scottish Naturalist* 1993; 105:271-82.

One lad even produced an anatomically explicit drawing of a long-necked, flippered animal labelled 'plesiosaur'.[9]

Descartes noted that 'Chance favours the prepared mind'. So does wishful thinking.

Waving, not drowning

Water can do extraordinary things. In August 1959, holidaymakers were entranced by a family of baby otters playing underwater below the sea wall at Oban. Eventually, someone noticed that the animals seemed to be stuck in a zone of turbulence, which turned out to be the outfall from a large pipe discharging clear water a few feet below the surface. The lithe black bodies of the 'otters' were just tricks of refraction.[10]

On Loch Ness, the wakes of boats are a common cause of misdiagnosed Monsters. The wave fronts thrown up by a boat's prow can be over a foot high and appear much bigger on flat-calm water. On Loch Ness, a particularly impressive wake was generated by the local steamer *Scot II*, which had been designed to bulldoze its way through ice in the Caledonian Canal.[11]

From far off and viewed from water level, a boat's wake can show as a dark hump or a series of humps which may appear to undulate, like the back of a moving caterpillar (Plate 42). Sometimes, the eye is drawn quickly from one hump to another, creating the illusion of something dashing across the water. Even at close range, a bow-wave can look surprisingly animate. Maurice Burton followed one in his boat as it travelled from mid-Loch towards the shore. At times, the wave looked like 'a brown glistening hump', and foam breaking at one end gave it the appearance of propelling itself.[12]

Interesting things can happen when a bow-wave reaches the edge of the Loch. If the shore shelves gradually, even a nondescript wake can be pushed up into a spectacular wave that can rush many yards up the beach – a small-scale version of the tsunami which towered 30 feet over coastal towns in Indonesia, but was only a foot high while travelling in mid-ocean. Rarely, a bow-wave meeting a sheer rocky surface may 'bounce' off, and the reflected wave can then fuse with an incoming wave front to produce a 'standing wave'. Depending on the geometry of the scene, a standing wave can appear to take on a life of its own. It may form a hump that seems anchored to the spot or moves in any direction, slowly or quickly and with or against the wind and current. The no-nonsense Richard Frere described one (created by the criss-crossing wake of his own boat) which he initially believed to be animate.[13]

Waves did not fool Andrew McAfee, who was surveying the shores of the Loch in July 1952. To his unaided eye, the three dark humps in the distance looked exactly like a large animal ploughing through the water; magnified through his theodolite, however, they were just waves about a foot high. This was evidence supporting the Natural History Museum's view, stated in their booklet *Scientific Research* (1956), that the phenomenon known as the Loch Ness Monster was simply 'one of waves and water currents'.[14]

Predictably, Monster believers were unimpressed by the wave theory, because, as Rupert Gould patiently pointed out, 'The presence of a boat will instantly "undeceive" the most casual observer'.[15] In fact, it is easy to be deceived by the Loch's indented shoreline. The bow-wave travels away from the boat's course at brisk walking pace and can take 20 minutes to reach shore from mid-Loch – by which time the boat responsible may be miles further on and out of sight.

Elemental forces

Wind can incite water to behave in a monstrous fashion. The mini-tornadoes known as dust devils have an aquatic counterpart, which can spring up, move across a stretch of water and then collapse. 'Water devils' can appear solid, especially when seaweed or floating debris are dragged up into the rotating column of water, and might explain reports of 'sea serpents' seen rising like pillars out of the ocean. The editor of the *Journal of Meteorology*, writing to Peter Scott in 1978, believed that water devils could account for some Monster sightings.[16]

Less dramatically, a tongue of wind can produce a localised patch of roughness where it licks the surface of the Loch. These 'windrows' may be static or mobile, and can look strikingly solid on mirror-calm water. A favoured site for the phenomenon is the northern end of the Loch between Abriachan and Aldourie – near where the Mackays witnessed the 'Strange Spectacle' which ushered in the modern age of Monster sightings in 1933.[17]

The wind can also cause a monstrous commotion deep in the bowels of Loch Ness, while leaving barely a ripple on the surface. To understand this, we need to make a brief diversion to a village called Paradise, on the two-mile stretch of the Rhine which connects the Lower and Upper Lakes Constance. On 23 February 1549, the townsfolk witnessed the extraordinary 'wonder of the rising waters'. In less than an hour, the river emptied and refilled itself repeatedly, sometimes even running uphill.[18] The 'wonder' was explained three and a half centuries later, when the same phenomenon was observed on a smaller scale in Lake Geneva. There, the water level can rise and fall like a tide, sometimes enough for normally submerged shorelines to dry out completely; this mass movement of water was called a seiche, from the Swiss-French word for 'dry'. A seiche forms when the sustained pressure of the wind pushes the surface layer of water towards the downwind end of the lake. When the wind relaxes, the displaced water flows back and can set up a series of oscillations, as witnessed in 1549 by the astonished inhabitants of Paradise.

Loch Ness has its own seiche, because the Great Glen acts like a wind tunnel that happens to point south-west, the predominant wind direction. A strong, steady wind pushes the water up into the north-eastern corner of the Loch. When the wind drops, the water runs back towards Fort Augustus. The amplitude of the seiche at the surface was measured in 1904 by the Bathymetrical Survey (a daily ritual continued afterwards by the monks at the Abbey), and amounted to only a few centimetres. Deep down, however, the Loch's internal seiche can be massive. An underwater wave up to 130 feet high sweeps to and fro along the length of the Loch with enough force to buffet submarines and generate sonar echoes.[19]

Reverberations of the seiche can also reach the surface, especially in places like Urquhart Bay. Odd currents can appear and melt away, sometimes carrying floating objects upwind or upstream.

Perhaps improbably, earthquakes might explain some Monster sightings. The Great Glen was carved out along a fault line several hundred miles long and thousands of feet deep. The fault remains seismically active, with a focus at the northern end of Loch Ness. It is much less irritable than the San Andreas Fault which bisects California, but significant tremors were felt in Inverness in 1890, 1901 (when the wall of the Caledonian Canal near Dochgarroch cracked) and 1934.

Earth tremors could cause transient but violent disturbances of the Loch's surface and could explain reports of 'a noise like a train', followed some time later by the crash of breakers on the shore. The notion that earthquakes could mimic a Monster was briskly dismissed by Constance Whyte in 1957, but Luigi Piccardi believes that he has found a historical correlation between earth tremors and Monster sightings.[20]

Root and branch

The long, dark object that glided out into the Loch early one morning in November 1933 caused great excitement among onlookers on Invermoriston Pier. Surveying the scene from his house above the Loch was the formidable Colonel W. H. Lane. Like everyone else, Lane was convinced that he was watching the Monster, until he fetched his binoculars. These revealed the Monster to be a large log, drifting with the wind.[21]

This was not the only log to masquerade as something more exotic. Several years earlier, a fabulous creature had been spotted in Nahuel Huapi Lake in northern Patagonia. Thanks to all the money raised to hunt down this 'living plesiosaur', it was captured and turned out to be a tree trunk.[12]

Floating logs or tree trunks could simulate a Monster, especially if magnified and distorted by refraction, or caught in the grip of the stray remnant of a seiche and seemingly swimming against the wind. Those fooled, at least momentarily, included observers on Peter Baker's 1960 Oxford–Cambridge expedition, and Elizabeth Montgomery-Campbell, who began the search for the water monster 'Morag'.[22]

Waterlogged tree trunks may travel long distances in mid-water. In 1934, the one that got away from an angler at Abriachan turned out to be a submerged log. After a marathon two-hour struggle to land what can only have been the Monster, the fishing line snapped. The lure was found weeks later, embedded in a waterlogged tree that had blocked the entry to Dochgarroch Lock.[23] It has also been suggested that waterlogged tree trunks rotting quietly in the depths could be lifted to the surface by gases; however, the cold, acidic water of Loch Ness may discourage such vigorous decomposition.[24]

Maurice Burton believed that many 'Monster' sightings were of masses of rotting vegetation, lifted to the surface and propelled across the water by methane produced by decay. To support the theory, Burton published a rather bad photograph of something thought to be a vegetable mat that popped up in Loch Lochy and promptly sank again. Vegetation mats are well described in Scandinavian lakes fed by rivers into which timber mills dump their debris, but Constance Whyte vehemently denied that such things ever occurred in Loch Ness.[25]

Empty vessels

Many types of boat – yachts, ocean-going trawlers, passenger ferries and rowing boats – use Loch Ness and the Caledonian Canal. A 12–15-foot open boat with a small outboard motor was the workhorse of the Loch during most of the twentieth century, and under some circumstances could be mistaken for a large aquatic animal. From a mile away, with or without a crew, the boat is just a blob to the naked eye.

Tim Dinsdale's 1960 film of the humped object zigzagging away across the Loch inspired a new generation of believers in the Monster. The benign but cynical Richard Frere always believed that Dinsdale had simply filmed a boat.[26] To head off that criticism, Dinsdale wisely went back to the same spot a few hours later and filmed a boat that followed a similar course to his Monster. He argued that the boat looked quite different and travelled more slowly than the object and, as triumphantly reported by the LNIB, the Joint Air Reconnaissance Intelligence Centre concluded that the object filmed was not a boat and was 'probably animate'.[18]

The list of inanimate items that could be mistaken for a Monster is completed by floating tar barrels. These were dumped in their thousands into the Loch during the extensive roadworks on the A82 in 1932-33 and were picked out by Alex Campbell as objects which optical illusion can transform into a 'grotesque shape'.[27]

Animal magic

And so to flesh and blood Monsters that are definitely, rather than 'probably', animate. Theorists in search of plausible Monsters have climbed out along several branches of the tree of animal life towards fish, amphibians, birds, mammals and even invertebrates. In some cases, the candidates come under the heading of 'life, but not as we know it'.

Queer fish

Fish are obvious candidates for a Monster which, if it needed to breathe air, ought to be seen more often at the surface. Several large fish are found in Scottish waters, but are either the wrong shape or, crucially, live in the sea. As well as the basking shark (as much as 30 feet long and four tons in weight) and conger eel (up to ten feet), we can exclude the rare oarfish (30 feet), which is so flat as to be almost two-dimensional, and the portly sunfish (up to a ton in weight) which looks like a potato stuck on a dagger.

Big fish that thrive in fresh water include salmon and eels, but a sole specimen of either is not big enough. The heaviest salmon ever recorded in Scotland (70 pounds, nearly twice the weight of the largest caught in Loch Ness) was only four feet long. However, salmon en masse could be more promising. While swarming, they can form long lines which snake across the surface and could create the illusion of a large underwater creature.[28]

Eels, once present in vast numbers in the Loch, hold greater potential. They can undulate their way across dry land as well as through mud and water, and sometimes lie on the surface, showing a series of humps and a projecting head. Unfortunately, size matters. Roy Mackal conducted detailed studies of the dimensions and chemical composition of eels caught at various depths in the Loch, only to conclude (after pages of mathematical formulae) that they were never going to exceed five feet.[29]

The eel hypothesis was enthusiastically endorsed for a time by Maurice Burton, but not by a specialist colleague at the Natural History Museum who was nicknamed 'Eel Man' by the popular press. Denys Tucker identified the humped creature which he saw in the Loch in 1959 as a plesiosaur, not an eel.

This leaves two fish which can grow to monstrous proportions and have an appearance to match. The European catfish is a shockingly ugly fresh-water bottom-dweller, which reaches nine feet in length and weighs up to 200 pounds. No part of it matches the Surgeon's Photograph, but it could have been the gigantic frog-like horror which was squatting on a ledge 30 feet down and scared a hardened diver witless in 1880.[30] However, catfish have never been reported in Scottish waters.

The sturgeon can grow to 18 feet and weigh a ton. With its long snout, serrated back and armour-plated flanks, it looks almost as prehistoric as the coelacanth. These fish are at home in both salt and fresh water, and can find their way far inland: the storm drain in Doncaster, where a six-foot specimen was harpooned in 1931, is 30 miles from the sea. In 1949, a five-foot sturgeon was caught in the locks of the Caledonian Canal. Adrian Shine has suggested that some Monster sightings could be explained by a large sturgeon swimming just below the surface with its jagged back and snout projecting.[8] This recalls the 'crocodile' which Miss MacDonald watched in the River Ness in 1933 – a report Gould discounted because it looked nothing like the common impression of 'X'.[31] Intriguingly, 'crocodile' was a common instant reaction of visitors on seeing the sturgeons which Shine kept in a pond at the Loch Ness Centre.[8]

Eye of newt

The identity of the Monster was no mystery to the intrepid Colonel W. H. Lane, the scholarly soldier who was as much at home excavating the ruins of Babylon as defending the Empire in Burma. The picture of his favoured candidate graced the cover of his pamphlet, *The Home of the Loch Ness Monster*, published in 1933 several months ahead of Gould's book.[32] This was the giant salamander.

Most salamander species are just a few inches long, such as the rough-skinned newt which an Oregon man swallowed in 1979 as a bet to prove that it was not poisonous (it is, and death prevented him from claiming his prize).[33] The giant salamander which Lane shot in the hills of Burma, where it had blundered into a quiet morning of trout fishing, was about five feet long. The animal is a fine example of a living fossil; remains from the Swiss Alps suggest that it has changed little in the last 20 million years. However, nothing else about the giant salamander fits the role. Non-fossilised ones have never come closer to Scotland than Vladivostok; its habitat and behaviour are all wrong; and although a giant among its own kind, five feet long is just not gigantic enough.

Birds of a feather

Cormorants were transformed by refraction into Alex Campbell's 30-foot Monster with its six-foot neck and agitated head movements. Similarly magnified, a cormorant fishing with its head below the surface can become a large humped creature. These birds can cover surprisingly long distances underwater and are masters of unexplained appearances and disappearances.

Other avian pseudo-Monsters have included geese, swans, ducks and divers, all resident or regular visitors to the Loch. Dick Raynor's 1967 film of a humped object trailing a hefty wake was one of the LNIB's pinnacles of achievement (and still is, according to some websites). Some 20 years later, Raynor spotted something that he believed to be identical, crossing calm water in Dores Bay. This time, powerful binoculars showed the 'hump' to be a female merganser (a small diving duck) swimming along with a brood of ducklings trying to keep up and climb on to her back. Other 'Monsters' have turned out to be lines of geese swimming in formation or flying close to the surface, a group of mergansers squabbling, and a pair of swans on the other side of a headland.[34]

The highest order

Various mammals could, in theory, adapt to life in the cold dark waters of Loch Ness – although, without a fundamental overhaul of their basic physiology, even the deepest divers would have to surface regularly to breathe.

Candidate Monsters have included marine mammals, notably whales, porpoises and seals. The word 'marine' is loaded with significance because of the need to explain how a sea-going species could enter Loch Ness and thrive in its fresh water. Very rarely, porpoises have been reported in the Loch, but their appearance and behaviour are obviously non-Monstrous.[35] Nonetheless, the dorsal fin of a bottle-nosed porpoise has been labelled as the Monster in several books, and someone with real imagination suggested that the Surgeon's Photograph shows the dorsal fin of a sick killer whale, swimming backwards.[36]

Various people have proposed that the Monster is a seal. The bearded seal shot in the Beauly Firth in 1904 was an extreme rarity, but common and grey seals are regular features of the Scottish seascape. They are adventurous, sometimes following rivers up to 50 miles inland. Constance Whyte and Maurice Burton were wrong to insist that no seal could ever reach Loch Ness. Hard evidence that they do includes photographs, a cine-film taken by Dick Raynor and the corpses of those unlucky enough to meet anglers.[37]

However, seals look nothing like the Monster, with their dog-like features, short, fat necks and habit of putting their heads out of the water to see what is going on. They are also too small. Even a dead bull elephant seal dumped on the Loch's shore in 1972 did not fool anyone for long – despite the post-mortem facelift and dental work to try to make it look the part.

'A large seal' was Marmaduke Wetherell's last-resort Monster diagnosis just before fleeing back to London in January 1934. This assertion was taken as seriously as his other utterances. The editor of the *Inverness Courier* wrote some months later, 'The seal hypothesis is now dead',[38] and nobody stepped forward to give it the kiss of life.

Occasional mammals resident in the Great Glen have made a convincing Monster. The ponies which masqueraded as water horses and terrified the

impressionable inhabitants of Lochend in 1852 are not the only large quadrupeds that can swim long distances.[39] In 1959, an experienced naturalist
was shaken to see a 'prehistoric' animal that he 'could have accepted as a
plesiosaur', crossing a lake in the Highlands. It proved to be an exhausted
roe deer, barely able to lift its head clear of the water.[40]

Young stags can cause confusion. Only mature males sport magnificent
'Monarch of the Glen' antlers; two-year-olds have short, unbranched horns
which at a distance look like stalks. These could explain the 'antennae'
reported on the Monster's head by some eyewitnesses, notably Mrs Greta
Finlay at Aldourie.[40]

And so to otters, on which Maurice Burton put his money in trying to
explain away the Monster as an unremarkable animal glimpsed under confusing conditions by the inexperienced or unprepared. Otters are common
in the Great Glen and nimble in water and on land. Slicing through the
water at up to ten feet per second, they can cross half the Loch's width on a
single dive. They can leave impressive trails of bubbles; shortly after D-Day
in June 1944, two otters in the Courtelles River in Normandy were misiden-
tified as a midget submarine and narrowly escaped being depth-charged.[41]
Otters do not always look like their photographs in wildlife books. They
can make a surprisingly long neck, especially when treading water to look
around, and may appear frightening to the uninitiated.

Otters are social animals, fond of family outings. Richard Elmhirst,
the canny Director of the Scottish Marine Biology Unit, was conned into
seeing a multi-humped water creature by a family of otters, swimming in
line behind the male (Figure 13) – until he trained his binoculars on them.
The same phenomenon has been recorded many times, from the Isle of
Mull and elsewhere.[42]

To Constance Whyte this was otter nonsense, mainly because of size. The
biggest otters are as long as a man, but still far short of a 20-foot Monster.
Reports of a 12-foot otter from South America did not impress Gould, who
wrote sarcastically: 'This seems very likely indeed.'[43]

Beyond the fringe

Conventional science has taken us as far as it can in our quest to solve the
mystery of the Loch Ness Monster. Arguably, the most exciting part of the
journey lies ahead, in a landscape that is brighter and more exotic than the
one behind us. We are about to enter the domain of cryptozoology – the
study of 'cryptids', animals that may or may not exist. In places, the path

is not paved with hard facts but with anecdote and legend. Depending on your viewpoint, you may find the going slippery and treacherous, or a strangely liberating experience that will put a spring in your step.

Figure 13 Family of otters swimming in line and simulating a large, multi-humped aquatic animal. Even experienced observers, such as Dr Richard Elmhirst of the Scottish Marine Biological Station, could be misled. Illustration redrawn from Gould *The Loch Ness Monster and Others* (1934) by Ray Loadman.

Cryptozoologists formally defined their field as a science in the 1950s, but the roots of cryptozoology run back much further. Man's fascination with fantastic animals, real and imagined, is evident throughout recorded history – Stone Age cave paintings of strange beasts; the bones of the sea dragon which Perseus killed while rescuing Andromeda, proudly displayed by showman Marcus Scaurus in ancient Rome; tall stories of 'tailed men', which persuaded Linnaeus to add *Homo caudatus* to his list of living species in 1758; the crowds awestruck by the concrete dinosaurs from the Great Exhibition of London, 1854; *Zoo Quest for a Dragon*, 1958; and *Jurassic Park*, 1993.

Cryptids and the people who study them are a mixed bag. At one end of the spectrum are conventional biologists and ecologists, who generate ideas, gather data and abandon a hypothesis if the evidence proves it wrong.

Their results – whether a mathematical prediction of how many unknown large marine species remain to be identified, or the discovery of a lungless frog living under a waterfall in Borneo – are presented and picked over at scientific meetings and published in peer-reviewed journals.

Cryptozoologists at the other extreme are just as dedicated, but see no need to observe the rules that bind ordinary researchers. They have no problem in stepping around whatever baggage the scientific establishment tries to dump in their way. The fruits of their labours might be found in *Fate* magazine, sandwiched between articles on UFOs and alien abduction.

The biodiversity of cryptids is apparent from a glance at the index of the many 'field guides' devoted to them. Supposedly extinct species that may still cling on somewhere sit cheek-by-jowl with creatures that, to date, exist only in anecdote or legend. For example, *Mysterious Creatures: A Guide to Cryptozoology* ('a reliable source of information') contains detailed entries on the ivory-billed woodpecker (painted by Audubon in 1829, declared extinct in 1944, possibly still surviving in forests in Florida); the green-scaled, three-fingered Lizard Man from the swamps of North Carolina; and the Mongolian Death Worm, a five-foot horror that can kill a man by both lethal injection and electrocution.[44]

In principle, cryptozoologists should be pushing at an open door, as mainstream scientists are so good at deluding themselves about how much they know. Georges Cuvier, the great French zoologist and anti-evolutionist, was quickly proved wrong when he declared in 1812 that no large animal species remained undiscovered. Another 220 large marine animals have been discovered since Linnaeus closed his list in 1758, 120 of them after 1800. The rate of discovery has flattened off, but Charles Paxton, an ecologist at the University of St Andrews, predicts that another 50 large species are still waiting to be found in the oceans.[45]

It is relatively rare for a cryptid to come in from the cold limbo of uncertainty and be 'welcomed into the ranks of known animals'. Nonetheless, cryptozoology has scored some palpable hits. Careful investigation of anecdotes and legends that would have been laughed away by mainstream scientists has turned up the giant squid (1893), okapi (1901), Komodo dragon (1910) and, in 2003, the Giant Terror Skink.[44]

The case of the giant squid illustrates the obstacles faced by cryptozoologists – the porous frontier between legend and reality, the lottery of serendipity and the prejudice of the scientific establishment. In 1857, the Danish zoologist Johan Steenstrup claimed the existence of a massive squid, ten times bigger than any ever seen before, on the basis of fishermen's legends and what appeared to be an oversized squid beak, retrieved in 1839 from a rotting carcass on a beach in Jutland. Steenstrup coined the name *Architeuthis* for this awe-inspiring but still hypothetical creature, and was promptly ridiculed by the French Academy of Sciences for having contradicted 'the great laws [governing] living Nature'. Steenstrup's reputation was restored in 1873, when a 60-foot giant squid was captured off the coast of Newfoundland.[46]

Large unknown animals that inhabit freshwater lakes figure prominently in the annals of cryptozoology. Like Loch Ness, many of the lakes lie in the northern hemisphere between latitudes 50° and 60° N. Notable examples are Okanagan (Canada), Storsjön (Sweden), Lögurinn (Iceland), Varota (Siberia), Duobuzha (Tibet) and the enigmatically named Dildo Pond (Newfoundland). In Great Britain, large lake-dwelling animals have been reported in three English lakes, with 13 in Wales, 55 in Ireland and 42 in Scotland other than Loch Ness. Most lake monsters are described as serpentine, with or without frills (e.g. Ogopogo, from Okanagan Lake) or worm-like (such as the Lagorfljótsomurinn which inhabits Lake Lögurinn). Others resemble horses or cattle; an identity crisis apparently afflicts the Tibetan 'hippoturtleox' in Lake Duobuzha.[47]

Of all these, one stands head, neck and shoulders above what Constance Whyte called 'ordinary run-of-the-mill lake monsters'. 'Bobby' is not just the best documented and most durable of the large unknown water creatures. It is at the very top of the cryptozoological pile: the 'Number One Cryptid'.

Nessie

The fact that 'Bobby' is an alternative name for the animal known scientifically as *Nessiteras rhombopteryx* is one of the revelations in its potted biography in *Mysterious Animals*.[44] The animal is described as 10–45 feet in length, with a blunt-ended 5–6-foot tail and a 4–8-foot neck topped by a small, flat head carrying two horn-like protrusions. It has oval eyes. The body is humped (usually one to three, sometimes up to eight), and its skin is rough like an elephant's, dark grey or black and possibly with white underparts. Sightings on land suggest that it has 'four short, thick flippers'. Most active during the day, the animal often swims just beneath the surface, raising a heavy V-shaped wake, but can also sink perpendicularly and sometimes 'lashes the water energetically'. It is believed to eat fish.

This thumbnail sketch is followed by a detailed list of the principal sightings, photographs and film records, backed up by a comprehensive bibliography that begins with Adomnán (circa AD 580), and ending with the link to the Loch Ness webcam.

Two pages are devoted to 23 broad categories for the possible nature of the Monster. All the usual suspects are there – seals, otters, stags, assorted birds, mirages, logs, vegetation mats – together with a couple of more imaginative suggestions from which readers of this book have been protected until now (a dragon popping in from a parallel universe, and 'an alien pet left by space travellers').

Most coverage is devoted to 'a surviving plesiosaur', as proposed in 1960 by Dr Denys Tucker and subsequently endorsed by countless others. However, the plesiosaur is not the only cryptozoological solution to the mystery of the Monster's identity. Ted Holiday's initial notion of a monstrous aquatic slug mutated into a 50-fold blow-up of *Tullimonstrum*, the odd fossil from the coal mines of Illinois which left no relatives anywhere when it died out 300 million years ago.[48] Roy Mackal also took thinking about candidate Monsters into new realms, with a 'hypothetical' amphibian based on the extinct *Eogyrinus*, and a relative of Steller's sea cow, which barely had time to be recorded before slipping into extinction. Unfortunately, both parent species had notoriously short necks and, as a learned zoologist explained to Mackal over lunch in London, sea cows 'really don't like to have their necks stretched'.[49]

Mammalia had one final trick up its sleeve – the long-necked seal (*Megophias megophias*) described by James Parsons FRS to the Royal Society in 1751.[50] This animal really did have a long neck (only slightly shorter than its body), and was believed by many to be the sea serpent of mariners' folklore. Sea serpents have been reported in every ocean, notably in Chesapeake Bay, the Indian Ocean, the Sound of Sleat off Skye, Liverpool Bay, and swimming quickly away from Butlin's holiday camp in Skegness.[51]

Sea serpents were made scientifically semi-respectable in Antonie Oudemans' book *The Great Sea-Serpent* (1892), in which he named the long-necked seal as his prime candidate. Seven decades later, Bernard Heuvelmans agreed in *In the Wake of the Sea-Serpent* (1965). Sandwiched between these two authorities, Rupert Gould laid out his belief in *The Sea-Serpent* (1925) and aimed to prove in *The Loch Ness Monster and Others* (1934) that the Monster and the 'Others' (his beloved sea serpents) were one and the same – although he was not sure what kind of animal it might be. Antonie Oudemans was more decisive, stating in 1933 that the Monster was indeed a long-necked seal.[52] However, the *Inverness Courier* gave

Oudemans' opinion short shrift and Roy Mackal did not even mention the species in his book.

Understanding of the long-necked seal has not advanced since Parson's paper in 1751. His only information was the description of a skin by Francis Grew, who had catalogued the Royal Society's collection of 'oddities' over 60 years earlier. Parsons had no opportunity to check Grew's measurements, as the skin – the only hard evidence that this extraordinary animal ever lived – had gone astray in the meantime.[50]

Which leaves . . .

. . . the plesiosaur-like animal that millions of people around the world would recognise in an instant as the Loch Ness Monster: the 'prehistoric' creature which hovered on the periphery of probability, then rushed on to centre stage in 1934 with the Surgeon's Photograph. The unlikely point of principle on which Denys Tucker bet his career in 1960. The template into which slotted neatly the new evidence that came through during the 1960s and 1970s and which Robert Rines's underwater photographs fitted like a glove. And the conservation icon that was immortalised in Sir Peter Scott's ethereal painting of two courting Monsters suspended in the amber waters of the Loch.

Not everyone agreed, of course. Back in 1934, Gould thought it 'unlikely' that a plesiosaur was in Loch Ness. Dr Richard Elmhirst, Director of the Scottish Aquatic Biology Unit, went further, stating that the odds of a plesiosaur surviving to live in Loch Ness were about 17 million to one against.[53] Throughout the 1960s, hardened scientists who had made their reputations as plesiosaur researchers poured devision on the notion. And in 1976, Roy Mackal ranked plesiosaurs below amphibians in his list of candidate Monsters.

But for the real opinion leaders – Constance Whyte, Tim Dinsdale, Peter Scott, Robert Rines and Nicholas Witchell – this was it. How it got there needed some explaining, but there was no doubt that Loch Ness harboured a large aquatic animal that was descended from the plesiosaurs.

Hindsight and some wisdom

Is minig a bha an fhìrinn searbh ri h-innse.
Truth is often harsh to tell

Gaelic proverb

Much of the evidence for the Monster is impossible to challenge or verify, because it consists of eyewitness reports. There is relatively little metaphorical meat in the form of photographs, cine-films and sonar tracings, and no actual meat or even a single bone. So how solid is the evidence that a large, unknown animal species inhabits Loch Ness?

What they saw is what you get

Roy Mackal estimated that roughly 10,000 sightings had been reported up to late 1969, of which some 3,000 were well enough documented to deserve further analysis. He cut these down to just 269 'valid' observations, beginning with St Columba in AD 565 and ending with the Craven family on 6 August 1969. Mackal's 'valid' sightings make an impressive graph (Figure 14).[1] This shows a massive spike in 1933–34 at the birth of the modern era of the Monster, and smaller peaks during the LNIB's golden decade. Sightings continue to be recorded to the present day. After the LNIB folded in 1975, Rip Hepple maintained his regular 'Nessletters', while Gary Campbell set up his Loch Ness Sightings website in 1996.

Mackal and others believe that 80–90 per cent of reported sightings can be discarded as error, misinterpretation or invention. However, none of the experts has ever laid out the criteria for 'validity' and they disagree over numerous reports, beginning with the first observation on Mackal's list – Rupert Gould had 'neither the time nor the inclination' for the tale of St Columba and the 'sacrilegious' water monster.[2] This raises the broader question of what persuades us that someone really has seen the Monster.

The essential qualifications of a top-quality eyewitness are good vision, correct interpretation of what was seen, accurate reporting of the facts and a sound memory which is also immune to peer pressure, media hype and the inducements of money and fame. This can be a tall order, as people tend to see what they expect or want. This was neatly demonstrated when Richard Frere induced a crowd of holidaymakers to conjure up a Monster (including a beautifully drawn plesiosaur) out of boat wakes at Lochend.[3]

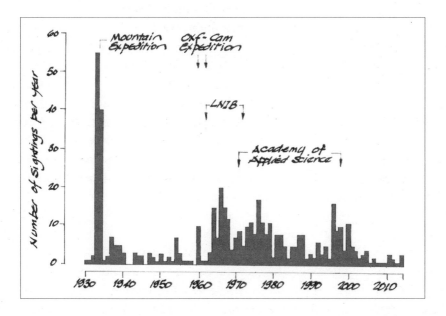

Figure 14 Graph of sightings of the Loch Ness Monster from 1933 to 2014. Data from R. Mackal, *The Monsters of Loch Ness* (1976), p. 87; H. Bauer, *The Enigma of Loch Ness* (1988) pp. 194–200; and G. Campbell, Loch Ness Sightings website (www.lochnesssightings.com). The timings of major expeditions are also shown. For comparison, between 30 and 60 people are struck by lightning each year in Britain.

Being a credible witness has nothing to do with intelligence, social standing or occupation – with the possible exception of those, like water bailiff Alex Campbell or steamer captain John Macdonald, who have spent decades on and around the Loch. Witnesses may look impressive because they are lords, ladies, counts, commanders, MPs, doctors, engineers or even a Nobel Prize-winner, but titles and qualifications are irrelevant when deciding whether to take them seriously or not.

Eyewitnesses may wave warning flags when telling their story. Seeing the Monster is such an extraordinary event that it should be seared on to the memory for ever, yet some observers seem remarkably vague about time, place or even the basic features of the creature itself. The bus driver who told his story to Philip Stalker in October 1933 dated his dramatic sighting as 'late August or the beginning of September'. Mr Spicer's prehistoric creature had grown from 5 to 6 feet to 25 feet by the time Gould interviewed him. R. K. Wilson could not 'place within a mile' where he took the Surgeon's Photograph. Even the Mackays' index sighting had slipped back from 28 April 1933 to 14 April when they told Gould about their 'Strange Spectacle'.[4]

Sceptics are bound to take an interest in anyone claiming to have seen the Monster more than once. The LNIB's decade of intensive surveillance reminds us that sightings are a vanishingly rare event – in fact, over twice as rare as the risk of being struck by lightning in Britain (about 30 cases per year; compare with Figure 14). A few lucky people have proved that being struck by lightning is not necessarily a once-in-a-lifetime experience, but what are the statistical odds against seeing the Monster twice? Or six times, as in the case of Hugh Gray? Or 12 times, for Winifred Cary, wife of the wing commander? Or 16 times for Alex Ross, harbourmaster at Temple Pier, and 18 times for Alex Campbell?[5]

Doubts may be raised about witnesses who stand to gain from reporting the Monster. The one-guinea reward which Sir Edward Mountain offered for each documented sighting may have been an irresistible temptation for some; this sum was roughly two weeks' pay for the ghillies and labourers hired as watchers, all of whom were all heading back to the dole queue after the expedition ended. And when anonymity was lifted from the couple who watched the 'Strange Spectacle', they stood revealed as Mr and Mrs Mackay, proprietors of the newly refurbished Drumnadrochit Hotel.

Finally, the witness's reliability or trustworthiness may not ring true. Miss Janet Fraser, proprietor of the Halfway House, described the Monster's 'glittering eye' and neck-frill 'like a pair of kippered herrings' from 1,000 yards away – details that could only be resolved at that range by a telescope powerful enough to make out the Rings of Saturn. Winifred Cary, veteran of 12 Monster sightings, was also a celebrated dowser, swinging a pendulum over a map to find her husband and other lost objects. Hugh Gray, who first photographed the Monster and subsequently saw it five times more, insisted that Royal Navy mines laid in 1918 were squashed 'as flat as pancakes' when pulled up from the bottom of the Loch.[6] The Benedictine Abbey at Fort Augustus, the source of many sightings, seemed unaware of the activities of a ring of paedophile monks who abused boys at the Abbey School. However, none of the above necessarily means that the witness did not see something extraordinary.

No matter how astute an observer of the human condition the inter-rogator believes himself to be, the decision to trust someone is entirely subjective. Peter Scott was prepared to stake 'all my knowledge of human relations' on the 'absolute honesty' of Bob Rines and his associates. A face-to-face meeting was enough to convince Sir Alister Hardy FRS that the photograph taken by bank manager Peter Macnab was genuine.[7]

Rupert Gould, the 'man who knew (almost) everything' and who was never contradicted on *Brains Trust*, was proud to report, 'I have never – I am glad to say – been compelled to discard the evidence of any witness be-cause I distrusted it'.[8] However, he occasionally harboured private doubts. In the margin of his proof copy of *The Loch Ness Monster and Others*, he wrote about one of the most striking sightings:

Were I re-writing the book, I should have dismissed this case. I think the Spicers saw a huddle of deer crossing the road. RTG.

But Gould made no changes to the proofs. The published version recounted his visit to interview the Spicers and concluded, 'I heard this story – and I became, and remain, convinced that it was entirely *bona fide*'.[9]

Gould's confidence was never dented in another celebrated sighting on land, by the earnest and sincere vet student Arthur Grant. 'A rumour reached me, from more than one quarter,' he wrote, 'that the story was a mere hoax . . . I am quite convinced that the rumour itself is groundless.'[10]

In 1975, 72-year-old garage owner Alec Menzies of Drumnadrochit told Dick Raynor that he had overheard Grant talking excitedly on the public telephone in the garage at around the time the story broke. Menzies alleged that Grant said, 'They've swallowed it!'[11] Even if true, the comment is open to many interpretations. As Grant was a veterinary student, 'it' could have been a horse pill or a yard of ale, rather than a cock-and-bull story about a moonlit night, a motorbike and a Monster.

Against the grain

In the absence of a whole specimen or even body parts, photographs and films of the Monster have been the next best thing to decisive evidence. Images such as the Surgeon's Photograph, the pictures taken by Hugh Gray, Lachlan Stuart, Peter Macnab and Bob Rines, and stills from Tim Dins-dale's cine-film are at the heart of all books and websites about the Monster and have converted more people into true believers than anything else.

Numerous other photographs have been passed off as the Monster, but so unconvincingly that they should never have been taken seriously. Early examples were two photographs that date from the first peak of Monster fever in 1934. They show a blunt, curved object and a dark triangular shape, both sticking out of the water (Plate 43). One was supposedly snapped near Fort Augustus, while the other is identified as a frame from the long-lost film taken by Captain James Fraser during Sir Edward Mountain's expedition. The first appeared in Witchell's *Loch Ness Story* (1974) and was 'positive evidence' for Roy Mackal (1976), although he struggled to explain which part of the Monster it might be. The second was reproduced in Henry Bauer's *The Enigma of Loch Ness* (1986), but with the warning that he regarded its provenance as 'doubtful in the extreme'.[12]

Each photograph shows only water and the object, and could have been taken anywhere, including the sea. In 2014, both were identified without hesitation by a professor of marine biology as the dorsal fins of a bottle-nosed porpoise and a basking shark, respectively.[13] How they eluded diagnosis at the time is a mystery, as both species are common in the seas around Britain. Porpoises have very rarely been seen in the Loch[14] and could conceivably have been photographed in honest error, but their unmistakable behaviour would immediately have given the game away. Basking sharks could never enter the Loch or survive in its fresh water.

Other obvious hoaxes included the handiwork of Tony 'Doc' Shiels and Frank Searle. Shiels, who probably tops the fraudsters' roll of honour, is best known for the 'Muppet' Monster which graces the cover of Tim Dinsdale's *Loch Ness Monster* (Plate 40). Frank Searle's 11-year vigil on the shores of the Loch yielded several front-page photographs of Monsters showing one or more humps and sometimes a long neck and a small, reptilian head (Plate 33). The more obvious fakes involved branches and a brontosaurus from postcards on sale in Inverness. In one instance where the site could be pinpointed, a post sticking out of the water had apparently acted as a support for Searle's do-it-yourself Monster kit.[15]

More controversial was Peter O'Connor's flashlit photo of the Monster's flank and long neck (Plate 20), taken while he sneaked up noiselessly through the shallows. This divided opinion from the start. Constance Whyte, Maurice Burton and Peter Scott instantly dismissed it as bogus, but Tim Dinsdale praised it as 'without doubt the closest and most revealing still photo ever taken – and therefore of unique importance'.[16] The photograph provided rich pickings for sceptics. The 'beast' gave no impression of ploughing through the water, as O'Connor claimed. Its flank appeared unnaturally smooth and the neck inanimate. The light was all wrong. At

6.30 a.m. on 19 May, the sun had already been up for two hours over Loch Ness – so why was flash necessary? O'Connor blamed his cheap camera, but even a Brownie Flash 20 could not turn the morning sky black.

Years later, Maurice Burton claimed that he had visited the spot shortly after O'Connor took his photograph and found the charred remains of white plastic sacks and a curved stick which looked uncannily like the neck of O'Connor's 'Monster'.[17] It would be hard to make O'Connor's Monster out of plastic sacks, but they could have acted as floats. Dick Raynor believes that the creature's flank resembles a folding canoe of the type owned by O'Connor. He speculates that the sky was blacked out to cover up an accomplice's hand, holding the canoe upside down on the surface behind the stick, which was anchored under the water.[18] O'Connor never admitted to faking his snapshot, and Dinsdale never retracted his praise of 'the closest and most revealing still photo ever taken'. However, the photograph did not appear in subsequent editions of Dinsdale's books.

Nicholas Witchell was bolder, condemning the photographs by Searle and O'Connor as 'fraudulent' in the first edition of his book (1974).[19] This was a courageous act for a 21-year-old, especially as Searle already had an unpleasant reputation; Searle's response included verbal abuse and explicit threats of violence serious enough for the police to become involved.[20] Witchell also faced criticism from another quarter, which suggested that exposés of eye-catching 'evidence' were not necessarily welcomed by the pro-Monster camp. He wrote, 'I have been strongly advised by many people simply to ignore these photographs',[19] but never revealed who they were.

All the above were relatively easy picking: low-hanging fruit, and some of it clearly rotten. However, the sceptics' assault on photographic evidence of the Monster did not end there.

Hugh Gray's photograph
Foyers, 12 November 1933 (Plate 1)

Gould gave this 'interesting and undoubtedly genuine' photograph star billing as 'The Loch Ness Monster' in the first plate of his book. Mackal classified it as 'positive evidence', echoing Whyte, Dinsdale and others. Believers have made much of the sinuous dark object which Gray said was 40 feet long and 'lashing about furiously'. In it, Rupert Gould saw his beloved sea serpent and Ted Holiday a gigantic slug. Recently, Roland Watson has argued that the open mouth of a massive fish or eel is visible at the right of the dark object.[21]

Others were less impressed. At the time, cynics wrote it off as a bottle-nosed porpoise or floating wreckage, while Maurice Burton later dismissed it as an otter (backing up his claim with a truly dreadful line drawing).[22] To some eyes, the 'spray' above the middle of the object is actually the out-of-focus head of a Labrador-type dog, showing its muzzle, both eyes and right ear. The dog is not only visible to sceptics. In 2012, a group of nine-year-old children quickly pointed it out when Tony Harmsworth showed them the picture with no other information.[23]

Curiously, the picture that would make Gray famous remained undeveloped inside the camera and in a drawer for three weeks. Cynics might argue that this would be expected if Gray assumed that he had only photographed a dog with a stick, but Gray had another explanation. While being grilled by local dignitaries, he insisted that he was afraid of being ridiculed by his friends for having photographed the 'notorious Monster' and thought he had missed it anyway.[24]

A Monster or a dog with a stick in its mouth? Eighty years on, sceptics and believers remain entrenched in their views, and some believe that the jury is still out.

Lachlan Stuart's photograph
Whitefield, 14 July 1951 (Plate 13)

Had he lived another four years, Rupert Gould would have been thrilled to learn that two men had watched a long-necked, three-humped Monster charging up the middle of the Loch – and had photographed the three humps just 50 yards offshore. Instead, it was Constance Whyte who went to interview the woodsman Lachlan Stuart, three days after the picture was published in the Scottish *Sunday Express*. What she learned persuaded her to put her neck on the block: 'I could not put forward this photograph with more confidence if I had taken it myself.'[25]

Two hardened hacks from the *Express* also tried to pick holes in Stuart's testimony, during a three-day assignment at Whitefield. Why had Stuart been so blasé about seeing the Monster so close? Answer: the animal had been seen several times recently. Why did he not rush to have the film developed? Answer: he wanted to finish off the roll with snaps of his wife and children. Could the picture have been faked? No. Stuart's simple box camera had a mechanical defect that made this impossible, and the film had been removed and processed under the gaze of a reporter. The journalists also left convinced that Stuart, a stalwart employee of the Forestry Commission for 15 years, was absolutely honest and that his photograph was a true likeness of the Monster.[26]

Nicholas Witchell (1974) and Roy Mackal (1976) both believed that the photograph was genuine, although Mackal thought it showed three separate Monsters. Even Maurice Burton hailed this as 'the most important photograph' (when he interviewed Stuart in September 1951, before his faith in the Monster disintegrated).[27]

However, there was a sting in the tale. In the 1989 edition of Witchell's book, Stuart's photo had become 'a hoax', the honest woodsman's story was 'a fabrication', and Constance Whyte had fallen prey to 'a most uncharacteristic lapse of judgement'. An anonymous local resident had told Witchell that, just three weeks after the photograph was published, Stuart confessed that the whole thing was a scam. Proof was provided by three bales of straw and tarpaulins, hidden in bushes on the shore. Stuart's confidant later turned out to be Richard Frere, lover of Loch Ness and Monster sceptic. Frere swore to keep Stuart's confession secret and did so until long after Stuart had disappeared from the area. Finally, 35 years later, Frere decided that it was time to set the record straight.[28]

Peter Macnab's photograph
Urquhart Bay, 29 July 1955 (Plate 14)

This is one of the brightest jewels in the believers' crown – and 'the most important piece of evidence yet' which converted Sir Alister Hardy FRS, the formidable Professor of Zoology at Oxford, into a believer.[7] It was one of only two photographs of the Monster which Peter Scott presented to his expert panel in 1960.

It is an arresting picture, with two long black humps that are not much shorter than the 55-foot tower of Urquhart Castle. The photograph's credibility comes from Peter Macnab's standing in society as well as proof of location. In Witchell's *Loch Ness Monster* (1989), Macnab is pictured as a pillar of the community in a smart suit, talking earnestly on the phone. 'He is a bank manager,' observed Professor Hardy, for whom this was a guarantee of probity.

There were, however, some curious aspects to the story. This photograph holds the record for the length of time it remained dormant before being awakened by the kiss of publicity. It was taken in mid-summer 1955 but did not figure in the first edition of Constance Whyte's *More Than a Legend*, published two years later. It was reproduced in the third edition (1961), with a lengthy caption which Whyte lifted from a letter which Macnab sent her.

Otherwise, 'Macnab' does not appear in the text or even in the index. 'The full story of what happened' was explained by Witchell (1974). Because of 'diffidence and fear of ridicule', Macnab sat on his photo for three years until October 1958, when he sent it to the *Weekly Scotsman* for publication.

Macnab's story, as reported verbatim by Witchell, reveals some further twists.[29] There were originally two photographs, one taken with an Exacta 127 single-lens reflex camera that had interchangeable lenses and the other with a child's-play Kodak Brownie. Macnab never explained why he dropped a high-quality camera with a telephoto lens in favour of one that would reduce Urquhart Castle to a small blob in the middle distance. His son saw nothing, because he had lifted the bonnet of the car and was checking the engine while his father, 'shaking with excitement' and shouting, took his photos. The lad was evidently not gifted with lightning reflexes. He did not look up while Macnab senior changed the Exacta's standard lens for the telephoto (both lenses had fiddly screw mountings); then took a photograph; then put down the Exacta; then lifted the Brownie and took the second photograph.

Later, Macnab revealed that he destroyed the second photograph from the Brownie, because his friends had made fun of him. The surviving photograph shows some unusual features. It is not blurred, as might be expected using a 135mm telephoto lens with shaking hands, no tripod and a relatively slow shutter speed. Also, the two-humped 'Monster' looks blacker and more clear-cut than everything else. The picture aroused the suspicions of Roy Mackal. In 1976, he asked for a copy of the original print from Macnab. This differed from the version in *More Than a Legend* (bushes in the foreground were missing from the print). Mackal wanted to know why; Macnab said that this must be another photograph which he had taken just after the first. However, the 'Monster' was identical and in exactly the same spot – and Macnab had told Witchell that he had destroyed the second photograph. Mackal also noticed that the tower of the Castle was plumb vertical, whereas its reflection was tilted a few degrees off true. He clearly suspected Macnab of faking the photograph, but even after a four-page, blow-by-blow analysis was unable to 'unravel the mystery'. Nonetheless, he dismissed the photograph as 'unacceptable as evidence'.[30]

By contrast, Nicholas Witchell saw no mystery to unravel. In the 1989 edition of *Loch Ness Monster*, published 13 years after Mackal's attempted hatchet job, the Macnab photograph was still a genuine 'Loch Ness classic' and Mackal's misgivings were not mentioned.[29]

Other interpretations of the Macnab photograph include interfering bow-waves from two or more trawlers that had passed out of sight beyond the Castle, and a pair of humps inked in over the wake. The tilted reflection of the Castle's tower is unexplained – as are the truncated patches of sunlight on the beach below the Castle, which appear to end along a sinuous line that could almost have been cut with a scalpel.

The Surgeon's Photograph, taken by Dr Robert K. Wilson
North of Invermoriston, 19 April 1934 (Plate 6)

Once seen, never forgotten. For hundreds of millions of people around the world, this was the ultimate proof that the Monster exists. Even non-believers knew what the picture showed. Indeed, the Surgeon's Photograph became one of the instantly recognisable images of the twentieth century.

The picture draws its power from its stark simplicity, the slightly sinister silhouette at its centre and the unsettling realisation that this is not like any animal currently alive on Earth. Predictably, Rupert Gould saw a sea serpent, and Ted Holiday his slug-like Great Orm. The majority view, endorsed by Constance Whyte, Tim Dinsdale, Denys Tucker, Nicholas Witchell and many others, was that this was a plesiosaur-like creature which, against astronomical odds, had cheated extinction and now provided a hotline back to the Age of the Dinosaurs.

Naturally, the cynics had their own theories, including the dorsal fin of a killer whale (perhaps sick and swimming backwards); the trunk of a submerged elephant which had said goodbye to the circus; and the limb of a submerged tree. Maurice Burton dismissed it as the tail of a diving otter, while Roy Mackal thought it was a water bird of unidentifiable species.[31]

A great deal hinged on the scale of the object. Everything happened too fast for Dr Wilson, fiddling with his unfamiliar camera, to estimate its size. The *Daily Mail* did not help, by blowing up the centre of the picture so that the Monster filled the front page. The relative size of the ripples around the silhouette provided clues that were read quite differently by various experts. Peter Scott believed that the ripples were small and that the neck stood only 12–18 inches out of the water. Tim Dinsdale, eyeballing the photograph at arm's length and invoking his statistical magic, made it six to eight feet high.[32] A detailed trigonometric analysis of the ripples produced a smaller estimate of 1.2 metres (four feet), which was almost twice as tall as the 0.7 metres (28 inches) calculated from exactly the same ripples by someone else.[33]

The authenticity of the photograph was widely taken for granted, largely because of the photographer's credentials. R. K. Wilson was a reserved man 'of unquestioned character and veracity', bound by the honourable traditions of the medical profession and apparently allergic to publicity. His reticence made a refreshing contrast to self-publicists such as Marmaduke Wetherell and he was distinctly unsensational when describing his 'strange animal': 'I am not able to describe what I saw. As I finished [taking photographs], the object moved a little and submerged.' Inevitably, some spoilsports suspected a hoax. In a letter to Tim Dinsdale, Peter Scott wondered if the ripples had been made by someone jiggling the object on the end of a stick.[32] And, on closer inspection, the Surgeon turned out to be a peculiarly inconsistent witness and his story began to wobble.

The photograph had probably been taken on Thursday, 19 April (not April Fool's Day, as some suggested) – but from where? Nowhere matching Wilson's description could be found by the road north of Invermoriston. Had he used a fancy quarter-plate camera and telephoto lens, loaned by his friend Maurice Chambers, or, as reported in the *Daily Mail*, his own reflex camera? Had he taken the photograph at 7.30 a.m., or around midday? Was he standing 100 feet or just three feet above the water, and had the 'commotion' been 100 yards or 300 yards offshore? Was the Monster really heading north towards Inverness, when the photograph showed it pointing right, i.e. south towards Fort Augustus? And who was Wilson's unnamed friend, who may or may not have shouted out, 'My God, it's the Monster!'[34]

It was almost as though Wilson had not been there.

Wilson provided some answers in a peculiar letter replying to questions from Constance Whyte, while she was drafting *More Than a Legend* in 1955. He admitted that 'there was a slight doubt and suspicion as to the authenticity of the photograph', but blamed any doctoring on either the pharmacist in Inverness or his own mysterious friend, who was evidently a woman but not Mrs Wilson.[35] Whyte's reaction was also strange. She destroyed Wilson's original letter, keeping only an edited version from which she had deleted names and personal details. In her book, Wilson was anonymised and referred to only as 'Dr -------' and 'a London surgeon' – even though anyone glancing at the *Daily Mail*, other newspapers of the day or Rupert Gould's book would have instantly found the name of Robert K. Wilson, gynaecological surgeon, of 42 Queen Anne Street, W1. For whatever reason, Wilson's disclaimer remained a secret between him and Constance Whyte.

Views about the Surgeon's Photograph should have changed forever on 7 December 1975. For various reasons, though, anybody who mattered was thinking about other things. The news was dominated by a dramatic siege in Balcombe Street, Chelsea, and anyone with interest to spare for the Monster was fully focused on the meeting to unveil Robert Rines's photographs at the House of Commons a couple of days later. The revelation that might otherwise have made the front pages provoked a brief letter from a journalist, O. D. Gallagher, and promptly sank into obscurity. It might never have been exhumed but for a huge coincidence: Gallagher's wife Betty became the curator of the Loch Ness Exhibition in Drumnadrochit, which Adrian Shine designed. After her husband died in 1991, Betty Gallagher showed his papers to Shine, who found the letter and set off down a trail that had gone cold for 16 years.[36]

Gallagher's letter to the *Sunday Telegraph* argued that the Loch Ness Monster existed and was a gigantic invertebrate – and that a recent article about it by 'Mandrake' (the television critic and writer Peter Purser) was irrelevant. Mandrake's piece turned out to be entitled 'The making of a monster' and its key sentence read: 'The man who took the most famous Loch Ness monster picture of all time comes clean at last: it was a fake'.[37] The man in question was Ian Wetherell, son of the film star and adventurer Marmaduke. Ian followed his father into acting, retiring in the late 1950s to run a pub in Chelsea which Purser happened to frequent.

Ian Wetherell never explained why he wanted to destroy an icon. He told Purser that his father came home angry after the *Daily Mail* expedition collapsed in January 1934 and said, 'We'll give them their monster'. So they made a model head and neck and stuck it on top of a clockwork toy submarine that cost half a crown. Both Wetherells and their amateur photographer friend Maurice Chambers took the Monster up to Loch Ness in Ian's car. They found an inlet where the ripples would look like full-size waves and set up the submarine with the neck and head sticking out of the water. Ian Wetherell took five photographs with his Leica camera – at which point a water bailiff appeared and Marmaduke Wetherell sank the Monster with his foot. Later, Chambers sent off the film to be developed.[37]

Purser did not say which photograph had been faked, but did not need to: 'The picture scoop of the century . . . a long reptilian neck and a little head', taken in 1934. It could only be the Surgeon's Photograph – but where was the Surgeon?

Shine followed up the story through David Martin, a friend and colleague from the Loch Ness Project who lived near London. Ian Wetherell had died in 1987 but Martin traced his son Peter to Wetherell's, an upper-crust estate agent's in Mayfair. In passing, Peter Wetherell mentioned that his father's stepbrother, Christian Spurling, lived in Worthing and might know something. Martin went to visit Spurling, then 87 years old and in poor health. There, Martin struck gold.[38]

Christian Spurling had earned his living as a maritime artist but was also an expert model-maker, specialising in ships and – in spring 1934 – monsters. His story filled in details missing from Mandrake's article. The toy submarine came from Woolworths in Richmond. It took him a week to build up layers of plastic wood on the conning tower into the head and neck, which he painted grey. Spurling confirmed his story by annotating a postcard of the Surgeon's Photograph. Above his signature, he wrote, 'This is a photograph of the "monster" made by me in 1933/34 for Mr. A. Wetherell' (Plate 44).

The sequence of events was probably completed as follows. Using his own quarter-plate camera, Maurice Chambers rephotographed two prints from the film taken with Ian Wetherell's Leica, preceding them with blanks to fill up the cassette. Wilson, a friend of Chambers, then entered the story as Wetherell's stooge and a medical man whose professional qualifications and West End reputation would make the scam credible. When he went up to the Loch, Wilson had no reason to stop – hence all the inconsistencies in his story. Instead, he drove straight to Ogston's chemist shop in Union Street, Inverness, and handed the plates over to George Morrison.

The first time Wilson set eyes on what he was supposed to have seen was when Morrison showed him the prints later that morning. The second picture which Chambers had rephotographed (Plate 7) was underexposed, blurred and so unspectacular that Wilson did not even bother to take it away from the chemist's. George Morrison kept a copy, which was later displayed in the shop window in Union Street and reprinted as a postcard 'of the Loch Ness monster', offered for sale on the premises.[34]

Robert K. Wilson FRCS then faded from the scene, as might be expected of a highly principled doctor with an upper-class clientele. David Martin continued to follow his trail with Alastair Boyd, a friend and Monster enthusiast. This took them to a battered suitcase that contained all of Constance Whyte's documents about the Monster. After she died in 1982, her daughter Jean passed the suitcase on to Nicholas Witchell, by then a rising star at the BBC and resting between editions of *The Loch Ness Story*.

Inside the suitcase, Martin and Boyd discovered Whyte's censored version of the 1955 letter in which Wilson admitted his 'slight doubt or suspicion about the authenticity of the photograph' and asked her to 'form your own opinions as to whether the photograph is likely to be genuine or not'.[35] Whyte evidently decided to bury the potentially bad news. In all three editions of *More Than a Legend* (1957 to 1961), the Surgeon's

Photograph was, quite simply, 'The Loch Ness Monster'. Witchell had presumably read the letter, too, and chose the same path as Whyte. In the 1989 edition of *The Loch Ness Story*, he described Wilson's 'most valuable' photographs in the same terms as in the first edition, 19 years earlier: 'They have been rigorously scrutinised for any evidence of faking, and no flaws have ever been found' – which, within the meaning of the act, was true at the time.

Whyte's suitcase contained a final clue, in a letter which Major Norman Eggington, owner of a second-hand bookshop in Hay-on-Wye, wrote to Witchell in November 1970 about trying to find some Monster books that were out of print. Eggington added that his 'Colonel in the Gunners' had confided 'quietly in the mess' that he and a friend, who was a keen amateur photographer, had 'hoaxed the local inhabitants of Loch Ness' with a trick photo of a model Monster. The mess in question belonged to the 85th Field Regiment of the Royal Artillery and was in the club of Gosforth Park Racecourse near Newcastle – and Eggington's Lieutenant Colonel was R. K. Wilson.[40]

Witchell chose not to act on the information, but Martin and Boyd did. They traced Eggington's widow, who had found Wilson 'rude, aggressive, unpleasant', and was happy to confirm the strange rumour of the faked photograph. 'Everybody knew,' she said. 'We all laughed about it . . . I find it incredible that the hoax lasted so long.'[41]

David Martin and Alastair Boyd revealed their findings in a *Sunday Telegraph* piece on 13 March 1994, a month short of the sixtieth anniversary of the publication of the Surgeon's Photograph.[42] The last surviving conspirator never saw their book, *Nessie: The Surgeon's Photograph Exposed*. Christian Spurling had died a few months earlier in his ninetieth year.

Martin and Boyd claimed that 'the image that had remained solid for virtually 60 years crumbled instantly' when their exposé was published (Plate 45). In fact, some Monster believers ridiculed the story and attacked this 'egregious violation of accepted scientific and journalistic protocols'.[43] Others, however, accepted that the Surgeon's Photograph had been faked without losing faith in the creature itself. One true believer had spent over 5,000 hours watching the Loch since the moment in 1979 when he saw a single-humped object gliding away at a range of 300 yards. The clever detective work by David Martin and Alastair Boyd might have demonstrated that one piece of evidence was fraudulent, but his own conviction that the Monster existed was undiminished.

And he should know. He is Alastair Boyd.

Tim Dinsdale's film
Foyers, 23 April 1960 (Plate 17)

This 50-foot strip of black-and-white cine-film had magical powers. In just four minutes it could transform apathy or even scepticism into fervent belief. Instant converts included thousands who watched *Panorama* on 20 June 1960 and some of the biggest names in the history of the Monster – David James, Roy Mackal, Robert Rines and Nicholas Witchell. Dinsdale's film was one of the prize exhibits in Peter Scott's bundle of evidence for his expert panel in 1960, and was shown alongside Rines's underwater photographs at the House of Commons meeting in December 1975. It also transformed Dinsdale's life, legitimising his crusade and bringing him respect, admiration and a place in history.

However, not everyone was persuaded. Richard Harrison, Professor of Anatomy and marine mammal expert, could not shake off the suspicion that 'someone was pulling Mr Dinsdale's leg'. Constance Whyte, Dinsdale's hero, mentioned the film only briefly in the 1961 edition of her book and made it plain in a private letter to Peter Scott that she believed Dinsdale to be honest but naive. Richard Frere, who grew up on the shores of the Loch, could not see what all the fuss was about. This was just a boat of the sort in which he had messed about since childhood.[44]

Dinsdale's sternest critic was Maurice Burton, his former 'scientific associate' who lent him the famous Bolex camera which took the film. From the footage, Burton claimed to see nine good reasons why the object was a boat, and none for it to be a Monster. Moreover, the 'Monster' had followed the course which a local man used regularly to cross the Loch, often setting off from near Foyers at around the time that Dinsdale spotted the object.[45]

Dinsdale had anticipated this criticism. He returned to the same spot later that morning and filmed a standard 15-foot fishing boat with a British Seagull outboard motor as it retraced the object's course with the hotel proprietor at the helm. The differences were pointed out to those who saw *Panorama*. Dinsdale's 'Monster' was a dark, triangular hump which disappeared in the dark water near the opposite shore, while the man in the boat could be nothing else. Also, the boat produced a clear propeller wake between the two bow-waves, which was absent from the Monster's trail.

To reassure those who mistrusted the human retina, David James sent the film to the Joint Air Reconnaissance Intelligence Centre (JARIC) to be analysed using the military's technical wizardry. In 1966, the LNPIB published JARIC's five-page report and triumphantly quoted its conclusion that Dinsdale's object was not a boat and was 'probably animate'.[46] Further powerful evidence emerged in 1972, when Alan Gillespie at the Jet Propulsion Laboratory in Pasadena, California, ran frames from the film through the same computerised image-enhancement process that he had applied to Rines's underwater photographs. This showed the original triangular hump, with a second hump that surfaced briefly behind the first.[47]

Game, set and match to Dinsdale?

The JARIC report published by David James was unsigned and did not mention JARIC or the RAF anywhere – possibly because a minor fuss had erupted over the misuse of military resources when there was a Cold War to be fought.

The report's analysis and conclusions were more guarded than the LNPIB's executive summary.[46] Even when magnified 20 times, the image resolution was poor because the object was nearly a mile away initially and ended up well over that distance. The viewing angle was very flat, meaning that all measurements could be wrong by 10 per cent or more if Dinsdale had been just ten feet lower than his estimated 300 feet above the water. JARIC refused to be drawn over whether Dinsdale's Monster had a unique wake: 'Further discussion of wake and wash patterns should be left to those familiar with fluid dynamics.' Neither could they confirm that it actually submerged, commenting that 'under certain conditions of light, reflectivity, aspect angle, etc . . . objects may NOT be visible'. In other words, the dark object might simply have blended into a dark background. This left JARIC with just two criteria for excluding a boat: the speed and appearance of the object itself. Its speed could be estimated accurately when it travelled straight across the field of view. This came out at 7 mph, exactly the same as the proprietor's boat going flat out. However, the 'Monster' must have moved much more rapidly, as its average speed, calculated from the time it took to cover its zigzag course, was 10 mph and too fast to be a boat.

In fact, this turn of speed was a miscalculation, because the Monster's journey lasted significantly longer than the film. The camera's clockwork drive mechanism ran for 20 seconds and then had to be rewound, which took about ten seconds. JARIC's experts had missed the gaps when Dinsdale broke off filming to rewind the camera. Allowing for this extra time, the object's average speed drops back to 7 mph, just the same as a Loch-standard boat. This undermines JARIC's conclusion that the object was 'probably animate', which was based on the fact that it moved too fast to be a boat (and the assumption that it was not a submarine).[48]

This left the object's distinctive appearance. It certainly did not look like the proprietor's boat which, like most on the Loch, was light in colour.

JARIC recognised the boat easily when it travelled straight across the frame, but noted that it was 'NOT identifiable as a boat' when moving away from the camera, and especially in the area where Dinsdale took most of the footage.[46] In retrospect, a light-coloured boat might not have been the best 'control' for Dinsdale's Monster.

In 1999, Richard Carter, Adrian Shine and Dick Raynor tried to recreate the sighting under similar lighting conditions, aided by a 16mm clockwork Bolex camera and a dark wooden 15-foot boat that followed the same zigzag course across the Loch. They believed that the result looked 'remarkably like' Dinsdale's 'Monster'.[48] This is in line with some alternative interpretations of the film. Shine and others have suggested that the 'Monster' can be changed into a man in a dark-hulled boat simply by watching the film on television with the contrast turned up. Shine later superimposed magnified images of the object taken from 170 frames using an 'image-stacking' method which removes the grain in the film and random artefacts. According to Shine, this reveals a 15-foot boat with one and possibly two passengers – and a lighter spot on the prow, where a circular licence-number plate would have been carried during the 1960s.[48] His conclusion has since been endorsed by commercial image processing experts, including some of the JARIC personnel who produced the original report in 1966.[49]

Others beg to differ. Angus, the Dinsdales' youngest son, has also seen a modern reanalysis of the original 1960 footage. No boat magically creeps out of the grain; instead, a dark 'shadow' is revealed beneath the surface, showing a body, tail and a distinctive diamond-shaped flipper.[50] To true believers, the situation has not changed since the heady days of 1960. The film remains 'the best moving film of a Nessie', and Tim Dinsdale still reigns supreme as the 'Man who filmed the Monster'.

Underwater photographs taken by the Academy of Applied Science
Urquhart Bay, 7 August 1972 and 20 June 1975 (Plates 31, 35 and 36)

These were the images that made people gasp out loud (and could have set fire to the carpet) in the Grand Committee Room at the House of Commons on 10 December 1975. The star of the show was the flipper picture that sliced through the prejudice of the scientific establishment and made the cover of *Nature*. It had also persuaded Peter Scott to end 15 years of indecision. He wrote, 'And then I saw Bob Rines' new photographic evidence',[51] almost as if about to pick up the refrain of a popular song from the Sixties, 'Now I'm a believer.'

Scott's belief was strong enough to survive Alan Gillespie's secret revelation that the computer-processed image of the flipper had been made 'more convincing' somewhere between JPL and *Nature*.[52] This was not the only potential skeleton lurking in the AAS's photographic cupboard. When Rines first mentioned the flipper, he claimed that it showed the pentadactyl (five-digit) structure common to reptiles, birds and mammals, ancient and modern. When the photograph was revealed, the pentadactyl had been ironed out and the flipper had acquired the central rib which looked more at home on a fish than a plesiosaur.[53]

There was also the enigma of the disappearing photographs. Rines tantalised his followers by describing other dramatic shots from the film that captured the flipper – an eight-foot tail and two whole animals – followed later by a massive object photographed at close range.[54] These pictures could not be released, however, as they were being analysed at JPL and/or the Smithsonian – and still were, three years later. They never materialised.

Like many others, Peter Scott found the 1975 photographs – the 'whole-body' and 'gargoyle head' – less convincing than the flippers of 1972. Rines claimed that the whole-body photograph showed a 20-foot plesiosaur-like creature with a body six feet across. He calculated its size using a complicated formula which assumed two mathematical constants, namely (A) the reflectivity of the animal's hide, and (B) a specific characteristic of the photographic film. (A) will never be known until a Monster is captured, while (B) was news to Kodak, the film's manufacturer.[55]

In 1977, G. E. Harwood reanalysed the image using mathematics that he believed to be more robust and allowing for the optical distortion introduced by the waterproof camera housing, which Rines had forgotten to include. Harwood's calculations shrank the 'whole-body structure from twenty feet to just under two feet'.[56] This brought it within the range of some of the establishment's alternative diagnoses, such as a puff of sediment lifted off the lake bed, or even a highly reflective swarm of phantom midge larvae.[57]

The riddle of the gargoyle head was finally solved in 1987. Adrian Desmond's comment that it 'looked like nothing in creation'[58] turned out to be wrong, because it closely resembled a water-nibbled tree stump which Operation Deepscan located on the floor of Urquhart Bay, not far from Temple Pier, in October 1987. The stump was recovered by Dick Raynor and is on display in the Loch Ness Exhibition at Drumnadrochit.

The verdict on the AAS's 1975 photographs: sloppy science and over-interpretation. But what about the flipper photographs of 1972?

*

The evolution of the first flipper photograph is a tribute to the power of photographic manipulation. The original, captured on colour transparency film by the underwater camera that Rines called 'Old Faithful', showed a vague diagonal line which is slightly brighter than the background, but no hint of a diamond-shaped structure. Enhanced by the JPL's computer, the diagonal structure appears stronger, with suggestions of irregular edges on either side of it (Plate 46). This was the image which Alan Gillespie sent to Rines late in 1972, and to Peter Scott just after Christmas 1975.[52] The version which Rines showed at the House of Commons and which appeared in *Nature* was a different beast altogether – not just 'more convincing', as Alan Gillespie wrote to Peter Scott, but as Adrian Shine said when he first saw it, 'a beautiful flipper'.[59]

The second flipper photograph, only made public in 1975, also has its mysteries. Taken 45 seconds before the first shot, it shows a diamond-shaped structure in a slightly different position but with its root at the same place in the frame. This means that the Monster had hovered in front of the camera, holding its position within inches for 45 seconds, just moving its flippers.

As a further complication, other unpublished flipper photographs existed, even though only two frames out of 2,000 had apparently captured the structure. A print which Rines sent to Steuart Campbell in 1984 showed a rhomboidal shape, but with a different outline and in a different position from both of the published flipper pictures. Challenged by Campbell, Rines made the bizarre statement that he had 'even better' photographs in his office. Rines and Wyckoff had mentioned that various composite versions had been derived by blending computer-enhanced images, but this does not explain how the process could have altered both the shape and the position of the flipper.[60]

Most people who found the flipper too beautiful or convincing to be true would instinctively have blamed the dark magic of the JPL's computer. Instead, Gillespie hinted to Scott of a deliberate act of man,[52] but it was several years before he went public. In 1984, Rikki Razdan and Alan Kielar accused the Academy of Applied Science of touching up the flipper photographs and quoted Gillespie as saying that 'the published pictures look a little suspicious around the edges'.[61]

Around the same time, Charles Wyckoff also broke ranks, at least according to Richard Demak at *Discover* magazine. In his 'Skeptical Eye' column, Demak quoted Wyckoff: 'After JPL finished with the photographs, they were retouched. Rines is the only one who could know how much they were retouched or who retouched them.'[61] When the article appeared, Wyckoff wrote an angry letter to the editor of *Discover*, claiming that he had been misquoted, creating 'false and seriously misleading impressions'. The AAS had never produced or released any image with 'the slightest bit of 'retouching' or change' and Robert H. Rines had 'no involvement with retouching of any kind of photograph, despite the outrageous innuendo to the contrary.'[62] *Discover* never published Wyckoff's rebuttal.

In 1989, Wyckoff had a further change of heart and made the full extent of the doctoring clear in a conversation with Adrian Shine. On the original flipper photograph, as reproduced in *Nature*, Wyckoff wrote: 'These portions were not in my original print. They appear to have been retouched. Charlie Wyckoff 7/7/89'. To indicate the 'portions', he drew all around the flipper's clear-cut edges.[63]

Verdict on the 1972 flipper photographs: a deliberate fraud, designed to deceive. Someone had taken a fine paintbrush and created an edge of darkness and a 'beautiful flipper' out of photographic chaos. And there appeared to be only one suspect – the President of the AAS, Dr Robert H. Rines.

What exactly did 'Old Faithful' photograph in the small hours of 8 August 1972? The revelation that the camera was not sitting on the bottom but dangling some feet above it provides a mundane but plausible answer. The camera could have swung back into shallow water and, still firing automatically, took a picture of a gouge mark and disturbed sediment where it had scraped along the lake bed, thus creating the illusion of a diagonal rib.[64]

As Rines said, the Academy of Applied Science were only 'instrumentation people'.

Echo chamber

How can you look for something in darkness that becomes absolute onetenth of the way to the bottom of Loch Ness? The obvious solution, and one which seemed to pull rabbits out of hats at strategic moments, was sonar. Sound waves are not troubled by the effluent from peat bogs and can 'see' a mile or more in Loch Ness, easily enough to plumb its deepest recesses. However, sonar can tell you things that you want to hear rather than what is actually there.

Like all clever technologies, sonar sometimes ran ahead of itself and was dragged down by the parasites of artefact. The Loch, particularly in areas such as Urquhart Bay, turned out to be strewn with elephant traps for those who simply threw sonar at the problem of finding the Monster, without realising the limitations of the equipment. In February 1976 the spring meeting of the British Underwater Association showed Peter Scott some of the false pictures that sonar could conjure up, and the Association's careful studies were later expanded by Adrian Shine and David Martin, and by Rikki Razdan and Alan Kielar.[61, 65]

An impressive range of sonar artefacts has been catalogued. Sonar echoes can bounce off underwater rocky outcrops, mooring chains, the wakes of boats and even the 'thermocline', the invisible boundary between layers of water at different temperatures. These spurious echoes can produce surprisingly strong false contacts, which may mimic movement and life. The thermocline can also bend sound waves as they cross the temperature gradient, just as light is refracted by layers of warm and cold air (see Figure 11). Above the surface, a boat can appear to hover over the horizon; beneath it, an acoustic 'mirage' can lift the Loch floor a hundred feet into mid-water.[65]

The steep rocky sides of the Loch can also throw back chaotic echoes which the unwary may embroider into an eye-catching tapestry of artefact. The best example was the wild terrain which the AAS's sonar mapping managed to create on the floor of Urquhart Bay. The AAS featured a stunning three-dimensional picture of the ridges and canyons on the cover of a monograph about Loch Ness, and Robert Rines showed the image at the House of Commons as a habitat where Monsters could easily hide.[66] It is unfortunate that he and his colleagues did not ponder the likelihood that canyons 200 feet deeper than the Admiralty Chart might exist within a couple of hundred yards of the shore. The high-resolution sonar surveys of Shine's Loch Ness Project and Witchell's Project Urquhart have confirmed that the floor of the Bay was canyon-free exactly as recorded 90 years earlier by Sir John Murray's Bathymetrical Survey.

Reflection artefacts of various types seem likely to explain the tantalising sonar contacts reported during the Oxford–Cambridge Expeditions of 1960 and 1962. Peter Baker, who led both trips, wrote in 1968 that 'echo-sounding has provided no evidence for the presence of a large unidentified object in Loch Ness'.[67] In the same year, Professor Gordon Tucker and Dr Hugh Braithwaite brought their novel fish-finding sonar equipment to the Loch and tracked large targets that seemed to swim and dive at high speeds.[68] Roy Mackal praised this as a 'rather spectacular success', but Tucker and Braithwaite were more cautious – and with good reason, as these 'contacts' also had the characteristics of thermocline artefacts.[69]

Similarly, the few contacts reported during Adrian Shine's Operation Deepscan (1987) appeared more like reflection artefacts than large animals.[65] Operation Deepscan was a bells and whistles version of Peter Baker's

1962 attempt to sweep a sonar 'curtain', hanging from a line of boats, from one end of the Loch to the other. Neither study cornered any Monsters or detected any evidence that they had pushed their way through the sonar curtain. Witchell's Project Urquhart (1994) mapped the Loch in minute detail and found only one anomalous contact, which is now thought to be due to reflection off the wake of a boat.

Shifting sounds

Sonar was at the heart of the AAS's expeditions to Urquhart Bay during the 1970s. Rines's pithy summary of what happened during the early hours of 8 August 1972 – 'something moved in and had its photograph taken' – was based on a heavy sonar tracing that appeared within range of the 'Old Faithful' camera just before it took the pictures that later yielded the diamond-shaped 'flippers'. The sonar experts who interpreted the record as showing large, moving objects had been told that the sonar unit was fixed on the Loch bottom. In fact, it was swinging freely in mid-water, which would make fixed targets appear to move. The smeared dots thought to represent salmon 'running away from a larger moving creature' were actually static; the sonar unit was travelling through the water, while the fish stayed where they were. The heavy trace that the experts labelled as animals 20–30 feet in length with protuberances up to ten feet long was probably the echo from the hull of *Nan*, from which 'Old Faithful' hung, moored about 30 yards further out.

In 1976, the AAS recorded another flurry of sonar excitement over heavy traces with a peculiar 'filamentous' appearance. These were so striking that Charles Wyckoff was emboldened to write 'Nessiteras Rhombopteryx?' beside them on the paper chart recording. In fact, the filamentous trace is characteristic of the echoes thrown back from the wake of a boat.[70]

Moral: when putting your reputation on the line over something truly extraordinary, make sure that you have excluded the common and banal.

Primary contact

This is a good point to revisit the first use of sonar to hunt the Monster. On 2 December 1954, Peter Anderson, the mate of the Peterhead trawler *Rival III*, produced a length of paper chart torn off the boat's echo sounder. This showed the 'truly amazing' object which had been detected on the approach to Urquhart Castle. With its tracing shaped something like a scorpion, this was estimated to be 50 feet long and lay 480 feet below the boat's keel and 120 feet above the Loch floor. An expert from Kelvin Hughes of Glasgow, the echo sounder's manufacturer, stated categorically that the tracing was impossible to fake – a claim that has been repeated down the decades.[71]

However, aspects of the tracing looked odd. Acoustically, the 'Monster' appeared semi-transparent: it strongly reflected the sonar signal, but the profile of the Loch's floor could be seen clearly below it. Also, it spent six minutes directly under the boat, which was travelling at 6 mph. Perhaps the creature was remarkably good at maintaining its position under the boat's keel, nearly 500 feet above it; alternatively, if it had been stationary in mid-water, it would have to be about 1,000 yards long.[72]

Despite the confident denial by the company's expert, the tracing could easily have been faked. As well as the stylus which automatically recorded the returning echo on the paper strip, the recording unit had a separate hand-operated stylus used to annotate the chart. The paper could be re-wound on its drum in the chart and the machine restarted while jiggling the second stylus to produce an irregular outline that apparently hung in mid-water. According to Stuart Campbell, Peter Anderson eventually owned up, telling staff at the Loch Ness Centre that the tracing had been faked with the connivance of the manufacturers, Kelvin Hughes.[72]

But by then this was ancient history. Back on 2 December 1954, *Rival III* cleared the Caledonian Canal and headed for Oban. By the time they got there, the press were waiting en masse and, as Nicholas Witchell reported, 'negotiations began for the exclusive rights to the chart'.[71]

And yet again, invention and the hand of man shaped history.

The human factor

The saga of Loch Ness is as much about people as the creature itself. The Monster has always been a powerful divisive force, leaving few indifferent while driving a wedge between believers and agnostics. At one extreme are those who gave up everything to chase the Monster; at the other are people who were just as determined to obstruct them.

Curiously, the Monster's most vociferous enemies were precisely those who should have been the most excited by the species whose existence they denied – scientists. The eight decades since the 1930s have been the most fertile period yet for scientific discovery, as is obvious from putting a cheap digital camera beside the Surgeon's quarter-plate contraption. From the Monster's viewpoint, though, the scientific establishment could just as well have been stuck in the Dark Ages, stamping down dangerous new ideas with the same pig-headedness that sent Galileo into exile in 1632.

Science in the dock

Many people sided with Dr Denys Tucker's view that the Trustees of the Natural History Museum were 'living fossils', petrified by their own narrow-mindedness. They typified the scientific mafia, tucked away in the fortified ivory towers that littered the Groves of Academe, who refused to consider the possibility that the Monster might exist. To Nicholas Witchell, this epitomised the 'pawky impotence of the establishment'. In the same vein, Gerald Durrell laid into the scientists who branded 'observant, intelligent witnesses' as 'drunk, insane, hoaxers or partially blind and all definitely mentally retarded', just because they reported what they had seen.[1]

Paradoxically, those same scientists – some with 'FRS', 'Professor' or the name of a great institution on their letterhead – prided themselves on thinking big in their everyday research. Why did they have a blind spot for the Monster? Constance Whyte put her finger on part of the answer: 'The Loch Ness story challenges beliefs which are integral to the makeup of most established zoologists.'[2] The Monster lay beyond the fringes of respectability, out in the no-go zone of pseudoscience, together with UFOs, ghosts and levitation. Billed as the 'number one cryptid', it shared its stable with Lizard Man and the Mongolian Death Worm. As well as having an image problem, the Monster was also potentially a threat to mainstream scientists. If it turned out to be an animal from prehistory, it might tear up the foundations of modern zoology and leave the superstructure tottering.

However, the problem that ran the Monster into the sand every time was the absence of incontrovertible evidence that it existed. In science, evidence is everything. It must be gathered and analysed with care and treated with the same respect whether it brings fame and fortune or consigns years of work to the dustbin. Professor Henry Bauer built up a database of 3,500 articles supporting the existence of the Monster. By conventional scientific criteria, most of the articles were junk and the few that had been subjected to the acid test of peer review did not provide acceptable proof. In the view of the establishment, even the Scott–Rines paper in *Nature* was so flimsy that it should never have been published.

Eyewitness reports, photographs or sonar traces – no matter how suggestive – can never be conclusive. As the Smithsonian Institution put it: 'We keep an open mind, as scientists should, and wait for concrete proof in the form of skeletal evidence or the actual capture of such a creature.'[3] A tiny shred of the Monster's skin would have done the trick, but Roy Mackal's biopsy darts never fulfilled their destiny.

Peter Scott argued that it was difficult to confirm the Monster's existence but impossible to prove its non-existence – because this could only be done by draining Loch Ness, which could never be achieved. His logic might have teased a philosopher, but brought the Monster no closer to being embraced by serious scientists. The impasse remained, waiting for a piece of evidence with enough magic to convert those without faith.

The potential rewards of finding the Monster were huge, but the risks for mainstream zoologists were even greater. Identifying a living link to the Age of the Dinosaurs would bring unimaginable riches: global recognition, publications to die for and an embarrassment of research funding. Realistically, though, the chances of success were tiny, as was the possibility that the establishment would ever regard Monster research as respectable. This left the Monster trapped in an unbreakable, Catch-22 vicious circle. As Peter Scott lamented in 1961: 'Without funding, we will never have investigations or results. Without results, we will never have funding.'[4]

Until that happy day, hunting the Monster could easily be the kiss of death for conventional scientists. They would be pulled away from the day job which kept them afloat in the academic rat race and paid their bills. Unless they came up with convincing results, they would soon be marked out as oddballs and then crackpots. From there, as Dr Denys Tucker discovered, it is a few short steps to the exit.

Many scientists appeared obstructive or dismissive when dealing with the Monster. In reality, many were just observing the basic rules of self-preservation. The sorry spectacle of Tucker being hung out to dry undoubtedly had a salutary effect, just as public executions in days gone by helped to encourage would-be felons to mend their ways.

All in a good cause

So much for the establishment, which put a huge amount of energy into doing the Monster down. What about the investigators on the other side of the divide?

Many were, in Constance Whyte's words, 'amateur zoological sleuths' with no formal scientific qualifications. That does not necessarily disqualify them from doing high-quality research, as evidenced by the four decades of systematic studies which Adrian Shine has conducted in the noble tradition of the Bathymetrical Survey. However, many other 'amateur zoological sleuths' jumped straight in without realising that they would soon be dangerously out of their depth – Rupert Gould, Tim Dinsdale, Ted Holiday and (even though she was highly critical of David James's flights of fantasy) Constance Whyte herself. Bob Rines, engineer turned lawyer, insisted that he and the AAS were only 'instrumentation people', but this did not deter him from confabulating about zoology with the best of them.

For various reasons, the few professional scientists who became card-carrying believers did not greatly advance the Monster's cause. Denys Tucker, world-famous expert on deep-sea fish and eels, defied the establishment by claiming that the Monster was a relict elasmosaur, but never lifted a finger to prove the case. His namesake Gordon Tucker, the professor of electronic engineering who brought his newly invented sonar equipment to find the Monster in 1974, reached the top of his field internationally but stayed well away from the Loch. Peter Baker, the enterprising PhD student who led the Oxford–Cambridge Expeditions in 1960 and 1962, went on to blaze an incandescent research trail which won him an FRS at the indecently young age of 37. Baker made his name thanks to an amazing aquatic creature – unfortunately, not a plesiosaur but the squid, whose giant nerve fibres were ideal for him to work out how nerves conduct electricity.

Scientific interest in the Monster would have been galvanised on the spot if Baker had found enough evidence for just one paper as good as his publications on squid nerves, but this never happened.

Maurice Burton and Roy Mackal were both professional scientists from contrasting backgrounds who, for different reasons, tied themselves in knots over the Monster. Burton was derided for switching sides from pro to anti, and for abandoning giant eels in favour of plesiosaurs, then otters and ultimately vegetation mats. In fact, he was not blowing in the wind of indecision, but weighing up new evidence as it came in. This is one of the duties of a good scientist; as John Maynard Keynes remarked: 'When the facts change, I change my mind. What do you do?' It is unfortunate, though, that Burton's train of reasoning eventually led him to a long-necked animal of indeterminate species sketched on a cave wall by a Stone Age artist.

Roy Mackal was knocked spectacularly off course by the Monster and became almost schizophrenic as a researcher. Back home in his molecular virology lab in Chicago, he was a methodical experimenter who published good work in high-quality journals. At Loch Ness, however, he behaved as though the water contained some mind-altering substance that made him throw away the basic principles of his research training. He bent facts, rewrote evolution, invented new species which had no grounding in zoology and covered pages with lengthy calculations that were obviously wrong. In the Great Glen, it was as safe to let him loose with a calculator as a toddler with a chainsaw. His grossly inflated estimate of 13 million salmon in the Loch was based on a single underwater photograph that showed four fish.[5] He did not seem to realise that the tens of thousands of salmon-free exposures were also trying to tell him something.

The most sustained scientific onslaught on the mysteries of the lake has been Adrian Shine's Loch Ness Project. For the last 30 years this has been billed not as a search for the Monster but a comprehensive ecological study of this unique habitat. This philosophy was later embraced by Nicholas Witchell, formerly a conspicuous believer, who also took pains to explain that the professional scientists whom he brought into Project Urquhart were not there to hunt Monsters.[6]

Extraordinary seduction

Writing to Peter Scott on 10 August 1960 from the Royal Yacht *Britannia*, Colonel the Honourable Martin Charteris remarked on the Monster's ability 'to make normally sane and balanced people behave in a highly emotional manner'. This may have been a gentle reality check for Scott, who had recently written to the Palace to support Tim Dinsdale's 'rather emotional' letters to the Queen – and had proposed naming the Monster after Her Majesty.[7]

Scott's head had already been turned, although he only became a true believer 15 years later when he saw Robert Rines's underwater flipper photograph. Most other devotees were converted much more quickly. Denys Tucker, Ted Holiday and Robert Rines were in the privileged minority who found faith on the spot, thanks to the bolt-from-the-blue experience of seeing the Monster with their own eyes. For almost all other believers, the Damascene moment was second-hand. Some fell under the spell of eyewitnesses such as Alex Campbell, while Tim Dinsdale and David James were among the many who were persuaded by reading *More Than a Legend*. For Roy Mackal, Dinsdale's film was 'the one thing which would not go away'.

In some cases, dalliance with the Monster was short and sweet, like a summer holiday romance. For others, it became an obsession that stalked them for the rest of their days. This was an extraordinary seduction that could change lives, and wreck them. Like a jealous mistress, the Monster could be demanding, destructive and dangerous, forcing previously rational people to abandon their livelihood, family ties and reputation. Denys Tucker was an obvious example of someone who sacrificed his career on the altar of his belief in the Monster, but he was by no means the only one.

What was the nature of the magnetism that pulled believers to the Monster?

Many fell for the romance of the vanishingly improbable – the glorious incongruity of a prehistoric Monster hidden in Loch Ness. The farthest-flung corners and depths of the planet were yielding up their greatest secrets, but the most astounding animal of all might live in a Scottish lake, just an overnight train journey from London.

The Monster was also a powerful antidote to the grim realities of the world. During the 1930s, it provided welcome relief from the Depression at home and the gathering storm in Europe, as evidenced by Inverness, lit up like a solitary beacon to welcome in the New Year of 1934 with street parties and fireworks. Three decades later, the feel-good Monster lifted spirits against the backdrop of the Cold War, Vietnam and terrorism on the streets of Britain.

The enthusiasts who chased the Monster came across as dedicated, while obviously having great fun. This was a magical mystery tour, complete with a yellow submarine, a flying machine lifted from James Bond and electronic wizardry straight out of *Tomorrow's World*. And what a cast! Television personalities, war heroes, clever Americans and a motley crew of volunteers, including some who had given up everything to chase the Monster. A chain of believers kept the faith alight, each handing it on to the next like the Olympic torch: Rupert Gould to Constance Whyte, Whyte to Peter Scott, Scott to David James, James to Robert Rines and Nicholas Witchell.

The Monster also provided a chance to kick against the traces. The 1960s were rebellious times, with draft-dodging protesters in America, students pelting police with cobblestones in Paris and CND marches at Aldermaston. In their own way, the Monster hunters also stuck two fingers up at authority, while helping to build the cult of the anti-establishment hero. Charismatic people like Tim Dinsdale were living proof that you did not need fancy academic qualifications to reinvent yourself as a guru, admired and envied around the world.

Above all, the Monster met a basic human need, so powerful that it underpins religion as well as belief in fabulous creatures. This is the insatiable hunger to be filled with awe by something so extraordinary that only those who have seen it for themselves could ever believe that it exists.

Dominant species

Loch Ness is not renowned for the richness of its biodiversity, but it has supported a unique and vulnerable species – the people attracted there by the seductive power of the Monster. Like the medieval pilgrims who converged on Canterbury, they have come from all walks of life. Some returned to their previous existence more or less unscathed, but others found their lives permanently changed, and some never left this holy place. The story of the Monster would be incomplete without knowing what became of them.

The Old Guard

Those who helped to put the Monster on the map in the early 1930s went their separate ways and ended up happy or otherwise. Colonel W. H. Lane, tireless promoter of the giant salamander, died content aged 72, after finally seeing the Monster with his own eyes some months earlier. Marmaduke Wetherell was only 51 when he died in Johannesburg in 1939, having squeezed in a couple of films after returning from England. In the same year, Captain John Macdonald, the skipper with 20,000 trips on Loch Ness under his belt and the first to publicly doubt the Monster's existence, died on the same day that his beloved *Gondolier* was scuttled in Scapa Flow to block the passage to enemy vessels.

Rupert Gould (1890–1948)

When Peter Scott launched *Look* in 1956, *Brains Trust* was still in full flow, but without 'the man who knows everything'. Rupert Gould had died eight years earlier from pneumonia and heart failure. His many obituaries celebrated his near-omniscience, his flair for exciting the man in the street with oddities and enigmas, and the place he had earned in history by resuscitating John Harrison's priceless chronometers.

The obituaries drew a discreet veil over a tragic alter ego which would never have been suspected from Gould's wit and wisdom on air. Outside the studio, Gould's life had been a mess.[8] His grand-sounding 'Lieutenant Commander' (often abbreviated to the even grander 'Commander') was merely the title which the Royal Navy gave to unpromotable lieutenants who remained in their employ, even if – as in Gould's case – they never saw action at sea. Gould was a fragile soul who could not face the reality of what the navy was all about. 'The shot that was heard around the world' in Sarajevo in June 1914 and that precipitated the Great War also tipped Gould into a mental breakdown, the first of several. He was invalided into a quiet office in the navy's Hydrography Office near Admiralty Arch, where he was tolerated rather than rehabilitated.

Just as wounding for Gould was an ugly divorce in 1927, with accusations of neglect and sexual depravity bandied about by his wife Muriel, and picked up with glee by the *Daily Mail* ('Husband's cruelty' ... 'Gagging story'). Muriel Gould had undoubtedly suffered, playing second fiddle to Harrison's chronometers and (it seems) her husband's fondness for orgies with prostitutes and the marmalade magnate Alexander Keiller. The outcomes of this sad affair were victory for the wronged wife, the perfect excuse for the navy to sack one of their oddballs and financial ruin for Gould.[8]

Broadcasting and writing brought back some meaning to Gould's life, but not enough money for him to live independently or even keep up his membership of the Royal Geographical Society (they expelled him for non-payment of dues). Commander Gould had lived the lie convincingly enough to win millions of admirers, but the real world steadily caught up with him. He died aged just 57, penniless and lodging in a friend's house, the victim of his own obsessions.

Robert K. Wilson (1899–1969)

After the war and the excitement of serving in the Special Operations Executive, 'RK' was unable to resume business as usual. He ran a trout farm on the Solway Firth for several years before returning to front-line medicine. By then, his smart suite in London's West End had lost its appeal. In 1950, he took up post as District Medical Officer of the Highlands – not in Scotland, but 9,000 miles away in New Guinea. There, he perfected all the skills of a surgeon, performing heroic operations and hurling instruments through the operating room's window if he thought they were too blunt.[9]

In 1964, he retired to a quiet life in Melbourne. After his indiscretions in the mess during the war, RK never said anything more about the Monster. He came back to England just once, in early 1969, so that specialists in Birmingham could have a last-ditch attempt at curing his oesophageal cancer. When that failed, he returned to Melbourne for his last few months.

The New Age

A few of those who joined the Loch Ness community from the 1960s onwards still live nearby, in some cases because whatever pulled them there all those years ago has not relaxed its grip.

A virtual Adrian Shine can be found through his Loch Ness Project website; the real thing, even more splendidly bearded than when he lifted his whisky glass to celebrate the launch of *Machan* in 1974, can be located at the Project's headquarters near his other grand design, the Loch Ness Exhibition in Drumnadrochit. He may not have found what he was looking for in the 1970s, but he has used his time wisely and has added more to the understanding of the Loch than anyone since Sir John Murray's Bathymetrical Survey.

If you take a cruise on the Loch, you may find Dick Raynor at the helm. His career as an engineer with the Merchant Navy carried him around the world, but ever since the magical day in June 1967 when he took the film that David James said was the best thing to come out of the LNIB, home has been in the northern end of the Great Glen. Chasing the Monster is now a spectator sport for him, enlivened by his spirited detective work to find out what really went on. Raynor also runs a website which, like the commentary on his boat, lays out the various twists and turns in the story of the Monster.

Down at Foyers is Holly Arnold, the girl from Chicago who dropped in to see what was happening in 1968; her strange premonition that she had come home to Loch Ness has turned out to be entirely accurate. Other members of the cast are scattered further afield. Nicholas Witchell has the Monster to thank for channelling him into journalism and the BBC, where he is Royal Correspondent and often to be heard on air from Riyadh, Washington or the Cenotaph. Professor Alan Gillespie's career has taken him to Highlands even more remote than those in New Guinea – the gigantic canyons of the Valles Marineris on Mars. At the age of 80, Ivor Newby still has a fine voice, with which he regales folk clubs in Worcestershire rather than the bar at the Drumnadrochit Hotel or a film crew on a bleak shore in Galway. Looking back nearly 40 years at himself, singing his lament by the lake where the legendary Beast of Connemara had kept its head down, Newby speaks for all of them: 'It was fun.'

Constance Whyte (1902–82)

Dr Whyte's achievements in the Great Glen were more conspicuous than her medical career. Single-handedly, she brought the Monster back from its post-war limbo and planted it firmly in the public consciousness. *More Than a Legend* inspired countless people to chase the Monster, and to many Monster enthusiasts during the 1960s she was a near-saint who fought the good fight and fended off charlatans.

After retiring with her husband to Ditchling in East Sussex, Whyte initially remained interested in the Monster, but from a safe distance. Her correspondence with Peter Scott shows her increasing frustration with the overblown claims of Dinsdale, James and others, culminating in her resignation from the LNIB in 1966. Disenchanted with the whole thing, she refused to update her book. It was her daughter who eventually handed her archives – including the battered suitcase containing her curiously censored correspondence with R. K. Wilson – over to Nicholas Witchell.

Constance Whyte died peacefully in Ditchling in 1982. By then, she might have begun to suspect that some of the 'ordinary people', whose confidence she had been so proud to win, had taken her for a ride. Mercifully, she did not live to see how much of the foundations that propped up *More Than a Legend* would crumble into dust.

Denys Tucker (1921–2009)

The newspaper headlines in spring 1961 which predicted 'Professional Death' for the Natural History Museum's 'Eel Man' were not that far from the truth. After being sacked from the Museum, Denys Tucker never found another full-time academic job, although he picked up occasional threads from his dazzling but truncated career. He edited the 1973 edition of *Freshwater Fishes of the World* and in 1989 wrote 'The Zoologist's Tale', a 24-page addendum for the new edition of Nicholas Witchell's *The Loch Ness Story*.[10] He dedicated his Tale to a French biologist who was sacked by the Musée National d'Histoire Naturelle in Paris in 1802 for believing in the giant squid, but was now 'wholly vindicated by Science'. In 2006, in his eighty-fifth year, Tucker could be heard talking about the European eel on the BBC Radio 4 series *Great Animal Migrations*. This was a reminder of what he had done during the 1950s to become 'one of the most brilliant zoologists connected with the Natural History Museum in recent times'.[11]

The most vivid picture of Denys Tucker is as a martyr for the Monster. Here was a principled scientist brave enough to stand up to the establishment, putting his own future on the line to defend the astonishing conclusion that the evidence had forced him to reach. The harshness of Tucker's fate seized the public's imagination and focused attention on his claims that the Monster was a plesiosaur, that the scientific establishment was determined to strangle any attempt to investigate it and that the Natural History Museum's Trustees were 'about 70 million years out of date'.

Internal documents about 'the Tucker business', recently released by the Natural History Museum, paint a more complex picture. Tucker was undoubtedly bright, but he was also contemptuous of his bosses, short-fused and easily goaded into 'intemperate' language and firing off abusive memos. The Museum's 'living fossil' Trustees were not all scientists, and included the Lord Chancellor and the Archbishop of Canterbury. As well as telling Tucker (a lifelong atheist) to forget eels and turn to God, the Archbishop kindly advised Tucker about the hazards of 'certain behaviours' – wise counsel which Tucker ignored. Tucker also failed to realise that he was being used as a pawn by his trade union in their vendetta against the Museum's new Director.[11]

The press coverage of the cruel fate of the brilliant and fearless 'Eel Man' missed one crucial point. Tucker's final act of defiance, which killed his job and wrecked his future, was to send yet another insulting memo to his boss. It had nothing at all to do with the Loch Ness Monster.

Denys Tucker spent his last years in a farmhouse near Bordeaux, lodging with a lady friend who took pity on him. He died at the age of 82, having satisfied himself that he had outlived his enemies.[12]

Maurice Burton (1898–1992)

Denys Tucker had been a thorn in the side of the Natural History Museum; Maurice Burton was (mostly) one of its darlings. As the Curator of Sponges, Burton had done 'zoology with the best of them' and he was well regarded as Deputy Keeper of Zoology, the post he held until his retirement in 1958.

His greatest impact was through his popular writing about natural history for the general public, especially children. The former colleague who wrote his obituary noted that 'some of the old school' sniffed at this, but pointed out that Burton had inspired 'a new generation of budding professional zoologists'. Like Gould, Burton was fascinated by natural enigmas and wrote the bestsellers *Animal Oddities: The Strangest Living Creatures* (1971) and *The Sixth Sense of Animals* (1973); unlike Gould, he had the background and authority to educate as well as to entertain. Decades before the term 'public engagement with science' was coined, Maurice Burton was a master of the art – and although the scientific establishment did not realise it at the time, they would reap the benefits for years to come.

Burton was never seduced by the Monster. He had a brief fling, broke off the relationship and never went back. He delighted in throwing the cold water of reality on to overheated speculation and crossing swords with (and if necessary spilling the blood of) outspoken believers such as Denys Tucker and Tim Dinsdale. In his later years, he returned to the solid foundations of zoology. His last book, the diverting but uncontroversial *The Life of Fishes*, was published in 1990, a couple of years before he died.

Maurice Burton was also remembered for his warmth, humour and generosity. The former colleague who wrote his obituary injected respect and affection into his account of how Burton, his old boss, had helped to get him a job at the Natural History Museum in 1949.[13]

The name of Burton's junior colleague? Denys Tucker.

F. W. 'Ted' Holiday (1920–79)

When Ted Holiday set off in his van for Loch Ness in August 1962, he was embarking on a fantastic journey that would take him far beyond Scotland – and, indeed, the confines of our universe.

In 1968, Holiday explained the Monster by inflating the extinct, 14-inch *Tullimonstrum* into the 50-foot Great Orm, but he soon became persuaded that this was not the full answer. Struck by the Monster's habit of appearing when observers were looking the wrong way, or cameras had been forgotten or malfunctioned, Holiday concluded that the Monster was telepathically aware of its hunters' intentions. He tried various tricks to outwit the Monster – turning around suddenly to catch it on the hop, or saying loudly (and sometimes in a Scottish accent) that he was setting off in the opposite direction to his true destination.[14] Perhaps the strategy worked; he saw the Monster twice more.

Holiday was also increasingly convinced that the Loch had a sinister alter ego, and that the Monster was at the heart of it all. On his first night by the Loch in 1962, he noted, 'After dark, Loch Ness was better left alone . . . [it is] not a water by which to linger.'[15] He believed that the Monster and its brethren were connected with paranormal phenomenon such as the UFOs which Wing Commander Basil Cary and his wife Freddie had seen hanging over the Loch, and the ghosts which haunted the house of Lionel Leslie, David James's eccentric Irish cousin. These links were explored in his new book, *The Dragon and the Disc*, published in 1973.[16]

In the same year, Holiday decided to purge the Loch of its evil. In June, he took Dr Donald Omand and his greyhound Twinkle out in a boat to exorcise the Monster. Dr Omand, the retired vicar of Chideock in Devon, was an experienced exorcist who had seen action in the Fjord of the Trolls in Norway and elsewhere. Twinkle was also an old hand at pinpointing bad places and confirming that the procedure had successfully banished the malign presence.[17]

The exorcism had no obvious results. Shortly after, however, Holiday had the unsettling experience of meeting a sinister Man in Black, who said nothing while watching him through motorcycle goggles.[18] A year later, and at exactly the same spot, Holiday suffered a heart attack. After he recovered, Holly Arnold drove him back to Pembrokeshire, where he continued to ponder the mystery of the Monster.

He wrote another book, *The Goblin Universe*,[18] reincarnating the Monster as an 'interdimensional being' which cruises through parallel universes and looks like a plesiosaur whenever it pops up in ours. *Goblin Universe*, edited by a friend, was published in 1986, seven years after Holiday's death from a second heart attack. It took several years more for death to interfere with Holiday's writing career. What was probably his last book, *Serpents of the Sky, Dragons of the Earth*, appeared in 1993.

Back in June 1973, Holiday and Dr Omand the exorcist agreed that the Monster was 'not a zoological specimen but a psychical manifestation'.[17] Someone who would have been fascinated to join them and Twinkle in their boat that day was Sir Arthur Keith FRS, forever reviled by those who believed in the Monster because of what he said in 1933:[19]

Professional zoologists [have been convinced] that the Loch Ness Monster is not a thing of flesh and blood . . . The only kind of being whose existence is testified to by scores of witnesses and which never reaches the dissection table, belongs to the world of spirits . . . the Loch Ness monster is not a problem for zoologists but for psychologists.

Frank Searle (1921–2005)

No saga is complete without its dark side. Frank Searle was 'likeable and sincere' when he fetched up on the shores of Loch Ness in June 1969, having cashed in his previous life as a soldier and greengrocer for a tent, a camera and the chance to photograph the Monster. When he disappeared from the Loch 15 years later, he had become the villain of the piece: a serial hoaxer with a malicious streak who threatened Adrian Shine, Nicholas Witchell and Dick Raynor with violence and threw a Molotov cocktail at a Loch Ness Project boat.[20]

As a fraudster, he achieved some notable successes on the front pages of the *Daily Mail* and *Daily Record*. His blue caravan on the shore at Lower Foyers, bearing the grandiose title of 'The Frank Searle Loch Ness Investigation Centre', pulled in up to 25,000 visitors a year – not to mention the string of Girls Friday who were happy to share 'the lot' with him.

Sometime in the spring of 1984, people realised that his caravan stood empty and that Searle had gone just as he had arrived, unannounced and unattached. He never returned to the Loch. In January 1984, Searle had written to an American friend, Mrs Harriet Ely, that he was going to search for a sunken Spanish galleon off the west coast of Scotland. Of course they would keep in touch; letters addressed to him, 'care of Loch Ness', would still reach him. If they did, he never replied. Mrs Ely continued to write until June 1991. Her last letter explained that she and her husband were now in their seventies and clearing out their house. She ended, 'We have always been proud of your Loch Ness findings and hope for your ultimate success, so let us know where you are', and signed off with 'Good Luck and Love'.[21]

Frank Searle's end would have remained a mystery without concerted detective work by a journalist, Andrew Tullis, who became intrigued by the story.[22] Tullis's last-ditch bid to find someone who knew him, through an advertisement in a metal-detectors' magazine, led to the Lancashire coast and a bedsit in Fleetwood. Searle had moved there in 1986, but to the end was a loner and good at covering his tracks. When Tullis reached Fleetwood in April 2005, Searle was not available for interview; he had died just six weeks earlier.

And nobody there knew that this odd, reclusive man who had appeared from nowhere and with no past worth talking about had spent the best 15 years of his life as the Monster Hunter Extraordinary.

David James (1919–86)

James's early years – before the mast on the tall ship to Australia, dodging the SS in Lübeck, Operation Tabarin in Antarctica – had been a Technicolor extravaganza. By comparison, the last decade of his life was placid and monochrome. He had already drifted away from his 'old wartime friend' Peter Scott when Scott and Rines shared the limelight in the chaotic run-up to Christmas 1975. The LNIB had done its best, and painful though it was to admit, hunting the Monster now had to be entrusted to the Americans. Also, James had been diverted once more by his second love: politics.

As the MP for Brighton Kemptown, he had been 'a young man with a future'. Returning to the House in 1970 as the MP for North Dorset, and reminded of allegations that he had put the Monster ahead of his constituents, he had become 'an ageing man with a past'.[23] *The Times* later described his parliamentary performance as 'forthright and highly individual'.[24] His stance as a Catholic Unionist was bold but failed to heal the sectarian rift in Northern Ireland. The sugar manufacturers of the Common Market went their own sweet way, despite his eloquent pleas for those in the Commonwealth who were in the same business. Finally, and to his surprise, the nasty business about neglecting his constituency bubbled up again – when all he had done was to tell Ted Heath that he should stand down in favour of Margaret Thatcher.

James had read the leadership runes with commendable clarity, but by the time Thatcher was poised to win the 1979 General Election his heart was no longer in politics. She wrote him a nice thank-you letter, and he took himself and his family north of the border, to be with his first love.[25]

This was not the Monster, but Torosay Castle on the Island of Mull, the ancestral home he had inherited in 1945. Since becoming Laird, he had done little other than live there. Now, he set about restoring the castle and its 9,000-acre estate to their former glory. The reborn Torosay became a notable tourist attraction as well as the James family home, complete with Scottish Gothic turrets, Italianate gardens, a miniature railway and memorabilia that included the forged papers of Lieutenant Ivan Bagerov of the Bulgarian navy.

Overlooking the gardens was a large, bright room which could almost have been a shrine. Displayed with reverence were newspaper cuttings going back to 1933, maps, charts, documents and dozens of photographs of the vanished community of the LNPIB. And looking down on the scene were two bulky plesiosaur-like creatures, caught in mid-courtship as they cruised through the amber waters of Loch Ness, painted by Sir Peter Scott for his friend and fellow Monster believer, David James MBE, DSC.

James always insisted that chasing the Monster did him no harm, but others were less convinced. Elspeth Huxley, Peter Scott's biographer, wrote that both Scott and James were 'very brave' to declare their belief in the Monster. The risks were particularly high for James and 'may well have damaged [his] chances of holding office' – although Huxley acknowledged that 'his independence of mind might also have been an impediment'.[25]

A couple of months before he died in December 1986, David James was asked if he had any doubts. He did not, adding that otherwise 'it would be the greatest con ever known to mankind'. He was talking about his Catholic faith, which was being sorely tested during his last pilgrimage to Lourdes with inoperable lung cancer.[26] His belief in the Monster also stayed with him to the end, but his obituary in *The Times* did not mention the place where, some claimed, he had spent more time than in his constituency.[24]

Roy Mackal (1925–2013)

Like Denys Tucker, Roy Mackal was a bright young scientist on a rising trajectory before he met the Monster. His research into 'bacteriophage' viruses at the five-star University of Chicago was going well and he was on course to become a leader in the field. Then, during his trip to London in the summer of 1965, he yielded to the impulse which took him to the Great Glen. From that moment, Mackal's promising career was history. His output flagged and soon he was just one of the crowd. The reason was obvious to all: his head had been turned by something more exciting than the day job.

Walking the tightrope between being Associate Professor at the University of Chicago and the Scientific Director of the Loch Ness Investigation Bureau was never going to be easy. In Chicago, Mackal endured 'ridicule and unprofessional treatment' and threats from above that things would turn sour unless he dropped his Monstrous obsession.[27] The publication of his book *The Monsters of Loch Ness* in 1975 (by which time his papers on bacteriophages were drying up) did not help. Neither did becoming Vice-President of the International Society of Cryptozoology, a title that meant nothing to the hard-nosed bean counters who paid his salary. Some believe that Mackal was 'booted out of the biology department'; an alternative view is that 'lateral promotion' landed him the post of Energy and Safety Coordinator for the campus.[28] Mackal's energy was never in doubt, but his fondness for carrying a loaded harpoon rifle might have raised concerns about his suitability as a safety adviser.

Losing his head and heart to the Monster was the first of Mackal's *liaisons dangereuses*. More hazardous were two expeditions to find Mokele-mbembe, the fabled creature said to inhabit swamps in the Congo. Kitted out in the same bush jacket that he sported on campus and supported by ten Pygmy bearers, Mackal found a couple of dodgy footprints which convinced him that Mokele-mbembe was 'a small sauropod dinosaur'.[29]

Later in life, Mackal became fascinated by longevity. Over a drink in the Drumnadrochit Hotel, he explained to Dick Raynor that death occurs when the 'telomeres', the ends of chromosomes which shrink progressively as an organism ages, reach a critical length. He believed that fish, whose telomeres remained long, would live forever if they could be protected from accidents, infections or getting eaten, and he intended to do the same.[30]

Roy Mackal died of heart failure at the age of 88. To the end, he believed in the Loch Ness Monster and Mokele-mbembe and had no regrets about the path down which curiosity had led him.

Tim Dinsdale (1924–87)

It is impossible to picture Tim Dinsdale behind a desk or standing on the doorstep trying to sell life insurance. Of all those whose lives were changed by the Monster, Dinsdale probably underwent the most dramatic metamorphosis. From an unknown with a dull office job at Heathrow, he was reborn as the man who, for millions around the world, personified adventure, passion and success against overwhelming odds. At Loch Ness, he was the public face of the Monster hunters: relaxed but always alert, cradling his trusty Bolex in the well of *Water Horse* (Plate 46). He was at peace with himself and the world at large, because he had found his niche and people believed in him, just as he believed in the Monster.

Back in 1960, Peter Scott regarded 'poor little Dinsdale' as an obsessive oddball who had to be kept away from the high-powered zoologists. Then, over Easter 1960, Dinsdale 'reached out through his magic lens and grasped the Monster by the tail', converting a whole new generation of Monster believers and outmanoeuvring Scott. His transformation from loose cannon into big gun was completed in 1966 with the triumphant announcement that JARIC had put his film under the microscope and confirmed its authenticity.

Dinsdale continued to live in Blewbury Drive, Tilehurst, but his soul was elsewhere. Between 1960 and 1987, he mounted 56 expeditions to Loch Ness, either alone or with his sons. He saw the Monster twice more but, with no camera to hand on either occasion, never matched the sighting that brought him fame. And famous he was, 'recognised internationally as the authority on the subject of Loch Ness' and the only author 'who has actually sighted the fabled Monster and who has filmed it'. He wrote five best-selling books; *Loch Ness Monster* went through four editions between 1961 and 1982, and was reprinted in 1989.

That last reprint was posthumous. Tim Dinsdale died suddenly on 1 December 1987, aged 62. The loss of a pioneer and visionary was widely mourned. The last photograph in the 1989 edition of Nicholas Witchell's *The Loch Ness Story* shows a smiling Dinsdale standing behind a camera rig that points out across the water. The caption reads simply, 'Tim Dinsdale, 1924–1987. Champion of Loch Ness'.

Dinsdale never had to confront Shine's evidence that his Monster was just a boat – and in any case, Angus Dinsdale is convinced that Shine is wrong and that 'The Man who filmed Nessie' did exactly that. It is generally agreed that Tim Dinsdale was characterised by a man of unshakeable faith and transparent honesty; even the arch cynic Richard Frere described him as 'an intelligent man of great integrity'.[31] It would seem that his worst crimes against science were wishful thinking and overinterpretation of ambiguous findings. Dick Raynor, a good friend for over 20 years, believes that Dinsdale filmed a boat, but wishes to pop the myth as humanely as possible:

This in no way detracts from Tim's tremendous contributions to the Loch Ness story. He was loved and respected by all who knew him and inspired many, myself included, to follow his guiding principle – the search for the Truth.[32]

Robert Rines (1922–2009)

This 'spectacularly polymathic' man was arguably the most complicated figure in the saga of the Monster. Rines was more than a successful patent lawyer in the family firm in Boston. He also lectured at MIT on inventions and patent law; his teaching sessions, held in a college dining room from 7 to 9 p.m., were popular and remembered with affection by his former students. He gave his first seminar at MIT in 1963 and his last over 45 years later, in his eighty-sixth year.

Along the way, Rines somehow acquired a doctorate from the National Chiao Tung University of Taiwan. He must have impressed them, as he still heads the University's list of Notable Alumni, ahead of three Nobel Laureates. He is named on 70 patents and, in 1994, joined Thomas Edison and Alexander Graham Bell in the American Inventors' Hall of Fame. His citation explains that his patents in microwave, high-resolution radar and sonar technologies were crucial in finding the wreck of the *Titanic*, refining medical ultrasound imaging and perfecting the Patriot defensive missile system used in the Gulf War.[33]

Rines's portrait in the Hall hangs beside the picture of a plesiosaur-like animal with large, rhomboidal flippers. Rines always said that the thrill of seeing the Monster himself, over tea at the Carys' house on 23 June 1971, stayed with him for the next 37 years. He made 30 visits to the Loch but never relived the heady days of 1972 and 1975. The fuel of unbridled optimism finally ran out in 2008, when he realised that he had lost faith in ever seeing the Monster again. Rines believed that the last members of the species had perished, probably the victims of global warming.[34]

This was the year in which Rines began to wind up his affairs. He gave his farewell seminar at MIT in May 2008, shortly before flying out to Scotland. This was the last time he looked down from Tychat on to the dark waters that had given him nearly half a lifetime of excitement and frustration. Now, with those waters devoid of any mystery, there was nothing to come back to. Robert Rines died at home in Boston on 1 November 2009, aged 87.

Rines brought a great deal to the chase: enthusiasm, money and friends with technological know-how. His legal training had programmed him to argue 'like a Baptist preacher' and win his case, whether or not this coincided with the truth. Scientific rigour was someone else's problem; Rines regarded evidence as an elastic commodity that he could stretch until it gave the right answer. One opinion on the 'flipper' images which only he could have faked is that 'Dr Rines' enthusiasm for the task in hand clouded his vision of what the Academy team had photographed'.[35] Some might regard that interpretation as unduly charitable.

Today, Dr Robert H. Rines, the venerable polymath, gazes out benignly from the website of the Academy of Applied Science. His achievements as lawyer, inventor and philanthropist (and a composer of Broadway musicals) were breathtaking; his widow was quoted as saying that 'people just don't have stories like that in their lives'.[36]

Rines's track record was so astounding that a cynic might wonder if it was too good to be true. Some of it was. His claim to have developed the 'advanced sonar technology' that hunted down the Monster[37] would have surprised his friend Martin Klein, whose company built and provided the devices that Rines used. The assertions that Rines's early inventions enabled the *Titanic* to be found and incoming Scud missiles to be shot down in Iraq have also been challenged. In November 2013, the Wikipedia entry about Rines included the statement that his patents looked nothing like the eventual technologies, were never tested and, if the laws of physics can be trusted, could never have worked. That commentary was later removed and does not appear in the current version of Wikipedia.

David James, with typical old school obliquity, once described Rines as 'not quite sixteen annas to the rupee'. Roy Mackal, who threatened to resign his Directorship of the LNIB if Rines joined the Board, was more direct in a letter he wrote to James in August 1972:[38]

He is first a lawyer [and shows] a disturbing lack of zoological knowledge and little appreciation of what constitutes scientific judgment and competence . . . an unscrupulous opportunist in the worst sense of the word.

Peter Scott (1909–89)

Towards the end of his autobiography, *Eye of the Wind*, Peter Scott wrote, 'I can think of nothing sadder than to live a happy and interesting life without recognising it.' Luckily, he did, and so did many others. Scott's 'happy and interesting' life was packed with achievement across an astonishingly broad spectrum. His output as 'naturalist and painter' was prodigious: on top of thousands of paintings and drawings, he wrote 18 books and illustrated 20 more. Keeping track of all the works which reproduced Scott's writing and artwork kept Paul Walkden, the bibliographer at Slimbridge, permanently busy during the 1980s.[39]

Along the way, Scott collected honorary degrees, grand titles such as Rector of Aberdeen University, Chancellor at Birmingham and President of the World Wildlife Fund, and much more (Plate 48). His Distinguished Service Cross and the Olympic Bronze Medal from Kiel were joined by a veritable cacophony of gongs, notably America's Getty Medal and the Gold Medal of the WWF. Iceland also offered him their greatest honour, the Order of the Falcon. This carried pleasant resonances with his father's middle name and the place where he had married Philippa Talbot-Ponsonby in 1951 – but Iceland's continuing slaughter of whales revolted him, and he turned it down.[40]

The period 1986–7 was an especially good one for Scott, with the fortieth anniversary of the Wildfowl Trust (by then renamed the Wildfowl and Wetland Trust) and two glittering prizes that he thought would be withheld from him. His botched attempt to name the Monster after the Queen had not caused lasting damage, as Her Majesty appointed him a Companion of Honour, the greatest accolade that the Monarch can bestow on a commoner. Then the Royal Society made him a Fellow under Statute 12, which recognises individuals who have done great things for science but not as a front-line researcher. His FRS raised a few eyebrows because he had walked away from the world of science as an undergraduate in Cambridge. However, nobody became apoplectic with rage, as had happened when Statute 12 was invoked in 1983. The person appointed FRS on that occasion began her career as Margaret Roberts, working as a research chemist with Richard Synge (FRS, Nobel Laureate and Monster believer), before defecting to politics and marrying Dennis Thatcher.

For Scott, the Loch Ness Monster was an obsession at times and merely a sideline at others. It was pushed off his list of priorities when the WWF devoured his time during the late 1960s – which is why he knew nothing about Rines's flipper photograph until 1975, over two years after David James saw it.

Once converted, he never lost faith. He was fully aware that he was playing with fire and that his entanglement with the Monster 'tarnished his reputation in some scientific circles'.[41] In an interview in 1977, he waved away the risk: 'Science is not a question of reputation, it is a question of truth; and if there is the chance of learning scientific truth, reputations have rather little importance.'[42] But behind the bravado lurked the anxiety that he had estranged the establishment, on which his conservation activities depended.

In 1975, Keith Shackleton warned Scott about the potential risks of 'a fetid bubble oozing up from your past'.[43] That particular bad smell came from the time that Scott had spent killing wildfowl; for some scientists, the Monster was just as malodorous.

Scott's best year, 1987, was also when he realised that youth and health had deserted him. 'I don't feel old,' he wrote to a friend, 'but I'm aware that I must seem so to those I meet.' By then, both prostatic cancer and angina had crossed his path. At first, angina only prevented him from scuba diving; before long, it forced him to pull out of opening the Peter Scott Walk along nine miles of Norfolk coastline, near the Nene Estuary.[44] Missing that event really hurt. The walk began at the East Lighthouse, newly restored and with a smart blue plaque that bore his name. Back in 1933, he had rented the lighthouse for five pounds a year, and this was where his complementary careers as ornithologist and painter really took off.

Despite worsening angina, Scott carried on much as before. During the spring of 1989, bibliographer Paul Walkden was planning a major exhibition of Scott's work to coincide with his eightieth birthday on 14 September.[39] The exhibition had to be cancelled at a fortnight's notice. Peter Scott was taken into the Bristol Royal Infirmary with a heart attack; he died peacefully, with Philippa and their family at his side, on 29 August.

There are many memorials to Sir Peter Scott CH, CBE, DSC, FRS. One that says a lot about him originated with one of the birds that he met at the East Lighthouse – Anabel, a young female pink-footed goose who sought him out each winter on touching down after her 1,100-mile flight from Iceland. American writer Paul Gallico heard Scott tell the story at the World Figure-Skating Championship in Berlin in 1935, and decided to write a book about it.

Gallico was liberal with artistic licence. Anabel lost her pink feet; Scott acquired a hunched back and sociopathic tendencies; and love interest and a sad ending were pasted on to the backdrop of Dunkirk. Published in 1946, *The Snow Goose*[45] became a bestseller but was not to everyone's liking. James Robertson Justice sent Scott a copy with an angry note: 'If properly handled, this story, which is clearly about you and your lighthouse, should be actionable.'[46] Instead of suing Gallico, Scott wrote to congratulate him and drew the illustrations for the second edition.

The greatest memorial to Peter Scott, the 'father of conservation', is intangible. Go to Slimbridge and stand on the Dumbles, the salt flats which pull in wintering geese every year and where Scott's ashes were scattered. Looking around, you can see what he created here beside the River Severn. Beyond that lies a world that would have been poorer without him, and that may yet be a better place because of him.

14

Last word

Constance Whyte maintained that the Monster was 'the perfect mystery for amateur zoological sleuths', but added that it was very complex and that 'every possible clue must be welcome'.[1]

In that spirit, this is your final briefing before you are pinned down, just 27 pages from now, to solve that perfect mystery. We begin with an update on the beast itself.

A twenty-first-century Monster

For millions of people around the world the Loch Ness Monster remains the Number One Cryptid. As with the premature obituary of Mark Twain, rumours of its demise in 2008 (*Daily Star*: 'Nessie dies aged 3 million'),[2] appear to have been exaggerated, as sightings continue to come in. However, these are less frequent than 20 years ago, and the absence of reports for almost a year in 2008–9 caused Gary Campbell, President of the Loch Ness Monster Fan Club, many anxious moments. Relieved beyond belief by news that the Monster was seen on 6 June 2009, Campbell reassured the fans that 'Nessie was just keeping her head down'.[3]

Monster-spotting has moved with the times and now embraces twenty-first-century technology, including mobile phones, webcams and satellite imaging, but the usual caveats apply. The object revealed by Google Earth at 37° 12' 52.13" North, 4° 34' 14.16" West on 26 August 2009 was just a boat, as was 'the shadowy form around 100 feet long with two flippers' picked up on Apple Maps in November 2014.

Today, the Monster has a solid presence on the Internet and social media. Out there, dozens of websites variously inform, disabuse, indoctrinate or simply entertain. You can find chat rooms, Nessie's Diary and galleries of images old and new ('This one [the Surgeon's Photograph] doesn't show me at my best'). A couple of clicks of the mouse will transport you into areas of overlapping interest, such as cryptozoology and paranormality. A handful of websites are excellent, and some are dreadful; visit them and decide for yourself. If it all gets too confusing, you can always contact Nessie directly.[4]

The Monster industry continues to thrive. Back in 1934, the Monster inspired the 'Loch Ness' fox-fur-trimmed women's outfit, together with novelty postcards, souvenirs and advertisements for porridge and chocolate bars. Since then, the Monster brand has gone global, with T-shirts, songs, horror and fantasy novels, films, a board game called 'Nessie Hunt' and encounters with Doctor Who and Astérix. Ever popular are postcards of the Surgeon's Photograph, some carrying a prominent health warning: 'Nessie has now learned to avoid photographers and lives on a diet of finest Scottish salmon and the occasional tourist'.

The Monster has always been a money-spinner for the Great Glen. In 1933–34, it rescued Inverness from the Depression that was closing its grip around the rest of the Highlands – so successfully that Inverness Town Council hacked back the region's tourism advertising budget. The Monster's powers of seduction still operate today: in 2014, it pulled in an estimated 300,000 visitors and £30 million.[5] Sadly, the Monster-related income stream has divided the local beneficiaries. Opposing camps, essentially believers and agnostics, periodically trade insults and accuse their rivals of filling tourists' heads with 'negative theories' or 'garbage', as appropriate. Like the surface of the Loch, the tranquil façade of Drumnadrochit conceals powerful undercurrents. The village is home to two rival Monster exhibitions, one more evidence-based and less sensationalist than the other, separated by 50 yards of no man's land.

For the Monster it is business as usual. On 11 November 2014, a 'long neck-like shape' was filmed on an iPhone as it moved across the Loch, some 500 feet offshore. The two-minute recording was made by a man who, according to his wife, was so technically incompetent that he could not possibly have faked it.[6]

But what *is* the Monster, and where did it come from?

Identity crisis

Workers on Chinese production lines churning out plastic Monsters know exactly what is in Loch Ness. So did Constance Whyte, Denys Tucker, Tim Dinsdale, whoever painted the LNIB logo on the Visitors' Centre at Achnahannet, and all those who received Peter Scott's hand-drawn 'Nessies' to celebrate Christmas 1975. Since 1934, the front-runner of all the candidate species has been a long-necked, small-headed reptile with flippers, which, as Scott remarked, 'looks rather like a plesiosaur.'

Scott admitted that massive odds were stacked against the survival of something that, according to much more powerful evidence, should have died out 65 million years ago and had left no traces in the interim. In 1933, Richard Elmhirst calculated those adverse odds at 17 million to one against – an impressively precise figure, but one which he plucked out of the air.[7]

That 65-million-year gap in the fossil record does not necessarily spell the death of an attractive idea, as proved by the coelacanth and other 'living fossils'. However, this is a long time to go missing, and the mass extinction event which saw off the dinosaurs was particularly thorough as it exterminated half of all the pre-existing genera. Pragmatists would argue that, as death goes, this was pretty well irreversible and that, like Godot, the plesiosaur is never going to turn up.

There is also the thorny problem of how a subtropical plesiosaur-like reptile ends up in a landlocked Scottish lake with water a few degrees above freezing-point. Creationists know the answer: God put the Monster there. They could be right, because the UK's school curriculum watchdog recently approved their Accelerated Christian Education programme.[8] For everyone else, though, access remains a problem, because the Loch's basin went directly (in geological terms) from being under thousands of feet of ice to sitting high above sea level. Admittedly, seals and perhaps even porpoises have been reported in the Loch, leaving open the possibility that a relatively nimble marine reptile might have managed to get in from the sea at some point since the last Ice Age.

It is generally assumed that Loch Ness would harbour a breeding colony of Monsters rather than one or two extremely elderly, lonely and frustrated individuals. The number of Monsters in the Loch has been estimated using various methods. On 30 September 1972, 16 were counted, including five off Dores and two in Urquhart Bay, one of which was directly below Bob Rines's house, Tychat. The population was mapped in real time by Freddie Cary, the former Special Operations Executive agent and wife of Wing Commander Basil Cary. She did this by 'dowsing', swinging a non-metallic pendulum over a map of the Loch.[9]

Others have used mathematical modelling, extrapolating from the population dynamics of living reptiles and marine mammals and making sweeping assumptions about the Monster's weight and diet and the availability of fish in well-studied habitats such as the Pacific Ocean.[10] The results, published in scientific journals, suggest that between 6 and 20 individuals could be present. These numbers are surprisingly similar to Mrs Cary's, but are inconsistent with Adrian Shine's ecological surveys of the Loch. Shine has shown that the Loch is nutritionally impoverished and contains

36. The 'gargoyle' photograph taken on the morning of 20 June 1975, near Temple Pier in Urquhart Bay. Robert Rines and others identified bilateral symmetry in the structure, consistent with the face of a living animal. The two slender stalked features at the top were suggested to be breathing tubes, and the heavily shadowed area its open mouth. (Academy of Applied Science, Belmont)

37. *Courtship in Loch Ness* (1975), painted by Peter Scott for David James. For many years, the painting hung in the Loch Ness Monster display at Torosay Castle, James's family residence on the Isle of Mull. (Dafila Scott)

38. Peter Scott and Robert Rines facing the media at the press conference organised by *Nature*, 10 December 1975. David Davies, editor of *Nature*, is standing at right. (Nicholas Witchell)

LOCH NESS MONSTER

TIM DINSDALE   FOURTH EDITION

39. Adrian Shine (born 1949) in *Machan*, the underwater camera hide which he built in 1973. After testing in the river at the foot of his uncle's garden in Surrey, *Machan* was deployed at depths of up to 30 feet in the search for 'Morag', the aquatic creature reported from Loch Morar on the west coast of Scotland. The observer sits between the ballast tanks and looks up towards the surface through the plate-glass windows in the vessel's lid. (Maralyn and Adrian Shine)

40. Cover of the fourth edition of Tim Dinsdale's *Loch Ness Monster*, featuring the 'Muppet' photograph taken by the Cornish wizard and psychic, Tony 'Doc' Shiels. (Routledge & Kegan Paul)

41. Operation Deepscan, organised by Adrian Shine in October 1987. Nineteen cruisers moved in formation down the Loch, sweeping the water for sonar contacts. One unexplained contact was described as 'smaller than a whale but larger than a shark'. (Derek Colclough and loch-ness.org)

42. Train of bow-waves from a trawler. These can appear large and solid, especially on flat-calm water, and may seem to move across the surface at speed or undulate like a caterpillar. (Dick Raynor)

43. Two photographs supposedly of the 'Monster', dating from 1934 and reproduced in N. Witchell, *The Loch Ness Story* (1974), p. 51 and H. Bauer, *The Enigmas of Loch Ness*, p. 62. TOP: dorsal fin of a bottle-nosed porpoise; BOTTOM: dorsal fin and snout (right-hand end) of a basking shark. Identification by Professor David Sims, Marine Biological Association, Plymouth.

44. Copy of the Surgeon's Photograph, supposedly taken by R.K. Wilson in April 1934, annotated by Marmaduke Wetherell's stepson, Christian Spurling, on 11 July 1992. (David Martin and Alastair Boyd)

This is a photograph of the "monster" made by me in 1933/34 for Mr. a. Wetherell

Capt. Spurling

45. Alastair Boyd holding a reconstruction of the model 'Monster' made by Christian Spurling in 1934.

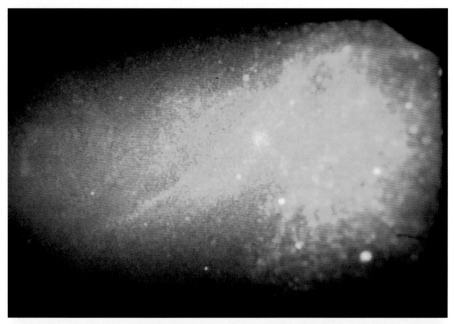

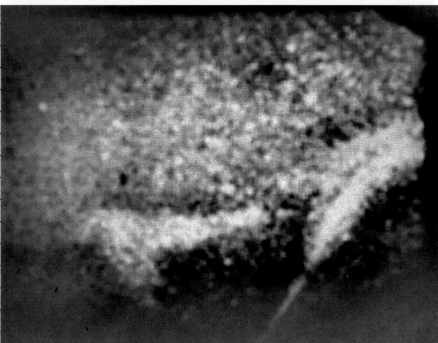

46. The origins of the underwater 'flipper' photograph published in *Nature* in December 1975. TOP: unenhanced colour transparency taken by the underwater camera. BOTTOM: the same image after enhancement by Alan Gillespie using the computer programme developed at the Jet Propulsion Laboratory for sharpening images from NASA space missions. This is the version which Gillespie sent to Robert Rines in late 1972. (Academy of Applied Science, Belmont and JPL, Pasadena)

47. Tim Dinsdale on *Water Horse*, photographed in the early 1980s. (Tony Healey)

48. Sir Peter Scott CH, CBE, DSC, FRS, photographed at Slimbridge by Lady Scott during the mid-1980s. (Dafila Scott)

49. Alex Campbell (1901–1983), water bailiff, Fort Augustus Correspondent for the *Inverness Courier* for 62 years, and veteran of 17 or 18 sightings of the Loch Ness Monster. Pictured in the early 1970s (left) and around 1980. (Nicholas Witchell; Tony Healey)

no more than 30 tons of fish, which would not be enough to support even a few plesiosaur-sized (one-ton) animals with reasonably healthy appetites.[11]

Blast from the past

Much more is known now about plesiosaurs than in Peter Scott's day; he would have been cheered by some, but not all, of that new understanding.

For most plesiosaur species, which frequented warm, shallow seas, the near-freezing waters of the Loch would be a no-go zone. Recently, however, the exotic *Umoonasaurus*, dug out of an opal mine in Australia, was shown to have flourished at a latitude of 70° South, which is equivalent to today's southern Antarctic Ocean. To survive, *Umoonasaurus* might somehow have broken the reptilian taboo on hot-bloodedness.[12] There is a precedent: dinosaur-like animals which pulled off the same trick and survived the mass extinction are still with us, as we are reminded every time we hear birdsong.

Unfortunately, the plesiosaur hypothesis runs into further trouble over the unhelpful way in which the animals' spinal column was articulated. The cervical spine of long-necked plesiosaurs such as the elasmosaurs contained 30–70 vertebrae and could exceed the combined length of body and tail. Detailed anatomical studies have shown that the neck could flex freely downwards, but not in other directions. This has transformed reconstructions of plesiosaurs and their lifestyle. Gone are the dramatic hunting scenes showing elasmosaurs hurtling after fish; instead, we should picture an unhurried creature that operated like an underwater hoover, sculling itself along above the sea bed and dipping its head down to grab shellfish off the bottom. The plesiosaur's backbone could never have bent itself into a series of humps, and a 30 mph sprint across the surface would have been at literally breakneck speed. Indeed, the plesiosaur's neck was so well designed for downward flexion that it could not even have lifted its head out of the water, let alone emulated the graceful curve of the Surgeon's Photograph.[13]

Putting together all the evidence, it now seems very unlikely that the Loch Ness Monster could be a surviving descendant of the plesiosaurs. So where did the idea come from?

Evolution or intelligent design?

On New Year's Day 1934, if various eyewitnesses could be believed, the large unknown animal in Loch Ness could have looked like a bog-standard water horse, a salamander, a crocodile, a prehistoric reptile, a huge frog or a sea serpent. The first feature film about the Monster pitted Seymour Hicks against a gigantic aquatic iguana. By the time *The Secret of the Loch* was released in summer 1934, the iguana was ancient history and the favoured candidate left in Loch Ness was the plesiosaur-like creature which, with minor modifications, has remained in the frame ever since.

Plesiosaurs have always pulled a good crowd. They first came to light in 1823 near Lyme Regis in Dorset, where 24-year-old Mary Anning scoured the Jurassic deposits with a hammer and her faithful rock hound, a black and white terrier named Tray.[14] Mary Anning's intuition for finding spectacular fossils matched the ability of Michelangelo to look into a virgin block of Carrara marble and see the angel trapped inside, waiting to be liberated by his chisel. Private collectors and museums fought over all her finds, but it was the long-necked plesiosaur which fired both the professional and the public imagination. Described much later as 'a snake threaded through the body of a turtle',[15] plesiosaurs rapidly took their place alongside the land-based dinosaurs, the creatures which have outstripped all others, living or dead, in their power to terrify and fascinate.

Dinosaurs and plesiosaurs starred in museums and the Great Exhibition of 1851, and clawed out ever-larger niches in popular natural history books for adults and children; *Science Fairy Tales* (1860) devoted a hefty chapter to 'The Age of Monsters'.[16] Prehistoric reptiles roamed through many fictitious landscapes, from hardy dinosaurs tramping across the Scottish Highlands[17] to the plesiosaurs that popped up in Jules Verne's *Journey to the Centre of the Earth* (1868) and *The Land that Time Forgot* by Edgar Rice Burroughs (1918). In the real world of the 1860s, sea serpents were suggested to be surviving plesiosaurs; 40 years later, the monster that had long inhabited the Patagonian lake of Nahuel Huapi was similarly identified.[18] Thanks to this popular obsession, there was nothing unusual in the posters of dinosaurs and plesiosaurs pinned up on the classroom wall at Fort Augustus primary school, or the illustrated book about prehistoric reptiles sitting on the bookshelf in Alex Campbell's cottage a few hundred yards away.

Until March 1933, prehistoric reptiles were frozen on the page and only came to life in the imagination. Then *King Kong* brought prehistoric reptiles back from the dead and, for the first time, dinosaurs and plesiosaurs were seen moving on the cinema screen. They only had supporting roles but the effect was electrifying. By the time *King Kong* reached England, it was a must-see blockbuster that everyone talked about. When Rupert Gould went to Golders Green in early 1934 to interview Mr and Mrs Spicer, they had all seen the film and been impressed by it. Significantly, Mr Spicer told Gould that the repulsive 'prehistoric' animal that had crossed the road in front of their car reminded him particularly of the Diplodocus.[19]

Gould's own undying loyalty to the sea serpent may explain why he ranked the plesiosaur fifteenth in his list of candidate species in *The Loch Ness Monster and Others*. By now, though, the Surgeon's Photograph had been published and Alex Campbell had run home and matched the 30-foot monster he had watched in Borlum Bay against the picture of the plesiosaur in his book of prehistoric animals. From then on, the Monster was a plesiosaur and it never looked back.

After the war, Constance Whyte picked up the plesiosaur theme, quietly dropping the sea serpent which Gould had espoused so devotedly. Tim Dinsdale's book, with its dramatic photograph of the fossilised elasmosaur skeleton from the clay pits near Peterborough, powerfully reinforced the case. So did the self-sacrificing Dr Denys Tucker; if this highly intelligent scientist believed in plesiosaurs so deeply that he was prepared to lose his job, there must be something in it. In his *Sunday Times* article of August 1960, Peter Scott was careful to sit on the fence, but the accompanying artist's impression of a plesiosaur will have convinced many readers that this was indeed the Monster. Next came Rines's underwater photographs of flippers, body, neck and head, which – depending on viewpoint – can be seen as either confirmatory evidence or a self-fulfilling prophecy.

The evolution of the Monster's identity was not a passive process, but the handiwork of people who were convinced that they knew what it looked like and did their best to make it fit their template.

Robert Gould was determined to squeeze the Monster ('X') into a sea serpent-shaped mould, and omitted observations that might have undermined his argument. He accepted Miss K. MacDonald's sighting of a serpent-like animal in July 1934, but not her earlier, closer observation of the 'crocodile-like' creature which she watched wading up the River Ness. He excluded the crocodile because he 'did not believe that she had seen 'X', even though Miss MacDonald was just as convinced by her 'crocodile'.[20]

Torquil MacLeod and Constance Whyte moulded perceptions of the Monster by drawing their own reconstructions. Later, Tim Dinsdale left out material that spoiled the picture he wanted to create. His statistically average Monster, derived from 100 sightings, had flippers rather than feet only because a small majority of witnesses believed they had seen flippers; he simply wrote off all those who were convinced that their monster had legs.[21]

And so the Monster became a plesiosaur. The final stages in its evolution can be seen in Peter Scott's drawings and paintings of *N. rhombopteryx* (Figure 15).

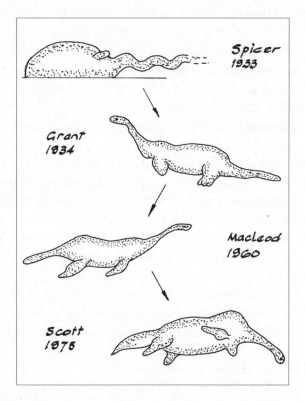

Figure 15   Stages in the evolution of the appearance of the Loch Ness Monster, as pictured successively by George Spicer (1933), Arthur Grant (1934), Torquil MacLeod (1960) and Peter Scott (1975). Illustration by Ray Loadman.

Taken for a ride?

In late 1978, Peter Scott received a polite letter from 12-year-old Nikki Hanley, inquiring if he was aware that his scientific name for the Monster was an anagram of 'Monster hoax by Sir Peter S'.[22] Scott wrote back, explaining that he did know and that – extraordinary though it might seem – this was just a coincidence. But was it?

Scott's notes show that he experimented with many potential names before coming up with *Nessiteras rhombopteryx* ('the wonder from Ness with the diamond-shaped fin'), which classicist Alan Wilkins judged 'excellent'.[23]

However, the reason which Scott gave for naming the Monster – that the animal could only be protected under international law if it had a proper scientific title – turned out to be invalid.[24] Could he have had an ulterior motive?

Scott was unconvinced that the 1975 pictures of the 'whole body' and 'gargoyle head' really showed an animal. He also had reservations about Rines, and had been firmly warned off him ('Tread very warily') by Sir Solly Zuckerman and the President of MIT.[25] Against that background, it is tempting to speculate that Scott deliberately concocted 'Nessiteras rhombopteryx' to protect his reputation. Should the tish ever hit the naf (anagr.), he could then reveal the message encrypted in the clever Greek name and so prove that he had known all along that Rines had faked his photographs.

However inviting this scenario, and however astronomical the odds against a coincidence, it seems that the anagram was just a pure fluke, as Scott always insisted. Dafila Scott well remembers her father being 'horrified' on hearing about it and has no reason to doubt the authenticity of that horror.

Nessiteras rhombopteryx turns out to be a gift on a plate for anagrammatists. As well as Rines's own riposte 'Yes, both pix are monsters. R.' and 'Extortion by Press shamer', almost 20 other reasonably relevant anagrams have been dug out of it.[26] Einstein once remarked that God does not play dice; perhaps word games are not excluded.

So much for a Monstrous non-hoax. It was greatly outnumbered by genuine hoaxes. Over the last 80-plus years, the Monster has acted as a magnet for tricksters, some of whom have shown commendable ingenuity. Constance Whyte was sensitised early to Monster hoaxers, as her brother happened to be working in the Natural History Museum when the plaster cast of the hippo's footprint arrived from Marmaduke Wetherell.[27] Her index of suspicion remained high thanks to Peter O'Connor's photograph and similar offerings, but it is unlikely that she had any inkling of how deeply the cancer of fraud had chewed into the evidence she presented in *More Than a Legend*.

Fake water creatures have a long and noble history. In 58 BC, Romans were thrilled to see the bones of the sea monster which Perseus killed while rescuing Andromeda, and asked no questions about their origin (probably from whales).[28] The celebrated 'Feejee Mermaid' which caused a sensation when it was displayed in London during the 1820s was constructed from the top half of a monkey, shamelessly but skilfully stitched to the back half of a fish.[29] The 114-foot fossilised skeleton of a sea serpent which mesmerised New Yorkers in 1845 was cobbled together from the bones of five *Basilosaurus* whales.[30] Numerous beached 'sea serpents', including the famous 'Marine Monster of Querqueville', found near Cherbourg in late February 1934, were the remains of sharks. These have a cartilaginous skeleton which rots down into bizarre structures that look monstrous, especially when wrapped up carefully in what is left of the skin and photographed from odd angles.[31]

The rogues' gallery of faked photographs of the Loch Ness Monster begins with Lachlan Stuart's three bales of hay and stretches via the Surgeon's Photograph and Frank Searle's handiwork to the flippers created by Bob Rines. To these can be added the fossilised plesiosaur vertebrae which tripped up a retired scrap-merchant on the shore of the Loch (nice try, but they were embedded in the wrong kind of rock) and a fibreglass hump left over from an American documentary.[32] Not forgetting the massive carcass found at Foyers in March 1972, which had its teeth filed, cheeks padded and whiskers shaved to conceal the fact that, in life, it had been an elephant seal in a zoo in Yorkshire.[33] And of course the silver ashtray that had belonged to the chain-smoking Marmaduke Wetherell, mounted in the dried foot of a hippopotamus, which was found by David Martin and Alastair Boyd while cracking the mystery of the Surgeon's Photograph. It is assumed that nobody reading this book would have been taken in by the American report in August 1995 that the Monster ('62-ft. beast was over 1,000 years old') had been harpooned and killed.[34]

These are all small fry beside something which, if true, would be the biggest and most outrageous fraud of all: the claim that there is not a shred of zoology in the Loch Ness Monster and that the entire story originated from a hoax.

In the beginning

Various people have claimed that they invented the Loch Ness Monster. The first is an old friend, well known for his misbehaviour and attention-seeking. While living at Boleskine House in the 1910s, Aleister 'The Great Beast' Crowley put up signs by the lochside which read, 'This way to the Kooloo Mavlick. Does not bite'. These were designed to shock the locals, and they succeeded. Word spread rapidly about Crowley's exotic monster and he was given a wide berth.

Writing more than 30 years later in 1947, Crowley recounted the story with great glee and speculated that 'this harmless invention of mine' had spawned the story of the Loch Ness Monster. Given the 20-year silence before the first reports of the Monster surfaced, this seems unlikely, but Crowley had the last laugh. This particular monster was not as terrifying as the demons which he tried to summon up, although it often hung around (metaphorically speaking) on such occasions. The 'Kooloo Mavlick' was the Great Beast's pet name for his procreative appendage.[35]

A less colourful pretender was Francesco Gasparini, an Italian journalist who supposedly dreamed up the Monster to fill a lean week in August 1933. Gasparini claimed paternity in an article in *Paris-Match* ('Je suis le père du Monstre')[36] but must have been elsewhere at the moment of conception; the Mackays had seen their Strange Spectacle three months earlier.

More plausible, but again lacking any contemporary validation, was the suggestion by English industrialist Lord Royden that hoteliers had dreamed up the Monster to attract visitors.[37] The comment was reported in passing by Sir Derek Bibby in his autobiography, *Glimpses* (1991), but went unnoticed at the time and was never followed up. Royden died in 1950 and Bibby in 2002, making his escape from terminal cancer with a compound so toxic that his post-mortem had to be deferred for several days.

Although impossible to verify, Bibby's suggestion points in the same direction as one final claim that the Monster was invented. Professor Henry Bauer, cryptozoologist and staunch believer in the Monster, deserves all the credit for hunting down this story.[38]

In 1980, Bauer remembered reading that the Monster had been concocted in the early 1930s as a publicity stunt. Chasing the source was complicated by the writer's multiple pseudonyms and prolific output of over 40 books. Eventually, Bauer found the quotation in *Marise* (1950) by 'Stephen Lister', a pen-name of D. G. Gerahty. *Marise* was one of a series of 'semi-autobiographical novels' about the adventures of an irascible and chauvinistic Englishman in a small town in Provence. In this episode, Lister is landed with an abandoned baby girl (Marise) and has to fend off rumours that he is hiding a lion on his property. He explains to the local gendarme that the non-existent lion is not the only large animal to have sprung from rumour:

. . . you may accept my assurance that the Loch Ness Monster was born in my presence, during a conversation which took place in a London public house, under the shadow of the monument erected to the great

Lord Nelson . . . The Loch Ness Monster . . . was invented for a fee of £150 by an ingenious publicity man employed by hotel keepers.[39]

Gerahty prided himself on writing about 'truth with trimmings', and, when Bauer contacted him for clarification, his reply began, 'The truth now follows.'

According to Gerahty, he had run a publicity business in London during the early 1930s and had been hired by 'a group of hotel owners in the area of Lossiemouth' to stump up trade. By coincidence, another customer at the time was an estate agent in the Okanagan Valley in British Colombia, who claimed to have invented Ogopogo, the indigenous lake monster, to attract tourists. Ogopogo had done so well that Gerahty decided to pull the same stunt in Loch Ness, where – even though he had not heard of it before – another legendary creature was supposed to dwell. And so –

Over several pints of beer we became midwives of the reborn Loch Ness Monster. All we had to do was to arrange for the monster to be sighted. This we did and the story snowballed. Thousands went north to see it, and they did. It was, of course, pure hokum. The unwitting parent was Ogopogo.[40]

Gerahty died a year later without revealing any more. Perhaps surprisingly, the pro-Monster Bauer believed him. Bauer reasoned that anything which brought in more potential observers could only increase the chances of witnessing a rare event – hence the flurry of sightings in 1933–34 of the elusive Monster that had always been there. But Gerahty's assertion was fundamentally different, namely that he had created the Monster *de novo*, where none had existed before. Conventional wisdom tells us that this is preposterous, because a Monster has been in the Loch for centuries.

Or has it?

Off the record

For Tim Dinsdale, the history of the Monster began in AD 565, with St Columba and the Pictiverous but God-fearing river creature which Adomnán described in his *Vita Columbae*. Dinsdale took the account seriously, because Adomnán was 'a pious and educated man and he would not have told lies on purpose'.[41] Nobody doubts Adomnán's integrity, but *Vita*

Columbae was written over a century after the event. Also, the Monster of Ness was not that unusual, being just one of many terrifying beasts on land and water that Columba met and vanquished, always in ways guaranteed to impress the Picts. So was this a positive sighting – which would immediately demolish Gerahty's claim – or a clever piece of early Christian propaganda?[42]

For the next 14-plus centuries, the Monster kept a low profile. Nicholas Witchell gave history the benefit of the doubt: 'the shreds of an historical record of strange animals in Loch Ness do exist, although the details are frequently clouded or temporarily forgotten.'[43] In fact, the bare handful of reports supposedly dating from the sixteenth to nineteenth centuries were all sketchy and had no contemporary validation.

The unkillable 'Monster of Loch Ness' of *c.* 1520 was allegedly described in 'an old book dealing with curiosities'. The two huge 'Leviathans' seen during the early 1730s by labourers building General Wade's Military Road were apparently recorded over 30 years later, in a book published in 1769. Sometime in 1896, *The Atlanta Constitution* is said to have run a newspaper article about Loch Ness and its Monster, complete with an artist's impression of a plesiosaur-like reptile. These accounts[44] are widely cited, but all were published only after the modern-era Monster had hit the headlines in 1933 – and none of the alleged sources has ever been found.

The same applies to verbal reports of large animals in Loch Ness: the 'very odd-looking beastie' like a huge frog which terrified diver Duncan MacDonald in 1880; Alexander MacDonald's gigantic 'salamander' of 1888; and Roderick Matheson's massive, maned eel from the same period.[45] None of these sightings was documented at the time, and they were not quoted until late 1933.

Believers would maintain that this is consistent with a rare creature inhabiting a remote area with a poorly literate population who lived in fear of a myth that actively discouraged its own telling. However, cynics would home in on the facts that the animals all looked different, and that no contemporary evidence of sightings could be found before 1933.

Through the centuries, there had also been a disconcerting silence from various parties which had also been expected to broadcast any hint of a Monster in the Loch.

In 1699, James Fraser sang the Loch's praises to the Royal Society: it was bottomless, surrounded by implausibly high mountains and rich in natural and archaeological curiosities.[46] Stories of a huge water creature would have been a gift on a plate for him, especially given his peers' fascination with monstrous exotica – such as the massive femur, supposedly from a race of giants who had been wiped out by the Flood, also reported in the Society's *Philosophical Transactions*.[47] Yet Fraser made no mention of any animals, big or small, in the Loch.

The scientific and popular literature of the eighteenth and nineteenth centuries also appeared to be Monster-free. In 1809, the influential *Scots Magazine* discussed Loch Ness and a monster in the same issue. However, the Loch excited interest because its bitterly cold yet non-freezing water could thaw out icicle-bound horses in the depths of winter, while the 'monster' was the 55-foot mystery carcass that had been washed up at Stronsay, in the Orkney Islands, 130 miles to the north.[48]

The *Encyclopaedia Londinensis* (1814), which drew on the knowledge of 'eminent scholars of the English, Scotch and Irish Universities', singled out Loch Ness for the 'most extraordinary' waves which followed the great earthquake that destroyed Lisbon, the waterfalls above Foyers and the Caledonian Canal, then under construction.[49] Elsewhere, the *Encyclopaedia* devotes many pages to 'sea serpents' and 30 different types of eel. The aquatic fauna of Loch Ness is not mentioned at all.

The famous Bathymetrical Survey (1899–1904) brought experts on aquatic zoology to the Loch for 15 months. Over 500 animal species were documented in and around its waters. Several new species were identified (the most exciting being *Ophryoxus gracilis*, an attractive one- to two-millimetre water flea) but no large water creatures were reported.[50] For sightseers of the time, the attractions of Loch Ness were laid out in David MacBrayne's comprehensive 'Royal Route' guide to the Highlands and Islands;[51] Monsters, legendary or otherwise, do not get a mention.

Rumours of a large unknown animal in Loch Ness would have been expected to reach the ears of the Inverness Scientific Society and Field Club, founded in 1875 to explore 'the chief natural phenomena of the neighbourhood'. The Club's interests were broad. In 1919, their Christmas Lecture was a lantern-slide presentation on 'The Marvels of the Deep' by Professor J. Arthur Thompson from the University of Aberdeen, while the Reverend Macfarlane of Dores covered 'Sea-myths of the Hebrides' in February 1923.[52] Members often descended on the Loch to satisfy their thirst for 'geology, botany, natural history, archaeology &c', but nowhere in their *Proceedings*, from 1875 to 1934, is there any reference to a Monster, real or mythical, in Loch Ness.

What about the newspapers, which had their fingers firmly on the pulse of life in the Great Glen? The *Inverness Courier* and the *Northern Chronicle* were always keen to report prodigious animals, such as a seven-foot sturgeon harpooned in the centre of Doncaster, and a 50-foot sea serpent seen 'about the South-West coast of Inverness-shire' in 1872.[53] Surely they would find space for sightings, or even rumours, of a Monster on their doorstep? During the nineteenth century there was just one newspaper report of a monster in the Loch. On 8 October 1868, a brief item in the *Inverness Courier* described the bizarre carcass of a six-foot animal – a flayed bottle-nosed whale, apparently dumped as a joke at Abriachan.[54]

After that, there was nothing for over 60 years until August 1930, when a two-foot wave swept across the Loch and rocked the boat under three young men fishing off Tor Point.[55] This was the first contemporary suggestion of any unusually large animal in Loch Ness.

The presence of a rare and timid species cannot be excluded in a lake of this size and depth but Monster agnostics could argue that, like the guard dog which aroused Sherlock Holmes' suspicions by failing to bark during the night, the centuries-long silence was significant.

Scotching a myth

The report of the three young anglers in their rocking boat heralded the advent of the modern-era Monster. This watershed – 1930 – was significant for another reason. It marks the start of the period when, if D. G. Gerahty was telling the truth, he invented the Monster.

Suspend any disbelief for a moment, and put yourself in Gerahty's shoes as he plans his great Monster scam to pull tourists into north-eastern Scotland. This should be a winner, because everyone is fascinated by monsters and would love to see one. With luck, the initial blaze of publicity will ignite the basic human urges to witness marvellous things and, in their absence, to create them from banal ingredients. Ideally, your Monster will quickly become a sitting duck for other hoaxers, who will keep the Monster in the public eye while distracting attention from your own efforts. More and more visitors will flood in, swept along by curiosity, wishful thinking and the determination to be the one who finally nails your Monster. Eventually, like a nuclear chain reaction, your hoax will become self-sustaining. You can then sit back, leave them all to get on with it, and congratulate yourself on a job well done.

It will not be plain sailing all the way. As with any successful hoax, you will need a convincing story, good publicity and an effective stooge to make it all happen. Your Monster must have a credible past; it cannot just suddenly appear in a land-locked lake. Apart from the ever-present kelpie, the only unknown animal reported in Loch Ness was the 'fish – or whatever it was', like an upturned boat, which was seen 'some years ago' by the unnamed keeper mentioned in the *Northern Chronicle*'s 1930 piece about the three young anglers. But a blank sheet has its advantages, and gives you free rein to create a fitting history into which you can plug your Monster. Its role model is its 'unwitting parent', Ogopogo, which took no time to weave itself seamlessly into the Okanagan Indian legends of the serpent-like creature they called N'ha-a-itk. However, your Monster and its concocted history may not convince those who regard themselves as experts on the Loch and its traditions. Be prepared for incredulous or even hostile reactions, and fight back by producing new 'evidence'. Fortunately, the masses prefer excitement to cynicism, and as long as you are not caught out, the naysayers will soon give up and be forgotten.

Effective publicity is crucial. In the Great Glen of the 1930s, newspapers – notably the twice-weekly *Inverness Courier* – were the best medium for spreading the word. Once the Monster has taken off, the national press and radio can be trusted to muscle in and give the story the proper coverage that only they can provide.

Ultimately, everything hangs on the stooge who introduces the Monster to the world and, if momentum flags, has to dream up ways of keeping it in the headlines. He must be trustworthy, resourceful and familiar with the area and its inhabitants. In an ideal world, he would also be respected in the community and have a hot line to the local press. Above all, he must be good at covering his tracks and at making whatever he invents look believable.

Finally, a word of caution. You should decide at the outset what sort of Monster you have planted in Loch Ness, and arrange for its description to be widely circulated. Otherwise, witnesses will not know what they are supposed to see, and you may end up with a menagerie of large, unusual species in the Loch.

Now imagine that you have switched sides and are investigating D. G. Gerahty's claim that the Loch Ness Monster began as a scam. The best time to look for evidence of the stooge and his fabrications would be early on, before the waters are muddied by other hoaxers.

Bearing all the above in mind, watch for clues as we revisit the first few months of the story.

Play it again

It all began on 2 May 1933 with the 'Strange Spectacle' witnessed by a businessman and his university graduate wife (both unnamed), as reported by an anonymous Correspondent of the *Inverness Courier*. The 'spectacle' was a patch of disturbed water, in which a creature 'resembling a whale ... many feet long' disported itself before disappearing. The Correspondent reminded readers that the Loch had 'for generations been credited with being the home of a fearsome monster' and that, some years previously, three anglers had encountered a large unknown creature which could not be a seal or a porpoise because these animals never get into the Loch.[56]

An immediate response came from steamer captain John Macdonald, who defied anyone to match his 50 years and 20,000 trips on Loch Ness. He had never heard of any Monster and had never seen any large animal there. The 'awful commotion' in the water was due simply to a swarm of salmon jumping, a phenomenon he had seen many times on the Loch.[57]

A month went by until the next report, by unnamed workmen, of a beast with an enormous head and a large, heavy body following a drifter up the Loch. Passengers on a bus (no details provided) were 'said' to have watched it too.[58]

Two months later, on 22 July, George Spicer saw his six- to eight-foot 'prehistoric' beast cross the road 'carrying a lamb or small animal'. His letter to the *Courier* was printed together with an unsigned introduction which rather downplayed the event, explaining that 'one who knows the ways of otters' had 'no doubt' that this species was Spicer's beast.[59]

Just four days later, the unnamed '*Courier* representative at Fort Augustus' reported three sightings near the town, detailing the witnesses' 'vivid descriptions': like a black horse with a knobbly back, an upturned boat with huge legs working beneath the surface and a big animal splashing itself with legs or flippers.[60]

The spoof article by another *Courier* Correspondent, 'A Mere Woman', followed on 11 August. This poked gentle fun at the jumbled identity of the 'Otterserpentdragonplesadaurus', which was grateful to the 'gentleman who writes to [the *Courier*] from Fort Augustus' for having dragged it into the public gaze.[61] On the same day, a woman tourist saw a black 'rock' in the water, but only realised it had been the Monster when shown a newspaper article by a shopkeeper in Fort Augustus.[62] There were no sightings for a month until 'our Fort-Augustus Correspondent' (anonymous) reported on

12 September that 'several people' in the town had seen a 30-foot creature travelling across the Loch at 30 mph.[63]

Two weeks later, the *Courier* devoted its editorial to the Monster, stating 'there is now no room for doubt about the existence of some abnormally large creature in the Loch'. This new certainty was cemented by 'numerous appearances during the previous weekend'. All the accounts were 'given exclusively to a *Courier* reporter' and described a long, multi-humped Monster that undulated like a giant caterpillar.[64]

For a month there were no new sightings. However, a steady trickle of Monsterabilia sustained interest, with letters about its possible identity (including Colonel Lane's giant salamander theory), the need for a public inquiry and the demand that the Ness Fishery Board should investigate. Newly discovered fragments from the Monster's past were also revealed. Years earlier, a nameless nun from the Abbey in Fort Augustus had told a woman in Cumbria about the 'uncanny beast' in the 'almost bottomless lake', which reportedly picked off animals grazing on the shore. The anonymous 'Fort Augustus correspondent' contributed an article about children of the town who had seen 'a queer-looking creature' near there 20 years earlier but had not mentioned it until now.[65]

In mid-October, the *Courier* printed a letter from a holidaymaker, reporting that two residents 'of undoubted honesty' had told him, back in May, that they had each seen a long-necked, 30-foot animal that looked like a plesiosaur.[66] A few days later, ace reporter Philip Stalker splashed the plesiosaur hypothesis across the *Scotsman*. Stalker was convinced by talking to a man who had refused to believe in the existence of unusual animals in the Loch until he saw a 30-foot creature raise its long neck and small head from the water, apparently listening to two drifters coming from the Caledonian Canal. When Stalker showed him a picture of a reconstructed plesiosaurus, the man stated that it was 'very like' the animal he had seen. This man was anonymous; all the other witnesses interviewed by Stalker were named.[67]

Subsequent landmarks on the road to Hogmanay 1933 included Hugh Gray's photograph of the 40-foot 'serpent' (12 November) and the sighting on 30 December by Mr W. U. Goodbody and his two daughters. They saw a multi-humped creature with a serpentine head and neck; a few days later, Mrs Goodbody and one of the daughters saw the creature again.[68] After that, the Marmaduke Wetherell debacle and the Monster's near-miss with Arthur Grant's motorcycle kept everyone talking well into January 1934, which otherwise was a lean month for sightings.

On 13 February 1934, a 'well-known Inverness gentleman who withheld his name because he 'did not wish to be pestered by the press' disclosed a sighting off Dores to 'a *Courier* representative', also anonymous.[69]

On the same day, the *Courier*'s 'own Correspondent' (anonymous) reported a sighting by 'several absolutely reliable witnesses', but omitted to give any names.[70] In addition, the monks at the Abbey in Fort Augustus revealed the long but previously unmentioned tradition of the Monster which lived near Horseshoe Crag, in a cave connected by a tunnel to the Loch.[71]

At the end of February, the Correspondent brought the exciting news that a serpentine Monster with a swan's neck and a small head had been seen by 'a well-known Ross-shire lady who does not desire her name to be published'.[72] By contrast, the 'well-known English lady hockey-player' who saw the head and tail of an animal that spanned at least 20 feet was happy to give her name to 'a *Courier* representative' in late March 1934.[73]

D. G. Gerahty claimed that the Monster was invented to pull in visitors to hotels 'in the area of Lossiemouth'. If so, the scam failed spectacularly. The Depression made Christmas 1933 thoroughly miserable in Lossiemouth, as across almost all of Scotland.

Forty miles away in Inverness it was a completely different story. In June 1933, the first Monster cruises took place on Captain Macdonald's trusty *Gondolier*. In October, twice as many buses as usual were laid on for the autumn holiday, and the roads beside the Loch were jammed with cars and motorcycles. In November, 'everyone in London' was talking about the Monster. On 19 December, the *Inverness Courier*'s editorial ('Monstrous!') predicted that Christmas would be exceptionally busy, and that 'if the present rate of progress continues, Loch Ness hotel keepers will have good reason to bless the Monster'. On Christmas Day, the city's streets were strangely deserted, but only because everyone was out by the Loch, looking for the creature that had lured them there.[74]

Hoteliers in Lossiemouth had been left out in the cold, but all the hotels in Inverness and around the Loch were fully booked. The beneficiaries included the Monster-spotting Mr W. U. Goodbody, who owned the Invergarry House Hotel south of Fort Augustus, and John and Aldie Mackay, the proprietors of the newly refurbished Drumnadrochit Hotel and the anonymous local couple who saw the 'Strange Spectacle' that started it all.

The veil of secrecy was never lifted from the well-known Inverness gentleman or Perthshire lady. However, the *Courier*'s secretive Correspondent at Fort Augustus finally revealed his identity some months after recording the piece that followed the King's Speech in the BBC's radio retrospective of 1933.

Alex Campbell (1901–83)

June 1983 had been a dreary month, but brighter times lay ahead if the Scottish National Party, about to celebrate its fiftieth birthday, could be believed.[75] Another fiftieth anniversary had recently passed: that of the 'Strange Spectacle on Loch Ness' reported in the *Inverness Courier* of 2 May 1933 by a nameless Correspondent. As it happened, the day which ended that dull June was also the Correspondent's last. The obituary in the *Courier* did Alex Campbell proud, praising his 62 years of service: 'Few newspapers have had so accurate and meticulous a correspondent for so long.'[76]

Campbell was a true son of the Glen. Born in Fort Augustus, he spent his life in the house beside the River Oich which his forebears had occupied since the 'Fifteen' (the Jacobite Uprising of 1715). For 47 years, his job with the Ness Fishery Board shackled him to the Loch, whether patrolling its southern stretches or raising millions of salmon fry. Campbell was much more than a skilled artisan. He read voraciously, had 'a great love for and knowledge of the wildlife of the region' and, although his origins were lowlier than Peter Scott's, he also turned out to be a 'natural' in front of the microphone and camera.[76]

Thanks to the Monster, the Great Glen became too small for Campbell. His anonymous radio piece at the end of 1933, in which he claimed credit for giving the Monster its name, was the first of hundreds of radio, television and film appearances. For the media, he was the obvious man on the spot: the softly spoken, salt-of-the-earth local who knew the Loch like the back of his hand, had seen the Monster himself and could tell the story as it really was. Campbell had trained in the Observers Corps and was ideally placed to see the creatures that inhabited the Loch. Three of his 17 or 18 sightings of the Monster became classics: the 30-foot, long-necked creature in Borlum Bay (22 September 1933), the pair of Monsters, also in Borlum Bay (16 July 1955), and whatever pushed his rowing boat out of the water in mid-Loch 'on a beautiful day in 1955 or 1956'.[77]

Campbell was the fount of knowledge for Whyte, Dinsdale, Witchell and others. Peter Scott brought him down to London to give vital evidence to his high-powered scientific experts in 1960. David James appointed him Local Coordinator for the LNPIB in 1962, and Campbell remained a cornerstone of the Bureau, as expert, eyewitness and the raconteur who enthralled the volunteers with his after-dinner talks each summer. In his posthumous tribute to Campbell, James praised him as one of the two people (the other being Constance Whyte), who originally inspired the research programme at Loch Ness. Latterly, James wrote, Campbell would have had the satisfaction of knowing that 'few scientists of repute would dream of questioning the validity of his evidence'.[78]

Campbell's integrity was legendary. Henry Bauer wrote that 'his honesty was apparent to the many people who asked him about the monster',[79] and Gould, Whyte, Dinsdale, Witchell and many others accepted his accounts of sightings as gospel. Yet inconsistencies riddled Campbell's reports from the start. According to his piece in the *Courier*, the Mackays saw their Strange Spectacle on 28 April 1933, whereas the couple told the meticulous Rupert Gould that the date was two weeks earlier, on 14 April. In his radio piece at the end of 1933, the date had slipped back to 'the end of May'.[80]

More significant questions surround his most famous sighting, which figures prominently in Whyte (1957), Dinsdale (1961), Witchell (1974) and Mackal (1975). His encounter with the 30-foot, long-necked animal that surfaced in Borlum Bay is instantly recognisable: the creature turned its head constantly as if nervous and submerged when two trawlers appeared from the Caledonian Canal. However, the dates and even the year of this unforgettable experience vary between these accounts, from 22 September 1933 (Whyte) to 'a June morning in 1934' (Dinsdale) and 'a calm May morning in 1934' (Witchell).[81]

Campbell's first version was carefully noted by Dom Cyril Dieckhoff, a monk at the Benedictine Abbey, Fort Augustus, who began collating reports of the Monster in autumn 1933. Initially, Campbell was willing 'to give his evidence to any scientific enquirer, but not to the press' but returned a couple of days later to retract his story. As Dieckhoff recorded:

[He] modified his statement to say that sun was in his eyes . . . visibility was poor or only moderate – light not too good – gave impression that for some reason was anxious to minimise what he had previously said and absolutely refused to allow name to be mentioned to anyone . . .[82]

What gave Campbell (who was still anonymous at that stage) cold feet is not clear, but there were hints of tension with his employers, the Ness Fishery Board. The Board had been criticised in the *Courier* for their inactivity over the Monster and they may well have worked out who the anonymous Correspondent was.

Exactly the same sighting is described in Gould's book (1934), but in a curious context: a letter which Campbell wrote to the Board on 28 October 1933. He explained that the 30-foot Monster which he saw in Borlum Bay on 7 September 1933 [sic] had apparently been a line of cormorants, magnified and distorted by optical illusion and poor visibility.[83]

Although prepared to admit to his employers that he had been fooled, Campbell never publicly retracted this sighting. Twenty years later, when Gould had been dead for several years and his book was out of print, Campbell reverted to the original version. Constance Whyte included the report by 'A.C.' in *More Than a Legend*, adding that his Monster had been magnified by optical illusion – which made the creature appear even bigger than its 30 feet. She cited Dieckhoff's account of Campbell's back-pedalling, but explained this away as a classic case of fear of the kelpie.[84]

This sighting became 'one of the most well-known, on account of Mr Campbell's obvious sincerity and his great experience of Loch Ness'.[85] Immortalised in Stalker's article in the *Scotsman*, it was instrumental in consolidating the Monster's identity as a plesiosaur-like animal.[67]

Campbell's job with the Fishery Board paid the bills, but the Monster became a profitable sideline which he exploited to the full. In his quiet way, he thrived on the celebrity status which the Monster brought him, savouring the high life beyond the Great Glen and the procession of exalted people like Peter Scott and David James who beat a path to his door.

The Monster was good for Alex Campbell and vice versa. After the initial report of the 'Strange Spectacle', the *Courier*'s anonymous Correspondent continued to drip-feed sightings and anecdotes (almost all unattributed or unverifiable) which kept the Monster's profile high. He also gave the Monster a history, beginning by reminding his readers about the experience of the three young anglers in 1930.[55] Most people would have forgotten the story because it was so unexciting; the men had watched a two-foot wave approaching across flat-calm water with 'a wriggling motion' but saw no sign of the 'fish or whatever it was' that might have caused it. When retold by Campbell in 1933, the invisible 'fish or whatever it was' had grown into an 'unknown creature' as big as a seal or a porpoise or – since neither of those species ever entered the Loch – 'the Monster itself!'[56]

This planted the Monster not just three years earlier but 'many years' before that, assuming that the unnamed 'keeper' mentioned in the piece had seen a 'fish or whatever it is' . . . like an upturned boat . . . and just as big'.[55] Campbell soon constructed a half-century of tradition with his stories of the 'queer-looking creature' that the schoolchildren of Fort Augustus allegedly watched during the 1910s,[65] and then MacDonald's 'salamander' of the 1880s.[86]

It is possible that Campbell was familiar with the 1930 report of the three anglers for another reason: he may have written it. He contributed to papers other than the *Courier* and there are similarities in the prose as well as the titles of the two items: 'What was it? Strange experience on Loch Ness' (1930) and 'Strange spectacle on Loch Ness. What was it?' (1933). Some also see Campbell's hand in two anonymous letters sent to the *Northern Chronicle* in response to the 1930 article, which claimed a tradition of large animals in the Loch.[87]

The saga of the Monster is like a long stick of rock, run through with lettering that spells out the names of those involved at each stage. You can snap the stick across at particular years and look to see who was there.

1980: Bob Rines, Adrian Shine, Tim Dinsdale, Nick Witchell, Alex Campbell. 1970: David James, Peter Scott, Dinsdale, Roy Mackal, Campbell. 1960: Constance Whyte, Dinsdale, Scott, Campbell. 1934: Rupert Gould, R. K. Wilson, Campbell. Midsummer 1933: the Mackays (anonymous), the Spicers, the unnamed Correspondent for the *Inverness Courier*. August 1930: three anonymous anglers and an unnamed reporter for the *Northern Chronicle*.

Before that, if we only accept contemporary reports, there is nothing.

Crunch time

The 'Loch Ness Monster' that thousands of people have seen, photographed, filmed and echo-sounded is not a single entity, but a ragbag stuffed with non-monstrous animals, tricks of nature and hoaxes. Your challenge now is to dig out all those sources of confusion and decide whether or not the ragbag also contains a large animal species unknown to zoology.

To misquote Sir Arthur Conan Doyle, the master of logical deduction who also wrote *The Lost World* and believed in fairies: when you have eliminated the possible, whatever remains – however improbable – must be the truth. If a hard core of Monster observations cannot be explained away by waves, mirages, otters, cormorants, touched-up photographs and sonar artefacts, then mainstream science has missed a massive trick and cryptozoology has scored its greatest victory ever.

When considering your verdict, try not to be swayed by prejudice or emotion. If your conclusion is different from what you believed almost 300 pages ago, so be it. As John Maynard Keynes said, 'When the facts change, I change my mind.' You might find yourself siding with Peter Scott, whose last letter about the Monster confirmed that, although exceedingly improbable, he still believed that it existed and was related to the plesio-saurs.[88] Or, like Robert Rines, you may decide that the Monster was real but died out around the close of the twentieth century. Perhaps you agree with Maurice Burton's view that, when everything else is eliminated, all that remains is the legend of the kelpie.[89] However, you owe no allegiance to these voices from the past, and we cannot tell what conclusion they would have reached if they had known then what you know now.

Finally, push to one side any sentimental attachment to your Monster. You might not feel the same anguish as the believer who said, 'You have no idea how much I want this thing to be there',[90] but it may still be a wrench. Almost all of us want the Monster to exist, whether or not we believe in it.

And this brings us at last to the Big Question: has Loch Ness ever harboured a large animal species unknown to science? Now that you have stripped away all the confounders, you can see what is left and can give your answer.

Yes or no?

Postscript

Somewhere along here (if you believe it) is where three tricksters photographed a cheap toy submarine that carried a stylised plesiosaur's head and neck instead of a conning-tower. Even if the spot were known, it would be easy to miss. The water is barely visible through the trees, and the 919 bus from Inverness to Fort Augustus makes no concessions for those who might want to savour the moment.

It's a mid-afternoon in early October 2013, and just over 50 years since I was here last. Back then, we were zigzagging across Scotland from Caithness in the far north-west to the Mull of Kintyre, where the Stranraer ferry would carry us home across the Irish Sea to Belfast. We went to Caithness because of the rocks. My father was the Professor of Geology at Queen's University, and a grand master at disguising field trips as family holidays. Luckily, Caithness was thrilling for an 11-year-old: peering straight down nine hundred dizzying feet of cliff into the grey-green ocean at Cape Wrath, six-foot conger eels, crabs as big as dinner plates and a white-knuckle ride in an open jeep along two miles of beach.

It was also a tough act to follow, even under ideal circumstances. Long before we reached Inverness, it had started to rain and tempers were fraying on the back seat of our Hillman Minx, where my sister was smug in the certainty that we would see nothing. When we stopped by the lochside for the first time, I still knew that my new 10 x 50 binoculars would show me something to beat the best in Tim Dinsdale's *Loch Ness Monster* and Constance Whyte's *More Than a Legend*, both of which were in my rucksack. It was only after we had left Fort Augustus that the cruel truth sank in. My sister had been right, and I had joined the millions who had failed to see the Monster.

Dinsdale and White remained on the shelf in my bedroom, alongside Gerald Durrell, David Attenborough and a low-rise, multicoloured terrace of the *Observer's* guides to nature. Above all, Peter Scott was there, with his *Eye of the Wind* and, in pride of place high on the wall, a print of ducks flying in at dusk. It was Scott who clinched my belief in the Monster; he was looking for it, therefore it existed.

In time, I was drawn from natural history into medicine. Studying anatomy in Cambridge in 1974 was enlivened by a visit to the basement where the bodies were prepared for dissection. On this occasion, the cadaver was bigger than usual: a three-ton killer whale from Dudley Zoo called Cuddles, parts of whom later appeared as a mystery slide in our end-of-year exam. Presiding over this formalin-tainted scene, resplendent in half-moon glasses and white wellingtons, was Richard Harrison FRS, Professor of Anatomy and a world authority on marine mammals. He taught us a lot and regaled us with many colourful yarns, but never mentioned that he had been in Scott's hand-picked scientific committee to review the evidence for the Loch Ness Monster.

The year 1974 was a memorable one for many people. Nicholas Witchell's book was published, Scott was so busy that he missed Bob Rines's flipper photographs and my father left Belfast for the Regius Chair in Geology at Birmingham University. He also regressed (as Denys Tucker would have viewed it) from palaeontologist to living fossil by becoming Chair of the Trustees of the Natural History Museum. In that role, and with Scott now Chancellor at Birmingham, it was inevitable that their orbits would eventually intersect. Nonetheless, while digging through the Peter Scott Archives in Cambridge to research this book, it gave me a jolt to see 'Professor Alwyn Williams FRS' in Scott's handwriting, on the invitation list for the ill-fated Edinburgh symposium of December 1975.

Today, Inverness remains well rooted in its past but is also moving with the times. A non-English conversation overheard on the street is just as likely to be in Polish as in Gaelic. The McDonalds who flaunt their insignia just a brick's throw from the Castle have nothing whatsoever to do with the Highlands. Most of the Nessie souvenirs are made in China. And Aleister Crowley would have approved of the lap-dancing club, but not of 'no touching, no full nudity'.

The once-noble Union Street has gone the way of most British high streets. Ogston's the Chemist, where Dr R. K. Wilson took his quarter-plates to be developed, is now a designer menswear shop. A few doors down was the fishing and shooting emporium once run by one of the three young anglers who were in a boat off Tor Point in late August 1930 at what might have been the very start of the story. The establishment is commemorated only by name, in the Gunsmiths Karaoke Bar which is to be found above an Indian restaurant.

The Inverness Library's Monster archive lies in a couple of cardboard boxes in a locked store-room, apparently undisturbed for years. In the reading room, the entire life history of the *Inverness Courier* (born 1817) has been boiled down to two drawers of microfilm spools, each contained in a square box of white card. The boxes labelled '1933' and '1934' are grubby and well used; almost all the rest, including '1983', wherein lies the obituary of part-time correspondent Alex Campbell, look untouched.

The 919 bus is now approaching Drumnadrochit. Out on the left, Urquhart Castle sits broken but proud on its headland; below the road is Temple Pier where, in early summer 1975, Dick Raynor photographed Bob Rines with the diving-suited Peter Scott.

Arriving at the Loch in 1967 as a 17-year-old volunteer, Raynor quickly became one of the LNPIB's stars by filming a wake that was ranked alongside Dinsdale's movie. He was inescapably hooked. After the LNIB folded, he spent a couple of years in the Merchant Navy, jumping ship when they refused him unpaid leave to spend a year chasing the Monster. Later, he worked as a diver on a wave energy project in Dores Bay, which is where he saw the family of mergansers that perfectly reproduced the object in his 'positive proof' film. His faith in his Monster was shattered, but the magic of the place never let him go. Three years from now, he will celebrate the fiftieth anniversary of coming here.

Yesterday, he took me out on the boat tour which he runs, complete with real-time sonar scanning of the Loch floor and a video that tells the story of the Monster, warts and all. The other passengers were two South Korean girls 'doing Scotland', who had to come here because 'the Monster is big at home'. They photographed everything, from the Castle to the unlucky boat that the sonar captured on the floor of the Loch, and the lake-wise ducks which clambered on board for lunch. Back at Temple Pier, the girls set off for Skye and their next photo opportunities, but the ducks remained while Raynor told me more. He made it sound huge fun, and, for the hundredth time, I wished that I'd been there.

Did he have any regrets? No. Not even when he realised that his film was not the Monster after all? No; that was just another piece of the puzzle.

What about the Monster? Raynor considered the Loch for a long moment before replying. 'People will still come here in twenty or thirty years' time,' he said. 'Maybe not to find it, but to see what all the fuss was about.'

The 919 has now passed the promontory at Achnahannet, where the LNPIB settlement stood in Fraser's Field. The next stop is Invermoriston, where I'm getting off. It's a five-minute walk down the hill, past the dank, fern-encrusted hollow containing St Columba's Well and over the bridge that carries the 'new' road across the River Moriston. My destination is Tigh-na-Bruach, 'The house on the edge'. Now a bed-and-breakfast establishment, this was formerly the home of Colonel W. H. Lane, the rifle-toting nemesis of giant salamanders and author of the first book about the Loch Ness Monster.

The back of the house looks across the water to the steep slopes of the eastern shore, a rugged tapestry of bracken and scrubby trees with grey rock showing through the bare patches. Here, in early 1934, Colonel Lane may well have chuckled as he trained his binoculars on the Monster-mimicking log that was causing such excitement as it floated past Invermoriston Pier. Here, too, in 1947, he finally saw his Monster, laying its torpedo-like trail across the water.

Down at the shore, the Loch is mirror-calm and a surprisingly rich blue. Colonel Lane's private jetty has been rebuilt, and the water beside its end is the colour of weak tea. I drop in a shiny new tenpence piece; it is silver when it hits the surface, and dull bronze when it settles on the bottom, just five feet down.

Nearly 20 miles up to the left is the beach at Dores, where I went to meet Steve Feltham a couple of days ago. His story began like Dick Raynor's, but a quarter of a century later and (so far) with a different ending. Feltham's childhood fascination with the Monster matured into an irresistible attraction. At the age of 28, he realised that he didn't want to reach the age of 70 only to look back on a life unfulfilled, so he abandoned a safe family job and his personal ties in Dorset and moved up to the Loch. That was in 1991. In 2012, *The Guinness Book of Records* contacted him to check that he really was the person who had maintained the longest uninterrupted vigil for the Monster. He didn't bother to reply.

During most of that time, home has been a mobile library van, which lost any capacity for mobility several years ago and is now rusted into its final resting place. Inside, there is still a library, but small and highly specialised (only about the Monster), a piano, a bed and a wood-burning stove. The stove bakes the clay Monster models he makes to stay above the financial waterline, and provides nearly enough heat in winter. Opening the van's roof-light at night creates his 'personal planetarium'; lying in bed, he can look straight up at the Milky Way and occasionally the Northern Lights. Other creature comforts are served by the pub at the end of the beach.

In front of the van, Feltham's chair sits behind a trestle table and a business-like binocular telescope on a tripod, pointing down the Loch towards Urquhart Bay. While we talk, his eyes constantly return to the water and his fingers work on autopilot, tearing off pieces of coloured modelling clay and forming them into his trademark Monsters. He knocks off three during our conversation.

You can find Feltham's website by typing into Google the question he is most commonly asked – haveyouseenityet? And he has, or at least a torpedo-like trail that raced away beneath the surface and that he couldn't otherwise explain. What has kept him here for 23 years is not *The Guinness Book of Records* (although he is in there now), but the hope of seeing the inexplicable once again and of capturing it on film.

Any regrets? Only that he'd not been better prepared when he saw the torpedo trail. His belief in the Monster? Just the same as always. It's out there, and, in time, he'll see it.

Back on Colonel Lane's jetty the light is fading and the wind has picked up. The Loch has turned grey and is now ruffled with waves. Beside the jetty, the water has darkened from weak tea to Guinness, and the silver coin that turned into bronze has vanished completely.

By the time I've climbed back up to Colonel Lane's lookout point beside the Scots pines, Loch Ness looks quite different. It's now vast and impenetrable, and a place where almost anything could be hiding.

B'amhail mura b'fior
Probable, if it were not true

Acknowledgements

Only an idiot would fail to put his wife at the top of any thank-you list, but mine deserves to be there, too. It's now thirty-four years since Caroline first read something I'd written, and her feedback has often been character-building; at least this book didn't put her to sleep on a train, as the last one did. I'm indebted to her, and to the equally tolerant Tim, Jo and Tessa (new dog, not another child) for their support, and for not changing the locks while I was away doing my research.

Ted Holiday, Monster-hunter and author, once admitted that he needed all the help he could get to write a book about such a strange subject. I can only agree. I was lucky to have the wisdom of several experts, all of whom have been extremely generous with their time and recollections, and all made me wish that I'd been there.

Top of the list are a triumvirate who, between them, have a century and a quarter of inside knowledge – Dick Raynor, Adrian Shine and Nicholas Witchell. All three kindly read various drafts of the manuscript and corrected my many errors of fact, chronology and interpretation, and also allowed me to reproduce photographs from their personal collections. Special mention goes to Dick, who was in attendance for almost half of the book's three-year gestation and has therefore suffered the longest. If I have managed to capture the fun and excitement of the hunt for the Monster during the late 1960s and early 1970s, it will be largely thanks to him. I'm profoundly grateful to Adrian for having guided me through the intricacies of life in deep Scottish lakes and of how to find large, unknown aquatic animals. Without Nick's help, I could never have made sense of one of the crucial episodes in the story, and there were many other occasions when I was glad of his wide-ranging knowledge. Invaluable help was also provided by Holly Arnold, Ivor Newby, Steve Feltham and Norman Newton.

Sir Peter Scott and David James MP played complicated roles, which would have been difficult to unravel without Dafila Scott, Chris James and the Honourable Jacquetta James. I would like to thank all of them for their time and for granting permission to reproduce material from the Peter Scott Archives and David James's memorabilia at Torosay Castle. Further insights into Peter Scott's life and works were kindly provided by Sheila Fullom, formerly of the BBC, and Peter Walkden, Scott's bibliographer at Slimbridge during the 1980s.

My thanks also go to Professor Alan Gillespie, of Washington University, US, and the Valles Marineris, Mars; Fiona Selkirk, Adrian Soar and Roger Woodham, formerly at *Nature*; Dr Tobias Churton, University of Exeter; Professor David Sims, Marine Biological Laboratory, Plymouth; Dr Adam Smith, Nottingham Natural History Museum; Sandi Shallcross, Thornbury & District Museum, Gloucestershire; Sir Michael Bibby and Mike and John Royden; and Claire Griffiths (Moray Heritage Memory Project) and Arthur Scott.

I am immensely grateful to the following for providing elusive photographs and images: Dawn Dinsdale and other members of her family; Richard Wilson and Professor David Watters; Tony Healey; and David Martin and Alastair Boyd.

I owe much to my 'Hanging Committee', a group of sharp-eyed and critical friends who have cheerfully read numerous drafts of the manuscript, and have been absolutely brilliant at spotting everything from one-letter typos to whole pages that were fit only for the shredder. They could tell you what the book would have been like without their input, but I hope that they won't. In alphabetical order, they are Dr Kathryn Atkins, Paul Beck, Moira Fozard, Dr John Lee, Jeanne and Ray Loadman, Jenifer Roberts, Tracy and Dr Robert Spencer, and Ernest Woolford. As in previous books, Ray Loadman has worn another hat together with his artist's smock, and has again brought beauty and style to the illustrations and maps.

One of the curses of the information highway is that it's too easy to find what you need while sitting behind your computer. This book gave me plenty of opportunities to get out and hunt through real archives. In the University Library, Cambridge (while fighting down the impulse to break Rule 26, 'No person shall go barefoot'), I was guided through the Peter Scott Archives by Frank Bowles, Louise Clarke and Adam Perkins (Curator of Scientific Manuscripts). Also exceedingly helpful were the staff of the Highland Archive Centre, Inverness, notably Lorna Steele, Debbie Potter, Peter Mennie and Colin Waller; Cait McCullough at High Life Highland, Inverness; Susan Skelton, Inverness Library; Dr Emily Goetsch, the National Library of Scotland, Edinburgh; Kate Tyte, John Rose and Yolande Ferreira, the Natural History Museum, London; Diane Manipud, the King's College Hospital Archives, London; Steven Dryden, the Sound Archives, British Library, London; and Lynda Brooks at the Royal Linnaean Society, London.

Moral support, far beyond the normal call of friendship, has also come from Tim and Julie Mann, Colin Gardner, Tim Jones, Alison Paton, Rob Bartlett, Gilbert Howe and Charlotte and Lesley Brockbank. For reasons too embarrassing to explain, this was the first book to land me in hospital. I'm grateful to Professor Jonathan Benger and his team in Bristol Royal Infirmary's excellent Accident and Emergency Department for sorting me out, and for kindly providing pen and paper so that I could catch up with the word count while they did so.

A top-class publisher does much more than turn a 1.3-megabyte email attachment into nearly 400 printed pages. The team at Orion have been superb in dealing with all aspects of this book. They have also been great fun to work with. I'm indebted to Alan Samson, Lucinda McNeile and Simon Wright for their expert guidance, good humour and encouragement throughout – and especially to Alan, for his enthusiasm from the start for what some might regard as a risky title. My sincere thanks also to Richard Collins, who has suffered as only a copy editor can; to Katie Horrocks and her excellent production team; to Sarah Jackson, for making this such a beautiful book; to Brian Roberts, for his brilliant jacket design; and to Margot Weale, for spreading the word so effectively.

Finally, this is a book that I've always wanted to write. It's a great story, and I hope that I've done it justice. If I haven't, the responsibility is entirely mine.

Gareth Williams

Rockhampton, Gloucestershire
June 2015

Bibliography

Websites

Loch Ness Investigation. Dick Raynor's 'factual website detailing personal study of Loch Ness phenomena since 1967'.
http://www.lochnessinvestigation.com/

Loch Ness & Morar Project. Adrian Shine's 'reference site for general scientific information concerning the exploration and investigation into Loch Ness and its famous monster controversy'. Many papers and documents, including the LNIB and Morar Project Annual Reports, available online through the Research & Archive Room.
http://www.lochnessproject.org/

Nessie Hunter. Website of Steve Feltham, 'full-time Loch Ness Monster hunter since 1991 on Dores Beach'.
http://www.nessiehunter.co.uk/

Loch Ness Information Website, by Tony Harmsworth. 'Devoted to the understanding of the Loch Ness Monster Mystery'.
http://www.loch-ness.org/

Loch Ness Mystery Blogspot, by Roland Watson. 'Reclaiming the Loch Ness Monster from the current tide of debunking and scepticism'.
http://lochnessmystery.blogspot.co.uk/

Loch Ness Sightings. Gary Campbell's 'official register of sightings of the Loch Ness Monster', from AD 565 to present.
http://www.lochnesssightings.com/

The man who filmed Nessie, by Angus Dinsdale. Devoted to 'Tim Dinsdale, whose incredible Nessie film continues to create conjecture the world over'.
http://www.themanwhofilmednessie.com/

Books

Adomnán. *Vita Sancti Columbae*, AD 690. British Library, London MS ADD.35110. Abridged version, Gregory J. *The Life of Columba.* Edinburgh, Floris Books, 1991.

Bauer H. H. *The Enigma of Loch Ness: Making Sense of a Mystery.* Urbana and Chicago: University of Illinois Press, 1968.

Betts J. *Time Restored: The Story of the Harrison Timekeepers and RT Gould, 'The Man Who Knew (Almost) Everything.'* Greenwich: National Maritime Museum & Oxford University Press, 2006.

Binns R. *The Loch Ness Mystery Solved.* Shepton Mallet, Somerset: Open Books, 1983.

Burton M. *The Elusive Monster.* London: Rupert Hart-Davis, 1961.

Campbell E. M., Solomon D. *The Search for Morag.* London: Tom Stacey, 1972.

Campbell S. *The Loch Ness Monster: The Evidence.* Wellingborough: The Aquarian Press, 1986. Revised edition, Aberdeen: Aberdeen University Press, 1991.

Carruth J. A. *Loch Ness and Its Monster.* Fort Augustus: Abbey Press, 1945.

Cassie R. L. *The Monsters of Achanalt.* D. Wyllie & Son, 1935.

Churton T. *Aleister Crowley: The Biography.* London: Watkins Publishing, 2011.

Coleman L. Huyge P. *Field Guide to Lake Monsters, Sea-Serpents and Other Mystery Denizens of the Deep.* New York: Jeremy P. Tarcher, 2003.

Costello P. *In Search of Lake Monsters.* London: Garnstone, 1991.

Dinsdale A. *The Man Who Filmed Nessie.* Surrey, BC, Canada: Hancock House, 2013.

Dinsdale T. *Loch Ness Monster.* London: Routledge & Kegan Paul, 1961; Philadelphia: Chilton, 1962; 2nd edition, Routledge & Kegan Paul, 1972; 3rd edition, 1976; 4th edition, 1982

Dinsdale T. *The Leviathans.* London: Routledge & Kegan Paul, 1966; 2nd edition, London: Futura, 1976.

Dinsdale T. *Project Water Horse.* London: Routledge & Kegan Paul, 1975.

Dinsdale T. *The Story of the Loch Ness Monster.* Allan Wingate, London, 1973.

Eberhart G. M. *Mysterious creatures. A Guide to Cryptozoology.* Santa Barbara, CA: ABC-CLIO Inc., 2002, pp. 375–84.

Frere R. *Loch Ness.* London: John Murray, 1988.

Gosse P. H. *The Romance of Natural History*, 1875. Republished with foreword by L. Coleman. New York: Cosimo, 2008.

Gould R. T. *The Case for the Sea-Serpent.* London: Phillip Allen & Co., 1930.

Gould R. T. *The Loch Ness Monster and Others.* London: Geoffrey Bles, 1934. Revised edition, New York: University Books, 1969.

Harmsworth T. *Loch Ness, Nessie and Me. The Truth Revealed*. Drumnadrochit: Harmsworth.net, 2010.

Harrison P. *The Encyclopaedia of the Loch Ness Monster*. London: Robert Hale, 1999.

Heuvelmans B. *In the Wake of the Sea-Serpents*. London: Rupert Hart-Davis, 1968.

Holiday F. W. *The Great Orm of Loch Ness*. London: Faber and Faber, 1968.

Holiday F. W. *The Dragon and the Disc*. London: Sidgwick & Jackson, 1973.

Holiday T., Wilson C. *The Goblin Universe*, revised edition. London: Xanadu Publishing, 1990.

Huxley E. *Peter Scott. Painter and Naturalist*. London: Faber and Faber, 1993.

Kallen S. A. *The Loch Ness Monster*. San Diego, CA: Reference Point Press, 2009.

Klein M., Rines R.H., Dinsdale T., Foster L. S. *Underwater Search at Loch Ness*. Monograph 1, Academy of Applied Science. Boston: Academy of Applied Science, 1972.

Lane, W. H. *The Home of the Loch Ness Monster*. Edinburgh: Grant & Murray Limited, 1934.

Lister S. (pseudonym of D. G. Gerahty). *Marise*. London: Peter Davies, 1950.

Mackal R. P. *The Monsters of Loch Ness*. The Swallow Press, Chicago, 1976.

Martin D. Boyd A. *Nessie – The Surgeon's Photo Exposed*. Martin & Boyd, East Barnet, 1999.

Meredith D. *The Search at Loch Ness. The Expedition of the New York Times and the Academy of Applied Science*. New York: Quadrangle, 1977.

Murray J., Pullar L. (editors). *Bathymetrical Survey of the Scottish Fresh-Water Lochs*, vols 1–6. Edinburgh: Challenger Office, 1910.

Oudemans A. C. The great sea-serpent. Luzac & Co., 1892.

Oudemans A. C. *The Loch Ness Monster*. Leyden: Late E.J. Brill, 1934.

Robson J. *One Man in His Time. The Biography of David James*. Staplehurst, Kent: Spellmount, 1998.

Searle F. *Nessie. Seven Years in Search of the Monster*. London: Coronet, 1976.

Searle F. *Monster Hunter Extraordinary: Ten Years in Search of the Loch Ness Monster*. Unpublished manuscript, undated (ca 1980). Highland Council Archives, Inverness D885/1.

Shuker K. P. N. *In Search of Prehistoric Survivors*. Blandford, London, 1995.

Whyte C. *More Than a Legend*. London: Hamish Hamilton, 1957; revised 3rd impression, 1961.

Witchell N. *The Loch Ness Story*. Lavenham, Suffolk: Terence Dalton, 1974; revised edition, Harmondsworth: Penguin, 1975; revised edition, London: Corgi, 1989.

Articles and papers

The Peter Scott Archives relating to the Loch Ness Monster are held at the University Library, Cambridge; the Frank Searle Archives are at the Highland Archive Centre, Inverness; and the archives of the Inverness Courier *and* Northern Chronicle *are at the Inverness Library.*

Anonymous. What was it? A strange experience on Loch Ness. *Northern Chronicle* 27 August 1930.

Anonymous (Alex Campbell). Strange spectacle on Loch Ness. What was it? *Inverness Courier* 2 May 1933, p. 5.

Anonymous. Loch Ness 'Monster': ship captain's views on occurrence. Letter by John Macdonald, *Inverness Courier* 12 May 1933.

Anonymous. Visitor's experience. Quoting letter by George Spicer, *Inverness Courier* 4 August 1933.

Anonymous. The Loch Ness 'Monster'. True story of his life. Told to A Mere Woman; Anonymous. Another sighting at Glen Urquhart. *Inverness Courier* 11 August 1933, p. 5.

Anonymous. Monster mystery deepens. Cast of spoor like a hippopotamus's. Zoology experts' report. *Daily Mail* 4 January 1934.

Anonymous. London Surgeon's photograph of the Monster. Monster yards from Lochside. *Daily Mail* 21 April 1934, p. 1.

Anonymous. News and Views. Monsters by sonar. *Nature* 1968; 220:1272.

Anonymous. Skeptical Eye. The (retouched) Loch Ness Monster. *Discover* September 1984, p. 6.

Anonymous. News and Views. Monsters by sonar. *Nature* 1968; 220:1272.

Anonymous. Obituary. Mr Alex M. Campbell, Fort Augustus. *Inverness Courier*, 1 July 1983.

Baker P. Echo-sounding as a method of searching underwater in Loch Ness. Appendix B in: Holiday F. W. *The Great Orm of Loch Ness*. London: Faber and Faber, 1968.

Baker P. F., Arnold R. The mystery of Loch Ness. *Scotsman* 12 September 1960.

Baker P. F. Westwood M. Underwater detective work. *Scotsman* 14 September 1960.

Baker P. F. Westwood M. Sounding out the Monster. *Observer* 26 August 1962.

Barker I. Evolution? I don't believe it. Haven't you heard of Nessie? *Times Educational Supplement Magazine* 6 July 2012.

Braithwaite H. Sonar picks up stirrings in Loch Ness. *New Scientist,* 19 December 1968; 40:664-666.

Burton M. The Loch Ness Monster. *Illustrated London News* 20 February 1960.

Burton M. Muck and monsters. *Illustrated London News* 1 October 1960, p. 568

Calman W. T. The evidence for monsters. *Spectator* 22 December 1933.

Campbell A. Letter to Ness Fishery Board, 28 October 1933. Quoted in: Gould R. T. *The Loch Ness Monster and Others*, p. 110.

Campbell E. M. Questing the Beast. *The Times* 2 October 1968.

Desmond A. Why Nessiteras rhombopteryx is such an unlikely monster. *The Times* 27 December 1975.

Dinsdale T. The Rines/Egerton picture. *Photographic Journal* April 1973, 162–165.

Ellis W. S. Loch Ness: the lake and the legend. *National Geographic* 1977; 151:758–79.

Fitter R. S. R. *The Loch Ness Monster: St. Columba to the Loch Ness Investigation Bureau.* Presentation to the International Society of Cryptozoology/Society for the History of Natural History Symposium on the Loch Ness Monster, Edinburgh, 25 July 1987. *Scottish Naturalist* 1988, Centenary edition, part 2: 47–51.

Fraser J. Part of a letter concerning the Lake Ness, etc. *Philosophical Transactions of the Royal Society* 1699; XXI:230–2; reproduced in Whyte C. *More Than a Legend*, 1961, p. 207–10.

Gould R. T. The Loch Ness 'Monster'. A survey of the evidence. *The Times* 9 December 1933.

Grimshaw, R. and Lester, P. (1976) *The Meaning of the Loch Ness Monster.* Occasional paper. Centre for Contemporary Cultural Studies, University of Birmingham. 42pp.

Halstead L. P., Goriup P. D., Middleton J. A. The Loch Ness Monster. *Nature* 1976; 259:75–6.

Harwood G. E. Interpretation of the 1975 Loch Ness pictures. *Progr in Underwater Sci* 1977; 2:83–90, 99–102.

Hepple R. Ness Information Service, Nessletter No. 1, January 1974. Nessletters from 1974 to 1995 are archived online at http://lochnessmystery.blogspot.co.uk

Holden J. C. The reconstruction of 'Nessie'. The Loch Ness Monster resolved. *J Irreproducible Results* 1974; 4:14–18.

Holiday F. W. The case for a spineless Monster. *The Field* 5 February 1976, 204–5.

House of Commons Debate, 3 February 1961. *Hansard*, vol 633, cc 1421–32 [Dr Denys Tucker, dismissal by the Trustees of the Natural History Museum].

House of Lords Debate, 16 July 1969. Hansard, vol. 304, cc 262–4 [Loch Ness Monster: submarine research].

Insight. Nessie, this is your best show yet. *Sunday Times* 30 November 1975.

James D. Time to meet the Monster. *The Field* 23 November 1961, pp. 951–3

James D. The Monster again. *The Field*, 14 June 1962, 1060.

James D. We find that there is some unidentified animate object in Loch Ness. (Report on the 1963 expedition). *Observer* 17 May 1964.

James D. Fine-weather monster. (Report on the 1964 expedition). *Observer* 27 December 1964.

James D. *Photographic Interpretation Report — Loch Ness. Report on a film taken by Tim Dinsdale.* Inverness: *Inverness Courier*, 1966.

James, D. Loch Ness Investigation. Annual Reports, 1965–9. London: The Loch Ness Phenomena Investigation Bureau. Available online at http://www.lochnessproject.org/adrian_shine_archiveroom/

Klein M., Finklelstein C. Sonar serendipity in Loch Ness. *Technology Review* December 1976: 44–57.

LeBlond, P. H. and Collins, M. J. 1987. The Wilson Nessie photo: a size determination based on physical principles. *Cryptozoology* 6, 55–64.

Lehn W. B. Atmospheric refraction and lake monsters. *Science* 1979 205: 183–5.

Macrae J. Letter to *Inverness Courier* 20 August 1872. Cited in Oudemans A. C. *The Great Sea-Serpent* (1892). Republished with foreword by Loren Coleman. New York: Cosimo, 2009, pp. 428–31.

'Mandrake' (Peter Purser). Making of a monster. *Sunday Telegraph* 7 December 1975.

Memory F. W. Loch monster's tracks found? *Daily Mail* 20 December 1933.

Mountain E. Solving the mystery of Loch Ness. *The Field* 22 September 1934, pp. 668–9.

Mountain E. Expedition to Loch Ness. *Proc Linnaean Society*, Session 1934-5, Part I, pp. 7-12.

Mulchrone V. The Loch Ness Monster does not exist. *Daily Mail* 27 September 1969.

Murray J., Pullar L. Mirages on Loch Ness. *Geogr J* 1908; 31:61–2.

Paxton C. A cumulative species description curve for large open water marine animals. *J Mar Biol Ass UK* 1998; 78:1389–91.

Presentation of Loch Ness Evidence to the Members of both Houses of Parliament, Scientists and Press, Grand Committee Room, House of Commons, 10 December 1975. Transcript, pp. 24–8. Peter Scott Archives C.725.

Proceedings of the Independent Panel on Unexplained Phenomena in Loch Ness. Inverness: *Inverness Courier* 1963. Reprinted in: Holiday F. W. *The Great Orm of Loch Ness*, 1968, pp. 205–12.

Razdan R, Kielar A. Sonar and photographic searches for the Loch Ness Monster: a reassessment. *Skeptical Inquirer* 1984; 9: 147–58.

Richardson E. S. Wormlike fossil from the Pennsylvanian of Illinois. *Science* 1966; 151: 75–6.

Rines, R. H. 1982. Summarizing a decade of underwater studies at Loch Ness. *Cryptozoology* 1, 24–32.

Rines R. H., Wyckoff C. W., Edgerton H. E., Klein M. Search for the Loch Ness Monster. *Technology Review* March–April 1976, pp. 25–40.

Scheider W., Wallis P. An alternate method of calculating the population density of Monsters in Loch Ness. *Limnology and Oceanography* 1973; 18:343.

Scott P. The Loch Ness Monster. Fact or fancy? *Sunday Times Magazine* 14 August 1960, p. 17.

Scott P. Why I believe in Nessie. *Wildlife* March 1976, p. 110–11.

Scott P, Rines R. (names not cited). Naming the Loch Ness monster. *Nature* 1975.; 258:466–8.

Sheldon R. W., Kerr S. R. The population density of Monsters in Loch Ness. *Limnology and Oceanography* 1972; 17:796–8.

Shine A. J. Loch Morar Expedition, Annual Reports, 1975–76; Loch Ness & Morar Project, Annual Reports, 1979–83. Available online at http://www.lochness-project.org/adrian_shine_archiveroom/

Shine A. J. The biology of Loch Ness. *New Scientist* 17 February 1983, p. 462–7.

Shine A. J. Postscript: surgeon or sturgeon? *Scottish Naturalist* 1993; 105:271–82.

Shine A. J, Martin D. S. Loch Ness habitats observed by sonar and underwater television. *Scottish Naturalist* 1988; 105:111–19.

Shine A. J., Minshull R. J., Shine M. M. Historical background and introduction to the recent work of the Loch Ness and Morar Project. *Scottish Naturalist* 1993; 105:7–22.

Sitwell N. The Loch Ness Monster evidence. *Wildlife* March 1976, p. 102–9.

Stalker P. Loch Ness Monster: a puzzled Highland community. *Scotsman* 16 October 1933, p. 11.

Stalker P. Loch Ness Monster: the plesiosaurus theory: old stories recalled. *Scotsman* 17 October 1933, p. 9.

Thorpe S. A., Hall A., Crofts I. The internal surge in Loch Ness. *Nature* 1972; 237:96–8.

Tucker D. W. The Zoologist's Tale. In: Witchell N. *The Loch Ness Story*, revised edition, London: Corgi, 1989, pp. 205–28

Watters D. A. Loch Ness, special operations executive and the first surgeon in Paradise: Robert Kenneth Wilson (26.1.1899–6.6.1969). *Aust New Z J Surg* 2007; 77:1053–7.

Wedderburn E. M. Seiches observed in Loch Ness. *Geogr J* 1904; 24: 441–12.

Wilkins A. The four vital sightings. *The Field* 29 November 1975.

Williams E. The 'Bulgarian' Naval Officer. In *Great Escape Stories*. London: The Heirloom Library, 1958, pp. 71–93.

Notes

The Peter Scott Archives relating to the Loch Ness Monster are held at the University Library, Cambridge; the Frank Searle Archives are at the Highland Archive Centre, Inverness; and the archives of the Inverness Courier *and* Northern Chronicle *are at the Inverness Library.*

Chapter 1

1 Scott P., Rines R. (names not cited). Naming the Loch Ness monster. *Nature* 1975; 258:466–8.

2 Stephenson D., Gould D. *British Regional Geology. The Grampian Highlands*, 4th edition. London: HMSO, British Geological Survey, 1995; Trewin N. H., ed. *The Geology of Scotland*. London: The Geological Society, London, 2002.

3 Murray J., Pullar L. *Bathymetrical survey of the Scottish fresh water lochs, during the years 1897 to 1909*. Part II. Report on the scientific results. Edinburgh: Challenger Office, 1910, pp. 381–7.

4 Shine A. J., Martin D. S. Loch Ness habitats observed by sonar and underwater television. *Scottish Naturalist* 1988; 105:111–19.

5 Fraser J. Letter to J. Wallace, Edinburgh, concerning the Lake Ness, etc. *Phil Trans Royal Soc* 1699; XXI:230–32.

6 Murray J., Renard A. F. *Report on the scientific results of the voyage of the HMS Challenger during the years 1873–1876*. London: HMSO, 1891.

7 Duck R. W. The charting of Scotland's lochs. *Forth Naturalist and Historian* 1990; 13:25–9.

8 Klein M., Rines R. H., Dinsdale T., Foster L. S. *Underwater Search at Loch Ness*. Monograph 1, Academy of Applied Science. Belmont, Mass.: Academy of Applied Science, 1972.

9 Blundell O. Notice of the examination, by means of a diving dress, of the artificial island, or crannog, of Eilean Muireach, in the south end of Loch Ness. *Proc Soc Antiq Scotland* 1909; 43:159–64.

10 Murray J., Pullar L. *Bathymetrical Survey* 1910, p. 381.

11 Mackenzie O. H. *A hundred years in the Highlands*. London: Edward Arnold, 1921, p. 16.

12 Mackinlay J. M. Folklore of Scottish Lochs and springs. Edinburgh: W. Hodge, 1893.

13 Briggs K. *An Encyclopedia of Fairies, Hobgoblins, Brownies, Bogies and Other Supernatural Creatures*. New York: Random House, 1997.

14 Witchell N. *The Loch Ness Story*. Lavenham: Terence Dalton, 1974, p. 30.

15 Anonymous. Monster sightings recalled. *Inverness Courier*, 6 November 1962.

16 Whyte C. *More Than a Legend*, 3rd edition. London: Hamish Hamilton, 1961, pp. 133–40.

17 Churton T. *Aleister Crowley: The Biography*. London: Watkins Publishing, 2011, p. 110.

Chapter 2

1 Carruth J. A. *Loch Ness and Its Monster*. Fort Augustus: Abbey Press, 1945.

2 Adomnán. *Vita Sancti Columbae*, AD 690. British Library, London MS ADD.35110. Abridged English translation, Gregory J. *The Life of Columba*. Edinburgh, Floris Books, 1991.

3 Mackal R. P. *The Monsters of Loch Ness*. The Swallow Press, Chicago, 1976, p. 224.

4 Gould R. T. *The Loch Ness Monster and Others*. London: Geoffrey Bles, 1934. Revised edition, New York: University Books, 1969, pp. 28–32.

5 News items in *Inverness Courier* 10 July 1858, p. 3; 8 October 1868, p. 5; 15 October 1868, p. 5; 26 May 1933, p. 6.

6 Anonymous. A scene at Lochend. *Inverness Courier* 1 July 1852, p. 3.

7 Anonymous. Strange fish at Abriachan. *Inverness Courier* 8 October 1868.

8 Macrae J. Letter to *Inverness Courier*, 20 August 1872. Cited in Oudemans AC. *The Great Sea-Serpent* (1892). Republished with foreword by Loren Coleman. New York: Cosimo, 2009, pp. 428–31.

9 Anonymous. What was it? A strange experience on Loch Ness. *Northern Chronicle* 27 August 1930.

10 Editor's Post Bag. Responses to 'What was it?' *Northern Chronicle* 3 September 1930, 10 September 1930.

11 'Piscator'. Letter to *Inverness Courier* 29 August 1930.

12 Anonymous. Germany and whisky. *Inverness Courier* 3 January 1930, p. 3.

13 Anonymous. Meeting of Rotary Club, Inverness. *Inverness Courier* 9 June 1933, p. 5.

14 Anonymous (Alex Campbell). Strange spectacle on Loch Ness. What was it? *Inverness Courier* 2 May 1933, p. 5.

15 Anonymous. Loch Ness 'Monster': ship captain's views on occurrence. Letter by John Macdonald, *Inverness Courier* 12 May 1933.

16 Anonymous. Angling Notes. *Inverness Courier* 23 May 1933, p. 3.

17 Anonymous. Loch Ness 'Monster'. Attempt to solve the mystery. *Inverness Courier* 30 May 1933, p. 4.

18 Anonymous. Loch Ness 'Monster' seen again. *Inverness Courier* 2 June 1933, p. 5.

19 Advertisement: Loch Ness Cruise. *Inverness Courier* 8 August 1933, p. 4.

20 Anonymous. Visitor's experience. Quoting letter by George Spicer, *Inverness Courier* 4 August 1933.

21 Anonymous. Loch Ness 'Monster'. Three appearances near Fort-Augustus. Eye-witnesses' vivid descriptions. *Inverness Courier* 8 August 1933, p. 5.

22 Spicer G. A visitor's experience: response from Mr Spicer; Anonymous. The Loch Ness 'Monster'. True story of his life. Told to A Mere Woman; Anonymous. Another sighting at Glen Urquhart. *Inverness Courier* 11 August 1933, p. 5.

23 Anonymous. Loch Ness 'Monster'. Numerous appearances during week-end. *Inverness Courier* 26 September 1933, p. 5.

24 Anonymous. Editorial: The Loch Ness Mystery. *Inverness Courier* 26 September 1933, p. 4.

25 Gould R. T. *The Loch Ness Monster and Others*. Revised edition, New York: University Books, 1969, pp. 62–5; Witchell N. *The Loch Ness Story*. Lavenham, Suffolk: Terence Dalton, 1974, pp. 48–9.

26 Anonymous. Loch Ness 'Monster'. Theory that it is an amphibian. *Inverness Courier* 3 October 1933, p. 5; Anonymous. Loch Ness 'Monster'. Problem for scientific society, with letters from W. H. Lane, A Russell-Smith. *Inverness Courier* 10 October 1933, p. 4.

27 Stalker P. Loch Ness Monster: a puzzled Highland community. *Scotsman* 16 October 1933, p. 11; Loch Ness Monster: the plesiosaurus theory: old stories recalled. *Scotsman* 17 October 1933, p. 9.

28 Anonymous. Editorial: The Loch Ness Monster. *Inverness Courier* 24 October 1933, p.4.

29 Green R. M. Letter to *Inverness Courier* 31 October 1933, p. 5; Views of Sir Murdoch Macdonald, *Inverness Courier* 3 November 1933, p. 5.

30 'A Correspondent' (Alex Campbell). A weird coincidence. *Inverness Courier* 31 October 1933, p. 7.

31 Whyte C. *More Than a Legend*. London: Hamish Hamilton, 1957. Revised 3rd impression, 1961, pp. 2–4.

32 Betts J. *Time Restored: The Story of the Harrison Timekeepers and RT Gould, 'The Man Who Knew (Almost) Everything.'* Greenwich: National Maritime Museum & Oxford University Press, 2006.

33 Gould R. T. *The Loch Ness Monster and Others*, 1969, pp. 2, 12.

34 Gould R. T. The Loch Ness 'Monster'. A survey of the evidence. *The Times* 9 December 1933.

35 Marmaduke Wetherell. Internet Movie Database, http://www.imdb.com/name/nm0923143/

36 Price H., Lambert R. The haunting of Cashen's Gap: a modern 'miracle' investigated. London: Methuen & Co. Ltd., 1936.

37 Memory F. W. Loch monster's tracks found? *Daily Mail* 20 December 1933.

38 Anonymous. Experts examine spoor cast at the Natural History Museum. *Daily Mail* 23 December 1933.

39 Anonymous. Editorial: A Monster Year. *Inverness Courier* 29 December 1933, p. 4.

40 BBC 1933 Year Round-Up, introduced by Lawrence Wager. Sound Archives, British Library, London, C1398/0144, C22–24.

Chapter 3

1 Witchell N. *The Loch Ness Story*. Lavenham, Suffolk: Terence Dalton, 1974, p. 59.

2 Anonymous. Loch Ness Mystery. Investigation by a Japanese journalist. *Inverness Courier* 26 January 1934, p. 5; Ishikawa, K. *A Book of Thoughts*. New York: Taplinger Publishing Co., 1958, p. 71.

3 Anonymous. Appearance on land twenty years ago. Fort-Augustus resident's description. *Inverness Courier* 3 October 1933, p. 5; Witchell N. *The Loch Ness Story*, p. 136.

4 Anonymous. Monster mystery deepens. Cast of spoor like a hippopotamus's. Zoology experts' report. *Daily Mail* 4 January 1934.

5 Witchell N. *The Loch Ness Story*, p. 63.

6 Memory F. W. More footprints. Amphibian with four toes and claws. *Daily Mail* 4 January 1934.

7 Calman W. T. Memo to Natural History Museum's Trustees, 18 January 1934. Natural History Museum Archives, DF ZOO/232/7/1/3

8 Memory F. W. 'Monster's' moonlight frolic on road. *Daily Mail* 6 January 1934.

9 Memory F. W. There is a seal in Loch Ness. *Daily Mail* 7 January 1934.

10 Anonymous. Editorial: 'Mud'. *Inverness Courier* 9 January 1934.

11 Gould R. T. *The Loch Ness Monster and Others.* Revised edition, New York: University Books, 1969, p. 87–9; Witchell N. *The Loch Ness Story.* Lavenham, Suffolk: Terence Dalton, 1974, p. 137–9.

12 Anonymous ('More Than One'). The Loch Ness 'Monsters'. *Inverness Courier* 13 February 1934, p. 3.

13 Anonymous. Loch Ness 'Monster' seen again. *Inverness Courier* 13 February 1934, p. 5.

14 Anonymous. Loch Ness 'Monster' seen at Fort-Augustus and Glen-Urquhart. Well-known Ross-shire lady's story. *Inverness Courier* 27 February 1934, p. 5.

15 Anonymous. The bearded seal. Well-known sportsman and Loch Ness Monster. *Inverness Courier* 26 January 1934, p. 4.

16 Anonymous. Loch Ness Mystery. Investigation by a Japanese journalist. *Inverness Courier* 26 January 1934, p. 5; Anonymous. Lecture on Loch Ness Monster, by Dr Richard Elmhirst. *Inverness Courier* 9 January 1934, p. 7; Anonymous. Loch Ness Mystery, American Professor's theory. *Inverness Courier* 23 January 1934, p. 5.

17 Anonymous. Is the Monster a giant squid? Mr Goodbody of Invergarry on American naturalist's theory. *Inverness Courier* 30 January 1934, p. 5.

18 Anonymous. Das Untier vom Loch Ness ist gefangen! *Berliner Illustrirte Zeitung* 31 March 1934.

19 Watters D. A. Loch Ness, Special Operations Executive and the first surgeon in Paradise: Robert Kenneth Wilson (26.1.1899–6.6.1969). *Aust New Z J Surg* 2007; 77:1053-7; Wilson R. K. *Textbook of automatic pistols, with a supplementing chapter on the light machine-gun, 1884–1935.* Plantersville, South Carolina: Small-Arms Technical Pub Co., 1943.

20 Maurice Augustus Chambers of Rosemount, Thornbury. http://theplain.thornburyroots.co.uk/HS%20n066%20Chambers.htm

21 Whyte C. *More Than a Legend*, revised 3rd impression. London: Hamish Hamilton, 1961, pp. 6–7.

22 Anonymous. London Surgeon's photograph of the Monster. Monster yards from Lochside. *Daily Mail* 21 April 1934, p. 1.

23 Anonymous. Loch Ness Monster. Seen by London surgeon yesterday. *Inverness Courier* 20 April 1934, p. 5.

24 Whyte C. *More Than a Legend* 1961, pp. 7–8.

25 Witchell N. *The Loch Ness Story*, p. 75.

26 *The Secret of the Loch* (1934). http://explore.bfi.org.uk/4ce2b6b55ac91

27 Whyte C. *More Than a Legend* 1961, pp. 84–5.

28 Witchell N. *The Loch Ness Story*, p. 83.

29 Witchell N. *The Loch Ness Story*, pp. 83–5.

30 Lane, W. H. *The Home of the Loch Ness Monster*. Edinburgh: Grant & Murray Limited, 1934.

31 Gould R. T. *The Loch Ness Monster and Others*. London: Geoffrey Bles, 1934. Revised edition, New York: University Books, 1969.

32 Gould R. T. *The Loch Ness Monster and Others*, 1969, p. 33.

33 Gould R. T. *The Loch Ness Monster and Others*, 1969, pp. 63–5

34 Gould R. T. *The Loch Ness Monster and Others*, 1969, pp. 44–6.

35 Anonymous. What was it? A strange experience on Loch Ness. *Northern Chronicle* 27 August 1930.

36 Gould R. T. *The Loch Ness Monster and Others*, 1969, pp. 36–8.

37 Gould R. T. *The Loch Ness Monster and Others*, 1969, p. 157.

38 Gould R. T. *The Loch Ness Monster and Others*, 1969, pp. 115–17.

39 Gould R. T. *The Loch Ness Monster and Others*, 1969, pp. 164–5.

40 Gould R. T. *The Loch Ness Monster and Others*, 1969, pp. 38–9 and 95–6.

41 Stalker P. Loch Ness Monster: the plesiosaurus theory: old stories recalled. *Scotsman* 17 October 1933, p. 9.

42 Campbell A. Letter to Ness Fishery Board, 28 October 1933. Quoted in: Gould R. T. *The Loch Ness Monster and Others*, p. 110.

43 Betts J. *Time Restored: The Story of the Harrison Timekeepers and R. T. Gould, 'The Man Who Knew (Almost) Everything.'* Greenwich: National Maritime Museum & Oxford University Press, 2006.

44 Pearson R. Sir Edward Mountain, first baronet (1872–1948). Oxford Dictionary

of National Biography, 2004. http://dx.doi.org/10.1093/ref:odnb/47692

45 Whyte C. *More Than a Legend* 1961, pp. 100–11.

46 Mountain E. Solving the mystery of Loch Ness. *The Field* 22 September 1934, pp. 668–9.

47 Whyte C. *More Than a Legend* 1961,

48 Mountain E. Expedition to Loch Ness. *Proc Linnaean Society*, Session 1934–5, Part I, pp. 7–12; reported in: Anonymous. The Loch Ness Monster. *Nature* 1934; 134:765.

49 Whyte C. *More Than a Legend* 1961, pp. 116–17.

50 Witchell N. *The Loch Ness Story*, p. 59.

51 Witchell N. *The Loch Ness Story*, p. 54.

52 Gould R. T. *The Loch Ness Monster and Others*, 1969, p. 165.

53 Letters from William Fraser, Chief Constable of Inverness-shire. Displayed in 'An Open Secret' Exhibition, National Archives of Scotland, Edinburgh, 2009 (NAS HH1/588). http://www.nas.gov.uk/documents/HH-1-588-31.pdf

54 Gould R. T. *The Loch Ness Monster and Others*, 1969, p. 22; Witchell N. *The Loch Ness Story*, p. 76.

55 Witchell N. *The Loch Ness Story*, p. 53.

56 Whyte C. *More Than a Legend* 1961, p. 54.

57 Whyte C. *More Than a Legend* 1961, p. 55.

58 Witchell N. *The Loch Ness Story*, p. 104.

59 Hardy A. C. Obituary. Stanley Wells Kemp, 1882–1945. *J Marine Biol Assoc UK* 1946; 26:219–34.

Chapter 4

1 Smith J. L. B. *Old Four Legs.* London: Longmans Green & Co., 1956

2 Attenborough D. *Zoo Quest for a Dragon.* London: Pan Books, 1961.

3 Colman J. G. A review of the biology and ecology of the whale shark. *J Fish Biol* 1997; 51:1219–34.

4 Edgerton H., Jussim E., Kayafas G. *Stopping Time: The Photographs of Harold Edgerton.* New York: HN Abrams, 1987.

5 Whyte C. *More Than a Legend: The Story of the Loch Ness Monster.* London: Hamish Hamilton, 1957; revised 3rd impression, 1961.

6 Whyte C. *More Than a Legend*, 1961, p. vxii.

7 *Ibid.* Chapter 3: Report without comment, pp. 37–71.

8 *Ibid.*, pp. 202–3.

9 *Ibid.*, pp. 203–4.

10 *Ibid.*, pp. 43, 94–5.

11 *Ibid.*, p. xx.

12 *Ibid.*, pp. 40–42, 47–8, 52, 60–61, 64–5.

13 *Ibid.*, pp. 5–10.

14 *Ibid.*, pp. 10–14.

15 *Ibid.*, pp. 14–18.

16 *Ibid.*, p. xxi.

17 *Ibid.*, pp. 4–5, 8–9.

18 *Ibid.*, pp. xviii, 9–10, 18–19, 103–7.

19 *Ibid.*, p. 89–94.

20 *Ibid.*, pp. 58-9.

21 *Ibid.*, p. 79.

22 *Ibid.*, p. 87.

23 *Ibid.*, p. 183.

24 *Ibid.*, pp. 171–6.

25 *Ibid.*, pp. 23–7.

26 *Ibid.*, p. 83.

27 *Ibid.*, pp. 4–5, 8–9, 101–2.

28 *Ibid.*, pp. 73, 87–8.

29 *Ibid.*, p. xi.

30 *Ibid.*, p. x.

31 *Ibid.*, p. 123.

32 *Ibid.*, pp. 7–8.

33 *Ibid.*, p. 29.

34 *Ibid.*, p. 214.

Chapter 5

1 Whyte C. *More Than a Legend: The Story of the Loch Ness Monster*, revised 3rd impression. London: Hamish Hamilton, 1961, p. ix.

2 Dinsdale T. *Loch Ness Monster*. London: Routledge & Kegan Paul, 1961, pp. 141–5; Witchell N. *The Loch Ness Story*. Lavenham, Suffolk: Terence Dalton, 1974, pp. 141–4; Letter from Denys Tucker to Peter Scott, 29 September 1960, Peter Scott Archives C.661.

3 Letter from Torquil MacLeod to Peter Scott, 12 September 1960, Peter Scott Archives C.669.

4 Witchell N. *The Loch Ness Story*. Lavenham, Suffolk: Terence Dalton, 1974, pp. 126–8.

5 Dinsdale T. *Loch Ness Monster*, 1961, pp. 11–28.

6 Dinsdale T. *Loch Ness Monster*, 1961, pp. 78–108.

7 Dinsdale T. *Loch Ness Monster,* 4th edition. London: Routledge, 1989, pp. 89–91.

8 Witchell N. *The Loch Ness Story*. London: Corgi, 1989, p. 112.

9 Huxley E. *Peter Scott. Painter and Naturalist.* London: Faber and Faber, 1993, p. 14.

10 *Ibid.*, pp. 42–50.

11 *Ibid.*, p. 205.

12 *Ibid.*, pp. 129–43.

13 *Ibid.*, p. 51.

14 *Ibid.*, p. 52.

15 Huxley E. *Peter Scott,* 1993, p. 264; Interview with François Gohier, 17 January 1977, transcript in Peter Scott Archives, M.2052.

16 Letter from Tim Dinsdale to Peter Scott, 12 March 1960, Peter Scott Archives C.658.

17 Burton M. The Loch Ness Monster. *Illustrated London News* 20 February 1960.

18 Tucker D. W. Obituary: Maurice Burton. *Independent* 23 September 1992. http://www.independent.co.uk/news/people/obituary-maurice-burton-1553125.html

19 Letter from Peter Scott to Tim Dinsdale, 22 March 1960, Peter Scott Archives C.658.

20 Letter from Tim Dinsdale to Peter Scott, 26 March 1960, Peter Scott Archives C.658

21 Letter from Michael Garside to Tim Dinsdale, 29 March 1960, Peter Scott Archives C.658.

22 Letter from Tim Dinsdale to Michael Garside, 30 March 1960, Peter Scott Archives C.658.

23 Letter from Tim Dinsdale to Peter Scott, undated, received at Slimbridge 30 April 1960, Peter Scott Archives C.658.

24 Letter from Tim Dinsdale to Peter Scott, 3 May 1960, Peter Scott Archives C.658.

25 Letter from Peter Scott to Colonel The Honourable Martin Charteris, 9 May 1960, Peter Scott Archive C.658.

26 Letter from Martin Charteris to Peter Scott, 11 May 1960, Peter Scott Archive C.658.

27 Draft agenda and list of attendees at Meeting at London Zoo, 11 May 1960, Peter Scott Archives C.659.

28 Letter (confidential) from Peter Scott to Sir Arthur Landsborough Thomson and others, 26 May, Peter Scott Archives C.659.

29 Letter from Tim Dinsdale to Peter Scott, 15 May 1960, Peter Scott Archives C.659.

30 Letter from Peter Scott to Tim Dinsdale, 22 May 1960, Peter Scott Archives C.659.

31 Letter from Tim Dinsdale to Peter Scott, 23 May 1960, Peter Scott Archives C.659.

32 Letter from Tim Dinsdale to Peter Scott, 31 May 1960, Peter Scott Archives C.659.

33 Letter from Maurice Burton to Peter Scott, 2 June 1960, Peter Scott Archives C.660.

34 Letter from Peter Scott to Maurice Burton, 11 June 1960, Peter Scott Archives C.660.

35 Letter from Maurice Burton to Peter Scott, 13 June 1960, Peter Scott Archives C.660.

36 Letter from Alex Campbell to Peter Scott, 16 May 1960, Peter Scott Archives C.659.

Chapter 6

1 Tucker D. W. On a rare deep-sea fish, *Notacanthus phasganorus* Goode (*Heterominotacanthidae*), from the Arctic Bear Isle fishing-grounds. London: British Museum (Natural History), 1951.

2 Tucker D. W. The Zoologist's Tale. In: Witchell N. *The Loch Ness Story*, revised edition, London: Corgi, 1989, pp. 205–28; Letter from Denys Tucker to Peter Scott, 15 June 1960, Peter Scott Archives C.661.

3 Witchell N. *The Loch Ness Story*. Lavenham, Suffolk: Terence Dalton, 1974, p. 151.

4 Letter from Terence Morrison-Scott to Peter Scott, 28 June 1960, Peter Scott Archive C.660.

5 Memo from A.V. Hill and Trustees of the Natural History Museum to Museum staff, 21 October 1959, Natural History Museum Archives DF ADM/1004/510.

6 Letter from Francis Fraser to Peter Scott, 24 June 1960, Peter Scott Archives C.660.

7 Letter from Peter Scott to Francis Fraser, 27 June 1960, Peter Scott Archives C. 660.

8 Letter from Peter Scott to Denys Tucker, 12 June 1960, Peter Scott Archives C.661.

9 Letter from Denys Tucker to Peter Scott, 15 June 1960, Peter Scott Archives C.661.

10 Letter from Denys Tucker to Peter Scott, 5 July 1960, Peter Scott Archives C.661.

11 Letter from Peter Scott to Denys Tucker, 6 July 1960, Peter Scott Archives C.661.

12 Letter from Denys Tucker to Peter Scott, 10 July 1960, Peter Scott Archives C.661.

13 Letter from Denys Tucker to Peter Scott, 29 September 1960, Peter Scott Archives C.661.

14 Letter from Peter Baker to Peter Scott, 15 June 1960, Peter Scott Archives C.660.

15 Letters from Lord Landsborough Thomson to Peter Scott, 8 June and 4 July 1960, Peter Scott Archives C.660.

16 Letter from Vero Wynne-Edwards to Peter Scott, 30 June 1960, Peter Scott Archives C.661.

17 Letter from Richard Harrison to Peter Scott, 22 June 1960, Peter Scott Archives C.660.

18 Letter from Sir Alister Hardy to Peter Scott, 2 July 1960, Peter Scott Archives C.660.

19 Letters to Peter Scott from Leo Harrison Matthews (7 July), Desmond Morris (15 July) and Carl Pantin (15 July 1960), Peter Scott Archives C.660.

20 Letter from Sir Alister Hardy to Peter Scott, 12 July 1960, Peter Scott Archives C.661.

21 Letter from Peter Scott to Tim Dinsdale, 24 June 1960, Peter Scott Archives C.661.

22 Letter from Tim Dinsdale to Peter Scott, 21 June 1960, Peter Scott Archives C.661.

23 Letters from Peter O'Connor to Peter Scott, 5 and 15 July 1960, Peter Scott Archives C.661; Witchell N. *The Loch Ness Story.* Lavenham, Suffolk: Terence Dalton, 1974, p. 182.

24 Letter from Tim Dinsdale to Peter Scott, 27 June 1960, Peter Scott Archives C.661.

25 Letter from Peter Scott to Tim Dinsdale, 6 July 1960, Peter Scott Archives C.661.

26 Letter from Martin Charteris to Peter Scott, 29 July 1960, Peter Scott Archives C.661.

27 Letter from Martin Charteris to Peter Scott, 10 August 1960, Peter Scott Archives C.661.

28 Scott P. The Loch Ness Monster – fact or fancy? *Sunday Times Magazine*, 14 August 1960, p. 17.

29 Scott P. Drafts of 'The Loch Ness Monster – fact or fancy? for *Sunday Times*. Peter Scott Archives, C661.

30 Letter from Peter Scott to Jack Lambert, 23 July 1960, Peter Scott Archives C.661.

31 Letter from Peter Scott to Dr R.S. (Dick) Scorer, 5 October 1960, Peter Scott Archives C.669.

32 Letter from Tim Dinsdale to Peter Scott, 14 August 1960, Peter Scott Archives C.665.

33 Letter from Peter Scott to Tim Dinsdale, 16 August 1960, Peter Scott Archives C.666.

34 Letter from Tim Dinsdale to Peter Scott, 18 August 1960, Peter Scott Archives C.666.

35 Miscellaneous letters to Peter Scott and responses, August 1960, Peter Scott Archives C.667, C.668.

36 Letter from Walter E. Astin to Peter Scott, 30 August 1960, Peter Scott Archives C.667.

37 Letter from Peter Du Cane to Peter Scott, 23 August 1960, Peter Scott Archives C.667.

38 Letter from Maurice Burton to Peter Scott, 5 October 1960, Peter Scott Archives C.669; Burton M. Muck and monsters. *Illustrated London News* 1 October 1960, p. 568.

39 Letter from Dr. L.B. Tarlo to Peter Scott, 23 October 1960, Peter Scott Archives C.670.

40 Letter from Commander J.C. Turnbull to Peter Scott, 23 August 1960, Peter Scott Archives C.667.

41 Gould R. T. *The Loch Ness Monster and Others*, revised edition. New York: University Books, 1969, pp. 115–16.

42 Letter from Torquil MacLeod to Peter Scott, 12 September 1960, Peter Scott Archives C.669.

43 Letter from Peter Scott to Torquil MacLeod, 18 September 1960, Peter Scott Archives C.669.

44 Letter from Peter Scott to R.S.R. Fitter, 25 September 1960, Peter Scott Archives C.669.

45 Letter from Sir Solly Zuckerman to Peter Scott, 17 October 1960, Peter Scott Archives C.670.

46 Letter from Tim Dinsdale to Peter Scott, 17 October 1960, Peter Scott Archives C.670.

47 Confidential minutes of Loch Ness Study Group, held on 19 October 1960. Peter Scott Archives C.670.

48 Letter from Constance Whyte to Peter Scott, 20 October 1960, Peter Scott Archives C.670.

49 Letter from Torquil MacLeod to Peter Scott, 9 November 1960, Peter Scott Archives C.671.

50 Baker P. F., Arnold R. The mystery of Loch Ness. *Scotsman* 12 September 1960; Baker P. F., Westwood M. Underwater detective work. *Scotsman* 14 September 1960.

51 Dinsdale T. *Loch Ness Monster*, 4th edition. London: Routledge & Kegan Paul, 1982, p. 104.

52 Letter from Tim Dinsdale to Peter Scott, 10 December 1960, Peter Scott Archives C.672.

53 Letter from Elisabeth MacLeod to Peter Scott, 14 December 1960, Peter Scott Archives C.672.

Chapter 7

1 Broadcast 12 February 1962.
http://www.bigredbook.info/david_pelham_james.html

2 Robson J. *One Man in His Time. The Biography of David James.* Staplehurst, Kent: Spellmount, 1998, pp. 38–90.

3 Ibid., pp. 102–62; Williams E. The 'Bulgarian' Naval Officer. In *Great Escape Stories.* London: The Heirloom Library, 1958, pp. 71–93.

4 Robson J. *One Man in His Time*, pp. 165–6.

5 *Ibid.*, pp. 187–202.

6 *Ibid.*, p. 204.

7 James D. Time to meet the Monster. *The Field* 23 November 1961, pp. 3–5.

8 Zuckerman S. Obituary: Dr Henry Gwynne Vevers. *J Zoology* 1989; 219:529–31.

9 Telegram from Peter Scott to Mr & Mrs Seddon-Smith, 5 April 1961, Peter Scott Archives C.673.

10 Letter from Revd W. L. Dobbs to Peter Scott, with copy of *St Michael Colehill Parish Magazine* (September 1960), 30 October 1960, Peter Scott Archives C.671.

11 Peter Scott. Notes from Loch Ness Monster Panel Meeting, 12 April 1961 at Burlington House, with various drafts of the minutes, Peter Scott Archives C.673.

12 Letters to Peter Scott from Carl Pantin (29 April 1961) and Sir Alister Hardy (9 May), Peter Scott Archives C.677.

13 Letter from Routledge & Kegan Paul to Peter Scott, 10 April 1961, Peter Scott Archives C.673.

14 Letter from Constance Whyte to Peter Scott, 8 May 1961, Peter Scott Archives C.677.

15 Letter from Tim Dinsdale to Peter Scott, 10 May 1961, Peter Scott Archives C.678.

16 Dinsdale T. *Loch Ness Monster.* London: Routledge & Kegan Paul, 1961.

17 *Ibid.*, p. 4.

18 *Ibid.*, pp. 6, 11–28, 125.

19 *Ibid.*, pp. 94–5.

20 *Ibid.*, pp. 43–51, 68, 152–8.

21 *Ibid.*, pp. 78–104; see Letter from Tim Dinsdale to Peter Scott, 15 May 1960, Peter Scott Archives C.659.

22 *Ibid.*, p. 203.

23 *Ibid.*, pp. 226–7.

24 *Ibid.*, p. 71.

25 *Ibid.*, p. 76.

26 *Ibid.*, p. 244.

27 Whyte C. *More Than a Legend*, revised 3rd impression. London: Hamish Hamilton, 1961.

28 Burton M. *The Elusive Monster*. London: Rupert Hart-Davis, 1961.

29 *Ibid.*, pp. 23–5.

30 *Ibid.*, pp. 62–4.

31 *Ibid.*, pp. 91–9, 118.

32 *Ibid.*, pp. 101–2, 106–7.

33 *Ibid.*, pp. 73–4.

34 *Ibid.*, pp. 174.

35 *Ibid.*, pp. 160–7.

36 *Ibid.*, pp. 154, 162.

37 *Ibid.*, p. 170.

38 D.W.T. (Denys Tucker). Review of *The Loch Ness Monster* (Gould), *More Than a Legend* (Whyte), *Loch Ness Monster* (Dinsdale) and *The Elusive Monster* (Burton). *Oryx* 1962; 6:241–2.

39 Burton M., Tucker D. W., Whyte C. The Loch Ness Monster. Letters to *New Scientist*, 17 November 1960, p. 1414.

40 Baker P. F., Tarlo L. B. The Loch Ness Monster. Letters to *New Scientist*, 24 November 1960, p. 1414.

41 Anonymous. Scientist tells of dismissal. Interview with Dr Fisher. *The Guardian*, 14 December 1960.

42 House of Commons Debate, 3 February 1961. Hansard, vol. 633, cc 1421–32.

43 Letter from Elisabeth MacLeod to Peter Scott, 14 May 1961, Peter Scott Archives C.678.

44 Letter from David James to Peter Scott, 18 July 1962, Peter Scott Archives C.682.

45 Letter from David James to Peter Scott, 25 July 1962, Peter Scott Archives C.685.

46 Letter from David James to Peter Scott, 1 September 1962, Peter Scott Archives C.685.

47 Letter from Peter Baker to David James, 4 September 1962, Peter Scott Archives C.685.

48 Dinsdale T. *Loch Ness Monster*, 4th edition. London: Routledge & Kegan Paul, 1982, p. 120; Holiday F. W. *The Great Orm of Loch Ness*. London: Faber and Faber, 1968, pp. 197–204.

49 Baker P. F., Westwood M. Sounding out the Monster. *Observer*, 26 August 1962.

50 Proceedings of the Independent Panel on Unexplained Phenomena in Loch Ness. Inverness: *Inverness Courier*, 1963. Reprinted in: Holiday F. W. *The Great Orm of Loch Ness*, 1968, pp. 205–12.

51 Robson J. *One Man in His Time*, p. 205; Holiday F. W. *The Great Orm of Loch Ness*, 1968, p. 207.

52 Holiday F. W. *The Great Orm of Loch Ness*, 1968, pp. 205–6.

53 Letter from David James to Peter Scott, 20 January 1963, with draft report of Operation Loch Ness from Independent Panel Meetings on 14 and 29 November 1962, Peter Scott Archives C.686.

54 Letter from Peter Scott to David James, 26 January 1963, Peter Scott Archives C.686.

55 James D. Script for ATV documentary *Report on the Loch Ness Monster*, broadcast 24 February 1963.

56 Letter from Pembroke Dutton to Peter Scott, 11 February 1963, Peter Scott Archives C.685.

Chapter 8

1 Holiday F. W. *The Great Orm of Loch Ness*. London: Faber and Faber, 1968, pp. 6–8.

2 *Ibid.*, pp. 12–17.

3 Letter from F. W. Holiday to Peter Scott, 28 March 1962, Peter Scott Archives C.683.

4 Letter from Peter Scott to David James, 20 October 1962, Peter Scott Archives C.683.

5 Holiday F. W. The case for a spineless Monster. *The Field*, 5 February 1976, pp. 204–5.

6 Holiday F. W. *The Great Orm of Loch Ness*. London: Faber and Faber, 1968, pp. 43–4.

7 Letters from David James to Peter Scott, 3–17 May 1963, Peter Scott Archives C.688.

8 Letter from Peter Scott to David James, 17 July 1963, Peter Scott Archives C.690.

9 James D. We find that there is some unidentified animate object in Loch Ness. *Observer*, 17 May 1964.

10 Letter from Maurice Burton to Peter Scott, 25 October 1963, Peter Scott Archives C.689.

11 Holiday F. W. *The Great Orm of Loch Ness*, 1968, pp. 64–7.

12 Letter from Constance Whyte to Peter Scott, 2 February 1963; Scott's reply, 7 February; Peter Scott Archives C.685.

13 Holiday F. W. *The Great Orm of Loch Ness*. London: Faber and Faber, 1968, pp. 70–72.

14 *Ibid.*, pp. 73–4, 86.

15 *Ibid.*, p. 85.

16 James D. Fine-weather monster. *Observer*, 27 December 1964.

17 Robson J. *One Man in His Time. The Biography of David James*. Staplehurst, Kent: Spellmount, 1998, p. 228.

18 Holiday F. W. *The Great Orm of Loch Ness*. London: Faber and Faber, 1968, pp. 98–100, 161.

19 *Ibid.*, pp. 104–10.

20 *Ibid.*, p. 161; Dinsdale T. *Loch Ness Monster*, 4th edition, 1982, p. 123.

21 Letter from David James to Peter Scott, 14 May 1965; Scott's reply, 23 May; Peter Scott Archives C.694.

22 Mackal R. P. *The Monsters of Loch Ness*. The Swallow Press, Chicago, 1976, pp. 1–2, 9–12.

23 *Ibid.*, pp. 14–17; David James's certificate of membership of the Adventurers' Club of Chicago, Torosay Castle collection of David James/LNIB memorabilia.

24 Mackal R. P. *The Monsters of Loch Ness*, 1976, pp. 26–7.

25 Letter from David James to Peter Scott, 4 February 1966, Peter Scott Archives C.696.

26 James D. *Photographic Interpretation Report — Loch Ness. Report on a film taken by Tim Dinsdale.* Inverness: *Inverness Courier*, 1966.

27 Dinsdale T. *Project Water Horse.* London: Routledge & Kegan Paul, 1975, p. 11; letters from Maurice Burton cited in Holiday F. W. *The Great Orm of Loch Ness*, 1968, pp. 56, 163.

28 Dinsdale T. *The Leviathans.* London: Routledge & Kegan Paul, 1966.

29 Holiday F. W. *The Great Orm of Loch Ness.* London: Faber and Faber, 1968.

30 *Ibid.*, pp. 162–3.

31 James D. The Loch Ness Investigation. Annual Report, 1966. Available online at http://www.lochnessproject.org/adrian_shine_archiveroom/paperspdfs

32 Mackal R. P. *The Monsters of Loch Ness*, 1976, p. 17.

33 *Ibid.*, pp. 18–20; Anonymous. The Monster is a giant sea-slug or squid. *Glasgow Herald* 20 September 1966; letter from David James to Peter Scott, 22 September 1966, Peter Scott Archives C.698.

34 Mackal R. P. *The Monsters of Loch Ness*, 1976, pp. 21–2.

35 Filming took place on 11–13 July 1966. Sheila Fullom, personal communication.

36 Letter from Peter Scott to Richard Fitter, 1 November 1966, Peter Scott Archives C.698.

37 Letter from Constance Whyte to Peter Scott, 22 March 1966, Peter Scott Archives C.696.

38 Holiday F. W. *The Great Orm of Loch Ness*, 1968, p. 164.

39 Mackal R. P. *The Monsters of Loch Ness*, 1976, p. 22.

40 Dick Raynor, personal communication; LNPIB 1967 brochure, Peter Scott Archives, C.701.

41 Mackal R. P. *The Monsters of Loch Ness*, 1976, pp. 120–21.

42 James D. The Loch Ness Investigation. Annual Report, 1967. Available at http://www.lochnessproject.org/adrian_shine_archiveroom/paperspdfs

43 Holly Arnold, personal communication.

44 Campbell E. M. Questing the Beast. *The Times* 2 October 1968; Anonymous. The great Monster hunt. *Weekend Scotsman* 6 July 1968; James D. The Loch Ness Investigation. Annual Report, 1968. Available online at http://www.lochnessproject.org/adrian_shine_archiveroom/paperspdfs.

45 Ivor Newby, personal communication.

46 Mackal R. P. *The Monsters of Loch Ness*, 1976, pp. 298–302.

47 Tucker D. G., Braithwaite H. Sonar picks up stirrings in Loch Ness. *New Scientist* 1968; 40:664–6.

48 Anonymous. News and Views. Monsters by sonar. *Nature* 1968; 220:1272.

49 James D. The Loch Ness Investigation. Annual Report, 1968. Available online at http://www.lochnessproject.org/adrian_shine_archiveroom/paperspdfs

50 Holiday F. W. *The Great Orm of Loch Ness*. London: Faber and Faber, 1968.

51 *Ibid.*, Appendix A: The Tully Monster, by Dr E. S. Richardson, pp. 191–5; Richardson E. S. Wormlike fossil from the Pennsylvanian of Illinois. *Science* 1966; 151:75–6.

52 Robson J. *One Man in His Time*, 1998, p. 209.

53 Frere R. *Loch Ness*. London: John Murray, 1988, p. 149; Mackal R. P. *The Monsters of Loch Ness*, 1976, pp. 42–8.

54 Anonymous. They're all hunting Nessie in a Yellow Submarine. *Daily Mail* 2 June 1969.

55 House of Lords Debate, Loch Ness Monster: submarine research. Hansard, 16 July 1969, vol. 304, cc 262–4.

56 Mackal R. P. *The Monsters of Loch Ness*, 1976, pp. 48–9, 305–6; Dick Raynor, personal communication.

57 *Ibid.*, pp. 296, 308.

58 Mulchrone V. The Loch Ness Monster does not exist. *Daily Mail* 27 September 1969.

59 Witchell, N. 1974. *The Loch Ness Story*. Terence Dalton, Lavenham, p. 168.

60 Mackal R. P. *The Monsters of Loch Ness*, 1976, pp. 51–2, 303–5.

61 Interview with David James. *TV Times* 31 July 1969, p. 31.

62 James D. The Loch Ness Investigation. Annual Report, 1969. Available at http://www.lochnessproject.org/adrian_shine_archiveroom/paperspdfs

63 Dinsdale T. *Loch Ness Monster* 4th edition, 1982, p. 133.

64 Anonymous. Obituary: Wing Commander Ken Wallis. *Daily Telegraph* 4 September 2013.

65 Mackal R. P. *The Monsters of Loch Ness*, 1976, p. 54.

66 *Ibid.*, pp. 54–6.

67 *Ibid.*, pp. 58, 382–7.

68 *Ibid.*, pp. 60–67.

69 *Ibid.*, pp. 59–67, 74; Dick Raynor, personal communication.

70 *Ibid.*, p. 67.

71 Dinsdale T. *Loch Ness Monster* 4th edition, 1982, p. 134.

72 Menu, undated, in Torosay Castle collection.

73 Mackal R. P. *The Monsters of Loch Ness*, 1976, p. 72; Dinsdale T. *Loch Ness Monster* 4th edition, 1982, p. 139; Rines R. H., Blonder I. S. Attractant investigation, Loch Ness, 1970. Academy of Applied Science, Belmont, Mass. Original document in possession of Dick Raynor; Anonymous. Loch Ness Monster pursued with scents. *Chemical & Engineering*, 31 May 1971, p. 25.

74 Klein M., Rines R. H., Dinsdale T., Foster L. S. *Underwater Search at Loch Ness*. Monograph 1, Academy of Applied Science. Boston: Academy of Applied Science, 1972.

75 Mackal R. P. *The Monsters of Loch Ness*, 1976, p. 73.

76 *Ibid.*, pp. 76–9.

77 Holiday F. W. *The Great Orm of Loch Ness*, 1968, dedication.

78 Robson J. *One Man in His Time*, 1998, pp. 221–2.

79 Mackal R. P. *The Monsters of Loch Ness*, 1976, pp. 73, 79.

Chapter 9

1 Letters from David James to Peter Scott 12 and 29 August 1971, Peter Scott Archives C.705.

2 Letter from David James to Peter Scott 20 September 1971, Peter Scott Archives C. 705.

3 Letter from Mrs D. Stubbs to Peter Scott 6 April 1973, Peter Scott Archives C.706.

4 Dick Raynor, personal communication.

5 Letter from Norman Collins to Peter Scott 16 November 1971, Peter Scott Archives C.705.

6 Witchell N. *The Loch Ness Story.* Lavenham, Suffolk: Terence Dalton, 1974.

7 Marquard B. Obituary: Robert H. Rines. *Boston Globe* 3 November 2009.

8 Insight. Nessie, this is your best show yet. *Sunday Times* 30 November 1975.

9 Grundberg A. Obituary: H. E. Edgerton. *New York Times* 5 January 1990.

10 Meredith D. *The Search at Loch Ness.* New York: Quadrangle, 1977, p. 57.

11 Nicholas Witchell, personal communication.

12 Telegram from Robert Rines to Harold Edgerton, June 1972. Cited in Naone E. The Nessie Quest. *MIT News Magazine* 15 October 2007.

13 Witchell N. *The Loch Ness Story*, 1974, pp. 187–8.

14 Sitwell N. The Loch Ness Monster evidence. *Wildlife* March 1976, p. 105.

15 Presentation of Loch Ness Evidence to the Members of both Houses of Parliament, Scientists and Press, Grand Committee Room, House of Commons, 10 December 1975. Transcript, pp. 24–8. Peter Scott Archives C.725.

16 Witchell N. *The Loch Ness Story*, 1974, pp. 188–9.

17 Campbell S. *The Loch Ness Monster: The Evidence.* Wellingborough: The Aquarian Press, 1986. Rev. edn Aberdeen: Aberdeen University Press, 1991, pp. 53–p5; Witchell N. *The Loch Ness Story*, 1974, p. 190.

18 Witchell N. *The Loch Ness Story*, 1974, p. 191–3.

19 Anoymous. A computer aids in search for 'Nessie'. *Caltech News*, December 1975.

20 Witchell N. *The Loch Ness Story*, 1974, p. 190.

21 Rines R. H. Search for the Loch Ness Monster. *Technology Review* March–April 1976, pp. 25–40; Lyman H., quoted in Witchell N. *The Loch Ness Story*, 1974, p. 194.

22 Telegram from Robert Rines to Holly Arnold and Dick Raynor, 12 September 1972, in possession of Dick Raynor; Holly Arnold, personal communcation.

23 Arnold H. LNIB Newsletter 29 October 1972. Original document in possession of Dick Raynor.

24 Dinsdale T. The Rines/Edgerton picture. *Photographic Journal* April 1973, pp. 162–5; *Time*, 20 November 1972.

25 Hepple R. Ness Information Service, Nessletter No. 1, January 1974; Dinsdale T. *Loch Ness Monster*, 4th edition. London: Routledge, 1989, p. 160.

26 Witchell N. *The Loch Ness Story*, 1974, p. 195; Dick Raynor, personal communication.

27 Searle F. *Monster Hunter Extraordinary: Ten Years in Search of the Loch Ness Monster*. Unpublished manuscript, undated (*c.* 1980), pp. 13 – 17. Highland Council Archives, Inverness D885/1.

28 Rines R. Lecture at MIT, April 1974. Recorded by Christopher E. Strangio. Available at http://www.cableeye.com/RH_Rines_Recordings.html.

29 Searle F. *Monster Hunter Extraordinary: Ten Years in Search of the Loch Ness Monster*, p p. 25–34

30 Hepple R. Nessletters from 1974-1995 are archived online at http://lochness-mystery.blogspot.co.uk

31 Letter from Peter Scott to David James 28 March 1975, Peter Scott Archives C.707.

32 Letter from David James to Peter Scott, 24 April 1975, Peter Scott Archives C.707.

33 Letter from Peter Scott to Robert Rines 17 May 1975, Peter Scott Archives C.707.

34 Letter from Robert Rines to Peter Scott, 2 June 1975, Peter Scott Archives C.707.

35 Postcard from Robert Rines to Peter Scott, 16 June 1975, Peter Scott Archives C.707.

36 Letter from Peter Scott to Solly Zuckerman 17 July 1975, Peter Scott Archives C.707.

37 Letters from Solly Zuckerman to Peter Scott 19 and 22 July 1975, Peter Scott Archives C.707.

38 Wilkins A. The four vital sightings. *Field* 29 November 1975, p p. 1047–8; The shapes on the Loch. *Ibid.,* 4 December 1975, p. 1086.

39 Sketches by Peter Scott, August 1975, Peter Scott Archives C.707.

40 *On the Track of Unknown Animals. Look*, no. 155. BBC Television, broadcast 9 December 1966.

41 Meredith D. *The Search at Loch Ness*. New York: Quadrangle, 1977, pp. 147–51.

42 Rines R. H., Wyckoff C. W., Edgerton H. E., Klein M. Search for the Loch Ness Monster. *Technology Review* March–April 1976, pp. 25–40.

43 Sitwell N. The Loch Ness Monster evidence. *Wildlife* March 1976, pp. 102–9.

44 Letter from Peter Scott to Russ Kinne, 27 November 1975, Peter Scott Archives C.731.

45 Witchell N. *The Loch Ness Story*, 1974, pp. 164–6.

46 Letter from George Zug to Peter Scott, 3 November 1975, Peter Scott Archives C.730.

47 Sheals G., Corbet G. B., Greenwood P. H., Ball H. W., Charig A. J. Report on Rines' photographs, 1972 & 1975, from the British Museum (Natural History), November 1975. Peter Scott Archives C.732; cited in Rines R. H. Search for the Loch Ness Monster. *Technology Review* March–April 1976, p. 37.

48 Drafts of guest list for symposium at Royal Society of Edinburgh, October 1975, Peter Scott Archives C.730.

49 Letter from Tim Dinsdale to Peter Scott, 23 October 1975, Peter Scott Archives C.729.

50 Scott P. Draft artwork for WWF's 'Save One Species' campaign, 30 November 1975, Peter Scott Archives C.738.

51 Letter from World Federation for the Protection of Animals, Zurich to Peter Scott, 28 November 1975, Peter Scott Archives C.735.

52 Scott P., rough notes, November 1975; letter from Alan Wilkins to Peter Scott, 6 November 1975, Peter Scott Archives C.731.

53 Letter from R. V. Melville to Peter Scott, 6 November 1975, Peter Scott Archives C.731.

54 Letter from Sir Arthur Landsborough Thomson to Peter Scott, 21 November 1975, Peter Scott Archives C.732.

55 Attenborough D. *New Life Stories*. London: Collins, 2011, pp. 117–18.

56 Letter from Nicholas Witchell to Peter Scott, 21 November 1975, Peter Scott Archives C.732.

57 Letter from Robert Rines to Gordon Sheals, 25 November 1975; Sheals's reply, 28 November 1975; Peter Scott Archives C.735, C.736.

58 Letter from Peter Scott to Maurice Burton, 2 December 1975, Peter Scott Archives C.740.

59 Letter from Peter Scott to Sir Frank Fraser Darling, 21 December 1975, Peter Scott archives C.740.

60 Notes from telephone call from Robert Rines to Peter Scott's secretary at Slimbridge, 4 December 1975, Peter Scott Archives C.740.

61 Witchell N. *The Loch Ness Story*. Revised edition, Harmondsworth: Penguin, 1975, p. 8.

62 Anonymous. The Loch Ness 'Monster'. *Nature* 1934; 132:921.

63 Scott P., Rines R. (names not cited). Naming the Loch Ness monster. *Nature* 1975; 258:466–8.

64 Fiona Selkirk, personal communication.

65 Adrian Soar, personal communication.

66 Letter from Peter Scott to Peter Newmark at *Nature*, returning proofs of *Nature* article, 21 November 1975. Peter Scott Archives C.732.

67 Deans J. Nessie picture show convinces more MPs. *Press and Journal (Aberdeen)* 11 December 1975, p. 1.

68 Cousteau J. Quoted in: Holden A. How can Nessie resist £44,000 bait? *Sunday Times* 6 June 1976.

69 Meredith D. *The Search at Loch Ness*, 1977, p. 15.

70 Wyckoff C, quoted in: Insight. Nessie, this is your best show yet. *Sunday Times* 30 November 1975.

71 Sitwell N. The Loch Ness Monster evidence. *Wildlife* March 1976, p. 102; Witchell N. *The Loch Ness Story*, 1974, pp. 170–73.

72 Presentation of Loch Ness Evidence to the Members of both Houses of Parliament, 10 December 1975. Transcript, pp. 2–4. Peter Scott Archives C.723.

73 *Ibid.*, pp. 9–43; Sitwell N. The Loch Ness Monster evidence. *Wildlife* March 1976, p. 102-9.

74 Presentation of Loch Ness Evidence to the Members of both Houses of Parliament, Scientists and Press, 10 December 1975. Transcript, pp. 45-7.

75 *Ibid.*, pp. 58–60; Meredith D. *The Search at Loch Ness*, 1977, p. 3.

76 Presentation of Loch Ness Evidence to the Members of both Houses of Parliament, Scientists and Press, 10 December 1975. Transcript, p. 74.

77 *Ibid.*, pp. 75–9, 83.

78 *Ibid.*, pp. 80–85.

79 Fairbairn N. Monster anagram. *The Times* 12 December 1975. Anonymous ('J.W.'). 'Nessie': what's in an anagram? *Science* 1976; 191:54.

80 Phillips P. Nessiteras absurdum. *Observer* 14 December 1975, p. 11.

81 Letter from David Davies to Peter Scott 15 December 1975, Peter Scott Archives C.740.

82 Notes written by Peter Scott during telephone conversation with Robert Rines, 18 December 1975, Peter Scott Archives C.738.

83 Letter from Peter O'Connor to Peter Scott, 15 December 1975 and Scott's reply, 20 December; Peter Scott Archives C.743.

84 Letters from Gordon Corbet and Humphrey Greenwood, and Alan Charig to Peter Scott, 19 and 21 December 1975, Peter Scott Archives C.743.

85 Letter from Peter Scott to Gordon Sheals 20 December 1975, Peter Scott Archives C.743.

86 Letter from Peter Scott to Martin Holdgate, 16 December 1975, Peter Scott Archives C.743.

87 Letter from Susan Young to Peter Scott, 15 December 1975 and Scott's reply, 19 December, Peter Scott Archives C.743.

Chapter 10

1 Desmond A. Why Nessiteras rhombopteryx is such an unlikely monster. *The Times* 27 December 1975.

2 Letter from Denys Tucker to Peter Scott, 31 December 1975, Peter Scott Archives C.745.

3 Letters from Dick Raynor and Holly Arnold to Peter Scott, 6 and 10 January 1975, Peter Scott Archives C.752, C.753.

4 Letter from Alan Gillespie to Peter Scott, 12 January 1976, Peter Scott Archives C.755.

5 Letter from David Attenborough to Peter Scott, 7 January 1976, Peter Scott Archives C.750.

6 Scott P. Why I believe in Nessie. *Wildlife* March 1976, pp. 110–11.

7 Corbett G. B. The Loch Ness Monster. *Nature* 1976; 259:75.

8 Halstead L. P., Goriup P. D., Middleton J. A. The Loch Ness Monster. *Nature* 1976; 259:75–6.

9 Scott P. Letters: Sir Peter Scott replies. *Nature* 1976; 259:75–6.

10 Letter from Constance Whyte to Peter Scott and Scott's reply, 2 and 8 January 1976, Peter Scott Archives C.754.

11 Holden J. C. The reconstruction of 'Nessie'. The Loch Ness Monster resolved. *Journal of Irreproducible Results* 1976; 4:15–18.

12 Letters from Adrian Desmond to Peter Scott, 1 and 28 January 1976, Peter Scott Archives C.754 and C.757.

13 Holiday F. W. The case for a spineless Monster. *The Field* February 1976, pp. 204–5.

14 Sitwell N. The Loch Ness Monster evidence. *Wildlife* March 1976, pp. 102–9.

15 Mackal R. P. *The Monsters of Loch Ness*. The Swallow Press, Chicago, 1976.

16 *Ibid.*, pp. 273–6.

17 *Ibid.*, pp. 224–64.

18 *Ibid.*, pp. 200–17.

19 *Ibid.*, pp. 138–9.

20 Letters from Roy Mackal to Peter Scott, 7 and 13 January and 1 June 1976, Peter Scott Archives C.756, C763.

21 Letter from Robert Rines to Peter Scott, 26 March 1976, Peter Scott Archives C.761.

22 British Underwater Association. Programme and Proceedings of Spring Conference, Natural History Museum, 26-7 March 1976. Peter Scott Archives C.761.

23 Searle F. *Monster Hunter Extraordinary: Ten Years in Search of the Loch Ness Monster*. Unpublished manuscript, undated (*c.* 1980), pp. 48–9. Highland Council Archives, Inverness D885/1.

24 *Ibid.*, pp. 54–8.

25 Searle F. *Nessie. Seven Years in Search of the Monster*. London: Coronet, 1976.

26 Carroll J. The man who tracked Nessie for seven years. *Argosy* November 1976, pp. 39–42.

27 Searle F. *Monster Hunter Extraordinary*, pp. 17, 44.

28 Rines R. H., Wyckoff C. W., Edgerton H. E., Klein M. Search for the Loch Ness Monster. *Technology Review* March–April 1976, pp. 25–40.

29 Meredith D. *The Search at Loch Ness. The Expedition of the New York Times and the Academy of Applied Science*. New York: Quadrangle, 1977, pp. 43–55.

30 *Ibid.*, p. 47.

31 *Ibid.*, p. 51.

32 *Ibid.*, p. 76.

33 Ellis W. S. Loch Ness: the lake and the legend. *National Geographic* 1977; 151:759–79.

34 Meredith D. *The Search at Loch Ness*, 1977, pp. 3–6.

35 *Ibid.*, pp. 132–7; Dinsdale T. *Loch Ness Monster*, 4th edition. London: Routledge & Kegan Paul, 1982, Fig. 16.

36 Klein M., Finkelstein C. Sonar serendipity in Loch Ness. *Technology Rev* December 1976, pp. 44–57; Meredith D. *The Search at Loch Ness*, 1977, pp. 138–46.

37 Dick Raynor, personal communication.

38 Anonymous. North Dorset MP to quit. *Evening Echo*, 19 October 1976.

39 Letter from Adrian Shine to Peter Scott, 5 January 1976, enclosing copy of Shine A, *Loch Morar Expedition 75 Report*. Peter Scott Archives C.755. Report available at http://www.lochnessproject.org/adrian_shine_archiveroom/

40 Campbell E.M., Solomon D. *The Search for Morag*. London: Tom Stacey, 1972, pp. 136–40.

41 Macrae J. Letter to *Inverness Courier*, 20 August 1872. Cited in Oudemans A. C. *The Great Sea-Serpent* (1892). Republished with foreword by Loren Coleman. New York: Cosimo, 2009, pp. 428–31.

42 Ivor Newby, personal communication.

43 Fitter R. S. R. Book reviews: *The Search for Morag*, E. M. Campbell and D. Solomon; *Loch Ness Monster*, 2nd edition, T. Dinsdale. *Oryx* 1973; 12:256–7.

44 Campbell E. M., Solomon D. *The Search for Morag*. London: Tom Stacey, 1972.

45 Adrian Shine, personal communication.

46 Shine A. *Loch Morar Expedition 76 Report*. Available at http://www.lochnessproject.org/adrian_shine_archiveroom/

47 Shiels T. *Monstrum! A Wizard's Tale*, revised edition. Bideford: CFZ Press, 2011.

48 http://hoaxes.org/archive/display/category/loch_ness_monster; Campbell S. *The Loch Ness Monster: the evidence*. Edinburgh: Birlinn, 2002, pp. 36–9.

49 Letter from Tim Dinsdale to Peter Scott, 11 October 1977, Peter Scott Archive C.772.

50 Dinsdale T. *Loch Ness Monster*, 1982, pp. 185–8.

51 Frazier K. UFOs, horoscopes, Bigfoot, psychics and other nonsense. *Smithsonian*, December 1978.

52 Letter from Robert Rines to Peter Scott, 11 January 1979, Peter Scott Archive C.779.

53 Anonymous. Ness hunter dies. *New Scientist* 5 July 1979.

54 Shine A. *Loch Ness Project 83 report*. Available at http://www.lochnessproject.org/adrian_shine_archiveroom/

55 Shine A. J., Martin D. S. Loch Ness habitats observed by sonar and underwater television. *Scottish Naturalist* 1988; 105:111–19.

56 Wedderburn E. M. Seiches observed in Loch Ness. *Geogr J* 1904; 24: 441–2.

57 Thorpe S. A., Hall A., Crofts I. The internal surge in Loch Ness. *Nature* 1972; 237:96–8.

58 Shine A. The biology of Loch Ness. *New Scientist* 17 February 1983, pp. 462–7.

59 Adrian Shine, personal communication; Ross P. Loch Ness Monster hunt continues 80 years on. *Scotsman* 24 February 2015.

60 Bauer H. H. *The Enigma of Loch Ness: Making Sense of a Mystery.* Urbana and Chicago: University of Illinois Press, 1968, pp. 82, 89, 90.

61 Symposium on the Loch Ness Monster, organised by International Society for Cryptozoology and Society for the History of Natural History, Royal Museum of Scotland, Edinburgh, 25 July 1987. Peter Scott Archives, C.793.

62 Shine A. S. Loch Ness habitats observed by sonar and underwater television. Presentation to the Symposium on the Loch Ness Monster, Edinburgh 25 July 1987; Shine A., Martin D. S. Loch Ness habitats observed by sonar and underwater television. *Scottish Naturalist* 1988; 105 (Centenary edition, part 2):111–9.

63 Fitter R. S. R. *The Loch Ness Monster: St. Columba to the Loch Ness Investigation Bureau.* Presentation to the Symposium on the Loch Ness Monster, Edinburgh, 25 July 1987; *Scottish Naturalist* 1988, 105 (Centenary edition, part 2):47–51.

64 Dinsdale T. Quoted by Dick Raynor, http://www.lochnessinvestigation.com/

65 Razdan R., Kielar A. Sonar and photographic searches for the Loch Ness Monster: a reassessment. *Skeptical Inquirer* 1984; 9: 147–58. See also Skeptical Eye. The (retouched) Loch Ness Monster. *Discover*, September 1984, p. 6.

66 Letter from Charles Wyckoff to Editor, *Discovery*, copied to Peter Scott, 27 August 1984, Peter Scott Archives C.784.

67 Shine A.J., Martin D.S. Loch Ness habitats observed by sonar and underwater television. *Scottish Naturalist* 1988; 105:111–9; Shine A. J., Minshull R. J., Shine M. M. Historical background and introduction to the recent work of the Loch Ness and Morar Project. *Scottish Naturalist* 1993; 105:7–22.

68 Shine A. J., Minshull R. J., Shine M. M. Historical background and introduction to the recent work of the Loch Ness and Morar Project. *Scottish Naturalist* 1993; 105:18.

69 Adrian Shine, personal communication.

70 Shine A. J., Minshull R. J., Shine M. M. Historical background and introduction to the recent work of the Loch Ness and Morar Project. *Scottish Naturalist* 1993; 105:14.

71 Letter from Tim Dinsdale to Peter Scott, 1 July 1987, Peter Scott Archives C.785.

72 Letter from Robert Rines to Peter Scott, 13 October 1987, Peter Scott Archives C.787.

73 Letter from Peter Scott to editor, *Western Morning News* 12 May 1988, Peter Scott Archives C.788.

74 Nicholas Witchell, personal communication.

75 Chalmers LN, in Project Urquhart brochure, 1992, p. 3. Natural History Museum Archives, DF704/16.

76 Hodges J. Witchell wary of monsters as survey begins. *Sunday Times* 12 July 1992; Ipsen E. Lowly worm to have its day. *International Herald Tribune* 13 July 1979.

77 *Project Urquhart. The definitive exporation of Loch Ness.* Press guide to participants. Natural History Museum Archives, DF704/16.

78 Cusick J. Sonar plumbs the depths in Loch Ness. *Independent* 14 July 1992.

79 Campbell S. *The Loch Ness Monster: The Evidence.* Edinburgh: Birlinn, 2002, p. 82.

80 Whyte C. *More Than a Legend.* London: Hamish Hamilton, 1957, p. 190.

Chapter 11

1 Lehn W. B. Atmospheric refraction and lake monsters. *Science* 1979 205: 183–185.

2 Murray J., Pullar L. Mirages on Loch Ness. *Geogr J* 1908; 31:61–2.

3 Campbell A. Letter to Ness Fishery Board, 28 October 1933. Cited in Gould R. T. *The Loch Ness Monster and Others.* London: Geoffrey Bles, 1934, pp. 110–12.

4 Ivor Newby, personal communication.

5 Gould R. T. *The Loch Ness Monster and Others* 1934, p. 105.

6 Boulenger E. G. The Monster of Loch Ness. *Observer* 29 October 1933. Cited in Gould R. T. *The Loch Ness Monster and Others* 1934, pp. 101–2.

7 Eberhart G. M. *Mysterious Creatures. A Guide to Cryptozoology.* Santa Barbara: ABC-CLIO Inc., 2002, p. 377.

8 Shine A. J. Postscript: surgeon or sturgeon? *Scottish Naturalist* 1993; 105:271–82.

9 Frere R. *Loch Ness.* London: John Murray, 1988, p. 101.

10 Burton M. *The Elusive Monster.* London: Rupert Hart-Davis, 1961, p. 117.

11 Campbell S. *The Loch Ness Monster: the Evidence.* Wellingborough: The Aquarian Press, 1986, p. 20.

12 Burton M. Letter to Steuart Campbell, 27 October 1984. Cited in Campbell S. *The Loch Ness Monster: The Evidence*, 1986, p. 19.

13 Baker P., Westwood M. Sounding out the Monster. *Observer* 26 August 1962; Frere R. *Loch Ness.* London: John Murray, 1988, p. 81.

14 Dinsdale T. *Loch Ness Monster.* London: Routledge & Kegan Paul, 1961, p. 53.

15 Gould R. T. *The Loch Ness Monster and Others* 1934, pp. 108–9.

16 Meaden G. T. Letter to Peter Scott, 7 June 1976. Peter Scott Archives, C763.

17 Dick Raynor, personal communication. See also http://www.lochnessinvestigation.com/Weather.html

18 Hollan E., Rao D. B., Bäuerle E. Free surface oscillations in Lake Constance with an interpretation of the 'Wonder of the Rising Water' at Kontanz in 1549. *Arch Met Geoph Biokl*, series A 1980; 29:301–2.

19 Wedderburn E. M. Seiches observed in Loch Ness, *Geogr J* 1904; 24:441–2; Thorpe S. A., Hall A., Crofts I. The internal surge in Loch Ness. *Nature* 1972; 237:96-98; Shine A. J., Martin D. S. Loch Ness habitats observed by sonar and underwater television. *Scottish Naturalist* 1988; 105:111–9.

20 Piccardi L. Seismotectonic origins of the Monster of Loch Ness (abstract). Presented at Earth System Processes Global Meeting, Edinburgh, 27 June 2001; Whyte C. *More Than a Legend.* London: Hamish Hamilton, 1957; revised 3rd impression, 1961, pp. 144, 164.

21 Gould R. T. *The Loch Ness Monster and Others* 1934, p. 107.

22 Campbell E. M., Solomon D. *The Search for Morag.* London: Tom Stacey, 1972, pp. 69–70.

23 Whitehead H. F. A monster catch. *Scotsman* 17 September 1960.

24 Seal D. T. Letter to Peter Scott, 17 November 1976. Peter Scott Archive, C732.

25 Burton M. Muck and monsters. *Illustrated London News* 1 October 1960, p. 568; Burton M. *The Elusive Monster.* London: Rupert Hart-Davis, 1961,

pp. 91–102; Whyte C. The Loch Ness Monster. Letter. *New Scientist* 17 November 1960, p. 1414.

26 Frere R. *Loch Ness*. London: John Murray, 1988, p. 102.

27 James D. *Photographic Interpretation Report — Loch Ness. Report on a film taken by Tim Dinsdale*. Inverness: Inverness Courier, 1966

28 'Clachnaharry'. The Loch Ness Monster. *Edinburgh Evening Dispatch* 27 November 1933.

29 Mackal R. P. Eel studies in Loch Ness. In: *The Monsters of Loch Ness*. The Swallow Press, Chicago, 1976, pp. 319–30.

30 Witchell N. *The Loch Ness Story*. Lavenham (Suffolk): Terence Dalton, 1974, p. 29.

31 Gould R. T. *The Loch Ness Monster and Others* 1934, pp. 38–9.

32 Lane, W. H. *The Home of the Loch Ness Monster*. Edinburgh: Grant & Murray Limited, 1934.

33 Bradley S. G., Klika L. J. A fatal poisoning from the Oregon rough-skinned newt (*Taricha granulosa*). *JAMA* 1981; 246:247.

34 Dick Raynor, personal communication; Campbell S. *The Loch Ness Monster: the Evidence*, 1991, p. 9; Gould R. T. *The Loch Ness Monster and Others* 1934, p. 114; Burton M. *The Elusive Monster*, 1961, p. 105.

35 Anonymous. Captured by Nature. *Daily Mail* 16 September 1914, p. 3

36 Mackal R. P. *The Monsters of Loch Ness*. The Swallow Press, Chicago, 1976. See Fig. P5, p. 100; Whyte C. *More Than a Legend*, revised 3rd impression, 1961, pp. 8–9.

37 Williamson G. R. Seals in Loch Ness. *Sci Rep Whales Research Inst* 1988; 39:151–7.

38 Anonymous. Editorial, 'The Loch Ness Monster'. *Inverness Courier* 8 May 1934.

39 Anonymous. A scene at Lochend. *Inverness Courier* 1 July 1852.

40 Burton M. *The Elusive Monster*. London: Rupert Hart-Davis, 1961, pp. 130–36.

41 Terence Morrison-Scott, letter to Peter Scott, 28 June 1960. Peter Scott Archives, C660.

42 Gould R. T. *The Loch Ness Monster and Others* 1934, pp. 115–17; Twelves J. Otters and monsters. *Oban Times* 11 May 1985.

43 Gould R. T. *The Loch Ness Monster and Others* 1934, p. 117.

44 Eberhart G. M. *Mysterious Creatures. A Guide to Cryptozoology*. Santa Barbara: ABC-CLIO Inc., 2002, pp. 375–84.

45 Paxton C. A cumulative species description curve for large open water marine animals. *J Mar Biol Ass UK* 1998; 78:1389–91; Paxton C. Predicting pelagic peculiarities: some thoughts on future discoveries in the open seas. *Dracontology* 2001; 1:20–25.

46 Roper C. F. E., Bosa K. J. The giant squid. *Scientific American* 1982; 246:96–105.

47 Coleman L., Huyge P. *Field guide to lake monsters, sea-serpents and other mystery denizens of the deep.* New York: Jeremy P. Tarcher, 2003, pp. 307–40.

48 Holiday F. W. *The Great Orm of Loch Ness.* London: Faber and Faber, 1968, pp. 140–44, 191–6.

49 Mackal R. P. *The Monsters of Loch Ness*, 1976, pp. 158–61, 212–15.

50 Parsons J. A dissertation upon the Class of the *Phocae marinae. Phil Trans Roy Soc* 1751; 47:111–12; Marshall M. The long-necked seal. *New Scientist* 4 February 2012, p. 44.

51 Anonymous. The Skegness Thing. *Skegness Standard* 6 November 1966.

52 Oudemans A. C. *The Loch Ness Monster.* Leyden: Late E.J. Brill, 1934.

53 Elmhirst R. Lecture at Royal Philosophical Society, Glasgow, January 1934, reported in *Inverness Courier* 9 January 1934, p. 4.

Chapter 12

1 Mackal R. P. *The Monsters of Loch Ness.* The Swallow Press, Chicago, 1976, pp. 87, 224–68.

2 Gould R. T. *The Loch Ness Monster and Others.* London: Geoffrey Bles, 1934. Revised edition, New York: University Books, 1969, p. 27.

3 Frere R. *Loch Ness.* London: John Murray, 1988, pp. 175–6.

4 Stalker P. Loch Ness Monster: a puzzled Highland community. *Scotsman* 16 October 1933, p. 11; Whyte C. *More Than a Legend.* London: Hamish Hamilton, 1957; revised 3rd impression, 1961, p. 6; Gould R. T. *The Loch Ness Monster and Others*, 1969, pp. 39, 46.

5 Whyte C. *More Than a Legend*, 1961, pp. 2, 43; MacFarquhar L. Letter from Scotland. Monster in the monitor. *New Yorker* 27 November 2000, 142–9; Dick Raynor, personal communication; Witchell N. *The Loch Ness Story.* Lavenham, Suffolk: Terence Dalton, 1974, p. 58.

6 Gould R. T. *The Loch Ness Monster and Others*, 1969, p. 63; Dick Raynor, personal communication; Witchell N. *The Loch Ness Story*, 1974, p. 108.

7 Scott P. Why I believe in Nessie. *Wildlife* March 1976, p. 110; Letter from Sir Alister Hardy to Peter Scott, 12 July 1960, Peter Scott Archives C.661.

8 Gould R. T. *The Loch Ness Monster and Others*, 1969, p. 33.

9 Betts J. *Time Restored. The Harrison Timekeepers and the Man who Knew (Almost) Everything.* Oxford: Oxford University Press, 2006, p. 259; Gould R. T. *The Loch Ness Monster and Others*, 1969, p. 44.

10 Gould R. T. *The Loch Ness Monster and Others*, 1969, p. 88.

11 Dick Raynor, personal communication.

12 Witchell N. *The Loch Ness Story*, 1974, p. 51; Mackal R. P. *The Monsters of Loch Ness*, 1976, p. 100; Bauer H. H. *The Enigma of Loch Ness: Making Sense of a Mystery.* Urbana and Chicago: University of Illinois Press, 1968, p. 62.

13 Professor David Sims, Marine Biological Laboratory, Plymouth, personal communication.

14 Anonymous. Captured by Nature. *Daily Mail* 16 September 1914, p. 3

15 Bauer H. H. *The Enigma of Loch Ness*, 1968, pp. 101–110, 113–14.

16 Letter from Tim Dinsdale to Peter Scott, 21 June 1960, Peter Scott Archive C.661; Dinsdale T. *Loch Ness Monster.* London: Routledge & Kegan Paul, 1961, Fig. 10.

17 Burton M. A ring of bright water? *New Scientist* 24 June 1982, p. 872; letter from Maurice Burton to Steuart Campbell, 27 October 1984, cited in Campbell S. *The Loch Ness Monster: The Evidence.* Wellingborough: The Aquarian Press, 1986. Revised edition, Aberdeen: Aberdeen University Press, 1991, p. 34.

18 Dick Raynor, personal communication; see lochnessinvestigation.com/o'connor.html

19 Witchell N. *The Loch Ness Story*, 1974, pp. 181–5.

20 Nicholas Witchell, personal communication.

21 Watson R. The Hugh Gray photograph revisited, 26 June 2011. http://lochnessmystery.blogspot.ca/2011/06/hugh-gray-photograph-revisited_26.html

22 Burton M. *The Elusive Monster.* London: Rupert Hart-Davis, 1961, p. 79.

23 Harmsworth T. *Loch Ness, Nessie and Me. The Truth Revealed.* Drumnadrochit: Harmsworth.net, 2010, p. 88.

24 Whyte C. *More Than a Legend*, 1961, pp. 2–5.

25 *Ibid.*, pp. 10–14.

26 Kemmet B. Is the monster picture a FAKE? *Scottish Sunday Express* 22 July 1951; Campbell S. *The Loch Ness Monster: The Evidence*, 1991, pp. 31–2.

27 Witchell N. *The Loch Ness Story*, 1974, pp. 108–10; Mackal RP. *The Monsters of Loch Ness*, 1976, pp. 100–102; Burton M. *The Elusive Monster*, 1961, pp. 74–6.

28 Witchell N. *The Loch Ness Story*, revised edition. London: Corgi, 1989, pp. 82–3; Adrian Shine, personal communication.

29 *Ibid.*, pp. 92–3.

30 Mackal R. P. *The Monsters of Loch Ness*, 1976, pp. 273–6.

31 Burton M. *The Elusive Monster*, 1961, pp. 159–60; Mackal R. P. *The Monsters of Loch Ness*, 1976, pp. 96–8; Burton M. A ring of bright water? *New Scientist* 24 June 1982, p. 872.

32 Letters from Tim Dinsdale to Peter Scott, 12 and 26 March 1960, and Scott's reply, 22 March, Peter Scott Archives C.658.

33 LeBlond P. H., Collins M. J. The Wilson Nessie photo: a size determination based on physical principles. *Cryptozoology* 1987; 6:55–64; Campbell S. The Surgeon's monster hoax. *Brit J Photogr* 20 April 1984, 407–11.

34 Martin D., Boyd A. *Nessie – the Surgeon's Photo Exposed*. Martin & Boyd, East Barnet, 1999, pp. 53–5.

35 *Ibid.*, pp. 63–5.

36 Adrian Shine, personal communication.

37 'Mandrake' (Peter Purser). Making of a monster. *Sunday Telegraph* 7 December 1975.

38 Martin D., Boyd A. *Nessie: The Surgeon's Photo Exposed*, 1999, pp. 19–20.

39 *Ibid.*, pp. 86–90.

40 *Ibid.*, pp. 71–3.

41 *Ibid.*, pp. 77–8.

42 Langton J. Revealed: the Loch Ness picture hoax. *Sunday Telegraph* 13 March 1994.

43 Smith R. D. The classic Wilson Nessie photo: is the hoax a hoax? *Fate* November 1995, pp. 42–4.

44 Letter from Richard Harrison to Peter Scott, 16 June 1960, Peter Scott Archives C.660; Letter from Constance Whyte to Peter Scott, 8 May 196, Peter Scott Archives C.661; Frere R. *Loch Ness*, 1988, p. 101.

45 Letter from Maurice Burton to Peter Scott, 2 June 1960, Peter Scott Archives C.660; Burton M. *The Elusive Monster*, 1961, pp. 73–4.

46 James D. *Photographic Interpretation Report — Loch Ness. Report on a film taken by Tim Dinsdale*. Inverness: Inverness Courier, 1966.

47 Dinsdale T. *Loch Ness Monster*, 4th edition. London: Routledge, 1989, Fig. 9a.

48 Dick Raynor and Adrian Shine, personal communications; Shine A. *The Dinsdale Loch Ness Film. An image analysis*. Loch Ness Project, unpublished, 2003; Campbell S. *The Loch Ness Monster: The Evidence*, 1991, pp. 47–8.

49 Terence Alan Patrick Dominic Foley, Kalagate Imagery Bureau, Cambs, signed statement, 16 July 2005. Document in possession of Adrian Shine.

50 Dinsdale A. *The Man Who Filmed Nessie*. Surrey, BC, Canada: Hancock House, 2013.

51 Scott P. Why I believe in Nessie. *Wildlife* March 1976, pp. 110–11.

52 Letter from Alan Gillespie to Peter Scott, 12 January 1976, Peter Scott Archives C.755.

53 Kallen S. A. *The Loch Ness Monster*. San Diego, CA: Reference Point Press, 2009, p. 44.

54 Witchell N. *The Loch Ness Story*, 1974, pp. 192, 210; Arnold H. *LNIB Newsletter* 29 October 1972. Original document in possession of Dick Raynor; letter from Robert Rines to Peter Scott, 11 January 1979, Peter Scott Archive C.779; Campbell S. *The Loch Ness Monster: The Evidence*, 1991, p. 53.

55 Rines R. H., Wyckoff C. W., Edgerton H. E., Klein M. Search for the Loch Ness Monster. *Technology Review* March–April 1976, pp. 25–40; Campbell S. *The Loch Ness Monster: the Evidence*, 1991, pp. 58–60.

56 Harwood G. E. Interpretation of the 1975 Loch Ness pictures. *Progr in Underwater Sci* 1977; 2:83–90, 99–102.

57 Sheals G., Corbet G. B., Greenwood P. H., Ball H. W., Charig A. J. Report on Rines' photographs, 1972 & 1975, from the British Museum (Natural History), November 1975. Peter Scott Archives C.732; cited in Rines R. H. Search for the Loch Ness Monster. *Technology Review* March–April 1976, p. 37.

58 Desmond A. Why Nessiteras rhombopteryx is such an unlikely monster. *The Times* 27 December 1975.

59 Shine A. J. The biology of Loch Ness. *New Scientist* 17 February 1983, pp. 462–7.

60 Campbell S. *The Loch Ness Monster: The Evidence*, 1991, pp. 55–7.

61 Razdan R., Kielar A. Sonar and photographic searches for the Loch Ness

Monster: a reassessment. *Skeptical Inquirer* 1984; 9:147–58. See also Anonymous. Skeptical Eye. The (retouched) Loch Ness Monster. *Discover* September 1984, p. 6.

62 Letter from Charles Wyckoff to Editor, *Discovery*, copied to Peter Scott, 27 August 1984, Peter Scott Archives C.784.

63 Adrian Shine, personal communication.

64 Dick Raynor and Adrian Shine, personal communications; Campbell S. *The Loch Ness Monster: The Evidence*, 1991, pp. 60–61.

65 Shine A. J., Martin D. S. Loch Ness habitats observed by sonar and underwater television. *Scottish Naturalist* 1988; 105:111–19.

66 Klein M., Rines R. H., Dinsdale T., Foster L. S. *Underwater Search at Loch Ness*. Monograph 1, Academy of Applied Science. Boston: Academy of Applied Science, 1972.

67 Baker P. Echo-sounding as a method of searching underwater in Loch Ness. Appendix B in: Holiday F. W. *The Great Orm of Loch Ness*. London: Faber and Faber, 1968, pp. 197–203.

68 Braithwaite H. Sonar picks up stirrings in Loch Ness. *New Scientist* 19 December 1968; 40:664–6.

69 Campbell S. *The Loch Ness Monster: the Evidence*, 1991, pp. 73–6.

70 *Ibid.,* pp. 77–80.

71 Witchell N. *The Loch Ness Story*, 1974, pp. 115–16; Harrison P. *The Encyclopaedia of the Loch Ness Monster*. London: Robert Hale, 1999, p. 170.

72 Campbell S. *The Loch Ness Monster: The Evidence*, 1991, pp. 65–8.

Chapter 13

1 Witchell N. *The Loch Ness Story*. Lavenham, Suffolk: Terence Dalton, 1974, p. 55; Foreword by Gerald Durrell, pp. 10–11.

2 Whyte C. *More Than a Legend*. London: Hamish Hamilton, 1957; revised 3ª impression, 1961, p. xi.

3 Nuwer R. Celebrating 1,447 years of the Loch Ness Monster. *SmartNews*, Smithsonian.com, 23 August 2012.

4 Letter from Peter Scott to Peter Baker, 11 October 1961, Peter Scott Archives C.664.

5 Mackal R. P. *The Monsters of Loch Ness*. The Swallow Press, Chicago, 1976, p. 333.

6 *Project Urquhart. The definitive exporation of Loch Ness.* Press guide to participants. Natural History Museum Archives, DF704/16.

7 Letter from Colonel the Hon. Martin Charteris to Peter Scott, 10 August 1960, Peter Scott Archive, C.663.

8 Betts J. Rupert Gould (1890–1948), horologist and broadcaster. *Oxford Dictionary of National Biography*, May 2013. Index no. 10104920.

9 Watters D. A. Loch Ness, Special Operations Executive and the first surgeon in Paradise: Robert Kenneth Wilson (26.1.1899–6.6.1969). *Aust NZ J Surg* 2007; 77:1053–7.

10 Tucker D. The Zoologist's Tale. In: Witchell N. *The Loch Ness Story*, revised edition. London: Corgi Books, 1989, pp. 205–28.

11 Professor George S. Myers, Stanford University, quoted at Press Conference called by Institution of Professional Civil Servants, 13 December 1960. Natural History Museum Archives, DF 5011/61; documents and correspondence relating to disciplinary action by Trustees of the Natural History Museum against Dr. Denys Tucker, 1959-1961. Natural History Museum Archives, DF 5011/61, 5011/64.

12 Anonymous. Obituary: Denys Tucker, zoologist. *The Times* 6 November 2009.

13 Tucker D. W. Obituary: Maurice Burton. *Independent* 23 September 1992.

14 Watson R. A story about Ted Holiday, posted 19 April 2012. Available at: http://lochnessmystery.blogspot.co.uk/2012_04_15_archive.html

15 Holiday F. W. *The Great Orm of Loch Ness.* London: Faber and Faber, 1968, p. 8.

16 Holiday F. W. *The Dragon and the Disc.* London: Sidgwick & Jackson, 1973.

17 Powell M. New theory about the Loch Ness Monster comes from East Devon. Pullman's Weekly, 5 June 1973, p. 1; Harrison P. *The Encyclopaedia of the Loch Ness Monster.* London: Robert Hale, 1999, p. 157.

18 Holiday T., Wilson C. *The Goblin Universe*, revised edition. London: Xanadu Publishing, 1990.

19 Keith A. Article in *Daily Express*, 1933, quoted in: Whyte C. *More Than a Legend.* London: Hamish Hamilton, 1957; revised 3rd impression, 1961, p. 87.

20 Dash M. Frank Searle's lost second book. Charles Fort Institute blogs, 27 December 2009. Available at: http://blogs.forteana.org/node/95

21 Letters between Frank Searle and Harriet Ely, from 14 March 1979 to 12 June 1991, Frank Searle Archives, Highland Archives, D 885/2/1.

22 Tullis A. Obituary: Frank Searle. Loch Ness monster hoaxer. *Independent*,

24 May 2005; Tullis A, producer. *The Man Who Captured Nessie*. Television documentary, broadcast Channel 4, 29 December 2005.

23 Robson J. *One Man in His Time. The Biography of David James*. Staplehurst, Kent: Spellmount, 1998, pp. 221–3.

24 Anonymous. Obituary: Mr David Guthrie-James. *The Times* 17 December 1986.

25 Huxley E. *Peter Scott. Painter and Naturalist*. London: Faber and Faber, 1993, p. 211.

26 Robson J. *One Man in His Time*, 1998, pp. 236–7.

27 Letters from Roy Mackal to Peter Scott, 7 and 13 January 1976, Peter Scott Archives, C.756.

28 Laurance W. The call of the weird: in praise of cryptobiologists. *New Scientist* 22 June 2011, p. 35.

29 Mackal R. P. *A living dinosaur? In search of Mokele-Mbembe*. New York: E. J. Brill, 1987.

30 Dick Raynor, personal communication.

31 Frere R. *Loch Ness*. London: John Murray, 1988, p. 169.

32 Raynor D. http://www.lochnessinvestigation.com/Remembered.html

33 Citation, American Inventors' Hall of Fame: Robert H. Rines. Available at http://www.invent.org/hall_of_fame/122.html

34 Dow B. Veteran Loch Ness Monster hunter gives up. *Daily Record* 13 February 2008.

35 Harrison P. *The Encyclopaedia of the Loch Ness Monster*, 1999, p. 15.

36 Marquard B. Obituary: Robert Rines, composer, professor, founder of law center, seeker of 'Nessie'. *Boston Globe* 3 November 2009.

37 Baker B. The scientist . . . and the Monster. *Boston Magazine* December 2008.

38 Letters from David James to Adrian Shine, and Roy Mackal to David James, in possession of Adrian Shine.

39 Paul Walkden, personal communication.

40 Huxley E. *Peter Scott. Painter and Naturalist*, 1993, pp. 327–9.

41 *Ibid.,* pp. 270, 335.

42 Peter Scott, interview with François Gohier, Slimbridge, January 1977. Peter Scott Archives, M.2052.

43 Huxley E. *Peter Scott. Painter and Naturalist*, 1993, p. 299.

44 *Ibid.*, p. 327.

45 Gallico P. *The Snow Goose: A Story of Dunkirk*. New York: Knopf, 1941.

46 Letter from James Robertson Justice to Peter Scott, cited in correspondence about *Snow Goose* Concert, Royal Festival Hall 19 February 1977, Peter Scott Archives, M.2052.

Chapter 14

1 Whyte C. *More Than a Legend*, revised 3rd impression. London: Hamish Hamilton, 1961, p. 123.

2 Quoted in: Davison P. Nessie hunter who helped save US Air Force (obituary, Robert H. Rines). *Financial Times* 14 November 2009.

3 Anonymous. Loch Ness monster death rumours denied. *Daily Telegraph* 6 January 2010.

4 http://www.nessie.co.uk/htm/nessies_diary/nessie.html, accessed 29 November 2014.

5 Witchell N. *The Loch Ness Story*. Lavenham, Suffolk: Terence Dalton, 1974, p. 65; Jonathan Brown. Loch Ness Monster: Nessie's back, just in time for Scotland's big year. *Independent* 27 April 2014.

6 Carter C. Has Nessie finally been caught on video? *Daily Mail Online* 10 November 2014.

7 Anonymous. Lecture on Loch Ness Monster, by Dr Richard Elmhirst. *Inverness Courier* 9 January 1934, p. 7.

8 Barker I. Evolution? I don't believe it. Haven't you heard of Nessie? *Times Educational Supplement Magazine*, 6 July 2012.

9 Cary W. ('Freddie'). Population distribution of unidentified large creatures in Loch Ness, 30 September 1972. Original document in possession of Dick Raynor.

10 Sheldon R. W., Kerr S. R. The population density of Monsters in Loch Ness. *Limnology and Oceanography* 1972; 17:796–8; Scheider W., Wallis P. An alternate method of calculating the population density of Monsters in Loch Ness. *Limnology and Oceanography* 1973; 18:343; Mortimer C. H. The Loch Ness Monster – limnology or paralimnology? *Limnology and Oceanography* 1973; 18:343–5.

11 Shine A. J., Minshull R. J., Shine M. M. Historical background and introduction to the recent work of the Loch Ness and Morar Project. *Scottish Naturalist* 1993; 105:7–22

12 Kear B., Schroeder N. I., Lee M. S. Y. An archaic crested plesiosaur in opal from the Lower Cretaceous high-latitude deposits of Australia. *Biology Letters* 2006; 2:615–19.

13 McHenry C. R., Cook A. G., Wroe S. Bottom-feeding plesiosaurs. *Science* 2005; 310:75.

14 Emling, S. *The Fossil Hunter.* New York: Palgrave Macmillan, 2009, pp. 77–85.

15 Owen, R. *Palaeontology, or a Systematic Summary of Extinct Animals and their Geological Relations.* Edinburgh: Adam and Charles Black, 1860, p. 230. [The origin of the simile is uncertain.]

16 Keene M. *Science in Wonderland: The Scientific Fairy Tales of Victorian Britain.* Oxford: Oxford University Press, 2015.

17 Cassie R. L. *The Monsters of Achanalt.* Aberdeen: D. Wyllie & Son, 1935.

18 Gosse P. H. *The Romance of Natural History*, 1875. Republished with foreword by L. Coleman. New York: Cosimo, 2008, p p. 357–8.

19 Gould R. T. *The Loch Ness Monster and Others.* London: Geoffrey Bles, 1934. Revised edition, New York: University Books, 1969, p. 46.

20 *Ibid.*, pp. 38–9, 95.

21 Dinsdale T. *Loch Ness Monster.* London: Routledge & Kegan Paul, 1961, pp. 14–27.

22 Letter from Nikki Stanley to Peter Scott, 11 November 1980, Peter Scott Archive C.778.

23 Letter from Alan Wilkins to Peter Scott, 6 November 1975; also Peter Scott's rough notes, Peter Scott Archives C. 731.

24 Letters from R. V. Melville and Sir Arthur Landsborough Thomson to Peter Scott, 6 November and 21 November 1975, Peter Scott Archive C.731 and C.732.

25 Letter from Sir Solly Zuckerman to Peter Scott, 22 July 1975, Peter Scott Archives C.707.

26 Meredith D. The Loch Ness press mess. *Technol Rev* March/April 1976, 10–12.

27 Whyte C. *More Than a Legend*, revised 3rd impression, 1961, p. xviii.

28 Mayor A. *The First Fossil Hunters. Palaeontology in Greek and Roman Times.* Princeton: Princeton University Press, 2000, pp. 138–9.

29 Bondeson J. *The Feejee Mermaid and Other Essays in Unnatural History.* New York: Cornell University Press, 1999, pp. 36–63.

30 Mayor A. *The First Fossil Hunters*, 2000, p. 139.

31 Gould R. T. *The Loch Ness Monster and Others*, 1969, pp. 196–204.

32 Owen J. Loch Ness sea monster fossil a hoax, say scientists. *National Geographic News* 29 July 2003; Munro A. Loch Ness Monster: George Edwards 'faked' photo. *Scotsman* 4 October 2013.

33 Witchell N. *The Loch Ness Story*, 1974, p. 187.

34 Anonymous. Loch Ness Monster is dead! *Weekly World News* 25 August 1995, p. 1.

35 Churton T. *Aleister Crowley: The Biography*. London: Watkins Publishing, 2011, pp. 110.

36 Gasparini F. Je suis le père du monstre du Loch Ness. *Paris-Match* 11 April 1959.

37 Bibby D. J. *Glimpses*. Liverpool: Bibby Bros & Co., 1991, pp. 14–15.

38 Bauer H. H. *The Enigma of Loch Ness: Making Sense of a Mystery*. Urbana and Chicago: University of Illinois Press, 1968, pp. 9, 155–7.

39 Lister S. (pseudonym of D. G. Gerahty). *Marise*. London: Peter Davies, 1950, p. 95.

40 Bauer H. H. *The Enigma of Loch Ness: Making Sense of a Mystery*, 1968, p. 3.

41 Dinsdale T. *Loch Ness Monster*, 1961.

42 Gregory J. *The Life of Columba, by Adamnan. An abridged translation*. Edinburgh: Floris Books, pp. 25–44; Grimshaw, R. and Lester, P. (1976) *The Meaning of the Loch Ness Monster*. Occasional paper. Centre for Contemporary Cultural Studies, University of Birmingham, pp. 3–9.

43 Witchell N. *The Loch Ness Story*, 1974, p. 30.

44 *Ibid.*, pp. 25–8.

45 *Ibid.*, pp. 28–30.

46 Fraser J. Part of a letter concerning the Lake Ness, etc. *Philosophical Transactions of the Royal Society* 1699; XXI:230–32; reproduced in Whyte C. *More Than a Legend*, 1961, pp. 207–10.

47 Mather C. An extract of several letters to John Woodward MD and Richard Waller, Esq. *Phil Trans Roy Soc* 1714; XXIX:61–2.

48 Anonymous. Letters to the Editor. Loch Ness; the Stronsa monster. *The Scots Magazine* 1809; 71:110–16.

49 Anonymous. Loch Ness. *Encyclopaedia Londinensis. Universal dictionary of arts, sciences and literature*. London: J. Adlard, 1814, pp. 864–5.

50 Murray J., Pullar L. (editors). *Bathymetrical Survey of the Scottish Fresh-Water Lochs*, vols 1–6. Edinburgh: Challenger Office, 1910, pp. 431–5.

51 David MacBrayne & Co. *Summer Tours in Scotland. The Royal Route.* Glasgow: Macbrayne & Co., 1898, pp. 31–46.

52 Proceedings of meetings of Inverness Scientific Society, February 1923 and 19 December 1919. *Transactions of Inverness Scientific Society*, 1918-25; IX:137, 360–70.

53 Macrae J. Letter to *Inverness Courier*, August 1872. Cited in Oudemans A. C. *The Great Sea-Serpent* (1892). Republished with foreword by Loren Coleman. New York: Cosimo, 2009; Gould R. T. *The Loch Ness Monster and Others*, 1969, pp. 216–17.

54 Anonymous. Strange fish at Abriachan. *Inverness Courier* 8 October 1868.

55 Anonymous. What was it? A strange experience on Loch Ness. *Northern Chronicle* 17 August 1930.

56 Anonymous (Alex Campbell). Strange spectacle on Loch Ness. What was it? *Inverness Courier* 2 May 1933.

57 Anonymous. Loch Ness 'Monster': ship captain's views on occurrence. Letter by John Macdonald, *Inverness Courier* 12 May 1933.

58 Anonymous. Loch Ness 'Monster' seen again. *Inverness Courier* 2 June 1933, p. 5.

59 Anonymous. Visitor's experience. Letter from George Spicer with editor's explanatory note, *Inverness Courier* 4 August 1933.

60 Anonymous. Loch Ness 'Monster'. Three appearances near Fort-Augustus. Eye-witnesses' vivid descriptions. *Inverness Courier* 8 August 1933, p. 5.

61 Anonymous. The Loch Ness 'Monster'. True story of his life. Told to A Mere Woman. *Inverness Courier* 11 August 1933, p. 5.

62 Anonymous. Loch Ness Monster seen again at Glen Urquhart. *Inverness Courier* 11 August 1933, p. 5.

63 Anonymous. Loch Ness 'Monster'. Numerous apearances during week-end. *Inverness Courier* 26 September 1933, p. 5.

64 Anonymous. Editorial: The Loch Ness Mystery. *Inverness Courier* 26 September 1933, p. 4.

65 Anonymous. Loch Ness 'Monster'. Theory that it is an amphibian. *Inverness Courier* 3 October 1933, p. 5; Anonymous. Loch Ness 'Monster'. Problem for scientific society, with letter from W. H. Lane, *Inverness Courier* 10 October 1933, p. 4.

66 Russell-Smith A. Problem for scientific society. *Inverness Courier* 10 October 1933, p. 4.

67 Stalker P. Loch Ness Monster: the plesiosaurus theory. *Scotsman* 17 October 1933, p. 9.

68 Anonymous. Loch Ness 'Monster'. Seen by a Highland Proprietor. *Inverness Courier* 2 January 1934.

69 Anonymous. Loch Ness Monster seen again. *Inverness Courier* 13 February 1934, p. 5.

70 Anonymous. Loch Ness Monster seen again. *Inverness Courier* 13 February 1934, p. 5.

71 Anonymous. Priests' views on the Monster. *Inverness Courier* 13 February 1934, p. 3.

72 Anonymous. Loch Ness 'Monster'. Well-known Ross-shire lady's story. *Inverness Courier* 27 February 1934, p. 3.

73 Anonymous. Loch Ness 'Monster'. English lady's description. *Inverness Courier* 3 April 1934, p. 5.

74 Articles in *Inverness Courier*, 16 June, 25 July, 30 October, 3 November, 12 December and 29 December 1933, 2 January 1934.

75 Choose Scotland. The SNP is good for Scotland. Advertisement, *Inverness Courier* 7 June 1983, p. 5.

76 Anonymous. Obituary. Mr Alex M. Campbell, Fort Augustus. *Inverness Courier*, 1 July 1983.

77 Harrison P. *The Encyclopaedia of the Loch Ness Monster*. London: Robert Hale, 1999, pp. 38–9.

78 James, D. Tribute: Mr Alex M. Campbell, Fort Augustus. *Inverness Courier* 8 July 1983.

79 Bauer H. H. *The Enigma of Loch Ness: Making Sense of a Mystery*, 1968, p. 5

80 Gould R. T. *The Loch Ness Monster and Others*, 1969, p. 39; BBC 1933 Year Round-Up, introduced by Lawrence Wager. Sound Archives, British Library, London, C1398/0144, C22–24.

81 Bauer H. H. *The Enigma of Loch Ness: Making Sense of a Mystery*, 1968, p. 5.

82 Whyte C. *More Than a Legend*, 1961, p. 203, footnote p. 75.

83 Gould R. T. *The Loch Ness Monster and Others*, 1969, pp. 110–13.

84 Whyte C. *More Than a Legend*, 1961, pp. 203–4.

85 Witchell N. *The Loch Ness Story*, 1974, pp. 80–81.

86 Witchell N. *The Loch Ness Story*, 1974, p. 28.

87 Anonymous. Monster fury over Nessie hoax claim. *Highland News* 1 March 2012.

88 Letter from Peter Scott to Editor, *Western Morning News* 12 May 1988, Peter Scott Archives C.788.

89 Holiday F. W. *The Great Orm of Loch Ness*. London: Faber and Faber, 1968, p. 53.

90 Grimshaw R., Lester P. (1976). *The Meaning of the Loch Ness Monster*. Occasional paper. Centre for Contemporary Cultural Studies, University of Birmingham, p. 53

Index

Page numbers in **bold** refer to figures.